Global Carbon Cycle and Climate Change

Springer
Berlin
Heidelberg
New York
Hong Kong
London
Milan
Paris
Tokyo

Kirill Ya. Kondratyev, Vladimir F. Krapivin and
Costas A. Varotsos

Global Carbon Cycle and Climate Change

Springer

Published in association with
Praxis Publishing
Chichester, UK

PRAXIS

Professor Kirill Ya. Kondratyev
Counsellor of the Russian Academy of Sciences
Scientific Research Centre for Ecological Safety
Nansen Foundation for Environment and
Remote Sensing
St Petersburg
Russia

Associate Professor Dr Costas A. Varotsos
University of Athens
Department of Physics
Laboratory of Meteorology
Athens
Greece

Professor Dr Vladimir F. Krapivin
Institute of Radioengineering and Electronics
Russian Academy of Sciences
Moscow
Russia

SPRINGER–PRAXIS BOOKS IN ENVIRONMENTAL SCIENCES
SUBJECT *ADVISORY EDITOR*: John Mason M.Sc., B.Sc., Ph.D.

ISBN 3-540-00809-8 Springer-Verlag Berlin Heidelberg New York

Bibliographic information published by Die Deutsche Bibliothek

Die Deutsche Bibliothek lists this publication in the Deutsche Nationalbibliografie; detailed bibliographic data are available from the Internet at http://dnb.ddb.de

Library of Congress Cataloging-in-Publication Data
Kondrat'ev, K. IA. (Kirill IAkovlevich)
Global carbon cycle and climate change / Kirill Ya. Kondratyev, Vladimir F. Krapivin, and Costas A. Varotsos.
p. cm.
Includes bibliographical references and index.
ISBN 3-540-00809-8 (acid-free paper)
1. Climatic changes. 2. Carbon cycle (Biogeochemistry) I. Krapivin, V. F. (Vladimir Fedorovich) II. Varotsos, Costas, 1956- III. Title.

QC981.8C5K6535 2003
363.738'74 – dc22 2003059053

Printed in Germany

Cover design: Jim Wilkie
Project Management: Originator Publishing Services, Gt Yarmouth, Norfolk, UK

Printed on acid-free paper

Contents

Preface

The growing interest in the problem of global climate change is due to the vital practical importance of this problem. Controversial assessments of anthropogenic contributions to climate change necessitates analysis of numerical modelling results and observational data on present climate changes and causes of these changes. Despite numerous efforts on varying scales to study the climatic trends, both in the past and present, the problem of reliable predictions of future climate change remains far from being solved. Emissions of greenhouse gases into the atmosphere (first of all – carbon dioxide) is considered as one of the principal causes of expected climate warming with its associated negative effects for humankind. Therefore the authors have tried to develop a formalized approach to assess the atmospheric greenhouse effect due to anthropogenic sources of carbon dioxide as well as to analyse the effect of various minor gas components on the formation of spatial distribution of carbon sinks in the biosphere.

To understand the factors determining the feedbacks in the global natural–technogenic system of the carbon cycle, a hierarchy of model units has been designed that parameterize a combination of physical and biogeochemical processes of the carbon transport. Components of these units have been developed on the basis of synthesis of individual models of budget relationship on the borders of various media.

The book contains six chapters. Chapter 1 deals with the present global ecodynamics, considers the concept of global warming due to the greenhouse effect, and discusses the problem of parameterization of anthropogenic processes. Chapter 2 gives data on both sources and sinks of carbon, considers the conceptual schemes of the global carbon dioxide cycle, and suggests a new approach to studies of the problem of the greenhouse effect. Chapter 3 assesses the role of 30 types of soil–plant formations in the assimilation of carbon dioxide from the atmosphere. Models of production processes in land biocoenoses have been substantiated, and characteristics of forest ecosystems have been analysed. Chapter 4 generalizes earlier models of the atmosphere–ocean gas exchange to simulate the carbon dioxide dynamics in the atmosphere–ocean system. Special attention has been paid to the role of the Arctic Basin in the assimilation of carbon dioxide from the atmosphere. Chapter 5 describes a parameterization technique to simulate the biospheric budget of carbon, nitrogen, sulphur, phosphorus, oxygen, ozone, and water within the framework of an integrated biogeochemical system. And, finally, Chapter 6 describes the global model of the nature–society system functioning, and with the use of this model, various scenarios of anthropogenic impact on the environment have been assessed.

Abbreviations

AADB	algorithmic adaptation of the database
APS	atmosphere–plant–soil
ASS	active surface substances
AVHRR	Advanced Very High Resolution Radiometer
CCAB	calculation of corrections for atmospheric balance (ozone)
CF	cloud feedback
CFC	chlorofluorocarbon
CFSU	control of the functions of the system units
CLaMS	Chemical Lagrangian Model of the Stratosphere
CTM	Chemical Transport Model
DIC	dissolved inorganic carbon
DIVERSITAS	biodiversity programme
DOM	dissolved organic matter
DU	Dobson units
EC	emission coefficient
EEC	European Economic Community
ENSO	El Niño Southern Oscillation
ERB	Earth radiation budget
EXPRESSO	Experiment for Regional Sources and Sinks of Oxidants
F	Forcings
FAO	Food and Agriculture Organization
FASIR	Fourier-adjusted, solar zenith angle-corrected, interpolated and reconstructed data
FBLO	formation of background level of ozone
FCCC	International Framework Convention on Climate Change
FPAR	fraction of PAR absorbed by the green portion of the vegetation canopy
FS	formation of scenarios (locations of flight corridors)

GCDC	global cycle of carbon dioxide
GCM	Global Circulation Model
GCP	Global Carbon Project
GHG	greenhouse gas
GIMS	Geoinformation Monitoring System
GIS	Geographical Information System
GOME	Global Ozone Monitoring Experiment
GSC	Global Sulphur Cycle
GSM	Global Simulation Model
GTE	Global Tropospheric Experiment
GWP	global warming potential
IAMAS	International Association of Meteorology and Atmospheric Sciences
IGBP	International Geosphere–Biosphere Programme
IGOS-P	Integrated Global Observing System
IHDP	International Human Dimensions Programme of Global Changes
IIASA	International Institute of Applied Systems Analysis
IPCC	Intergovernmental Panel on Climate Change
IRDB	information renewal of database
ISY	International Space Year
ITCZ	Intertropical Convergence Zone
JGOFS	US Joint Global Ocean Flux Study
LAI	leaf area index
LME	large marine ecosystem
LUCC	land use/land cover change
LWRF	longwave radiative forcing
MBWB	Model of the Biosphere Water Balance
MEA	millennium ecosystems assessment
MFDO	Model of Formation and Destruction of Ozone
MGC	minor gaseous component
MGNC	modelling the global nitrogen cycle
MGOC	modelling the global oxygen cycle
MGPC	modelling the global phosphorus cycle
MGSC	modelling the global sulphur cycle
MIF	management of information fluxes
MOSE	Model of the Okhotsk Sea Ecosystem
MOZART	Model for Ozone and Related Chemical Tracers
MPB	Mean Plant Biomass
MPCE	Model of the Peruvian Current Ecosystem
MPP	mean pure primary production
MPR	modelling photochemical reactions (flight corridors)
MPTO	modelling the processing of propagation and transformation (ozone)
MUE	Model of Upwelling Ecosystem

MUII	management of user informaton interface
NASA	National Aeronautics and Space Administration
NBP	net biome production
NDVI	normalized difference vegetation index
NIR	near-infrared range
NMHC	non-methane hydrocarbons
NO_x	nitrogen oxides
NPP	net primary production
OClO	chloroperoxy radical
P/B coefficient	produce/biomass coefficient
PADB	parametric agreement with the database
PAGES	Pilot Analysis of Global Ecosystems
PAME	Protection of the Arctic Marine Environment
PAR	photosynthetic active radiation
PBP	pure biome productivity
PCL	permissible concentration level
PEM	Pacific Exploratory Mission
PIC	products of incomplete combustion
PIK	Potsdam Institute for Climate Impact Studies
POAM	polar ozone and aerosol measurement
PPE	pure production of ecosystems
PPP	pure primary production
PSC	polar stratospheric cloud
RF	radiative forcing
RS	remote sensing
SAS	surface active substances
SAT	surface air temperature
SBUV	solar backscatter ultraviolet
SiB2	Simple Biosphere Model-2
SR	synthesizing reconstruction
SRVI	simple ratio vegetation index
SSROC	simulation system for regional ozonosphere control
SST	sea surface temperature
SSW	sudden stratospheric warming
SVI	spectral vegetation index
SWRF	shortwave radiative forcing
TCO	terrestrial carbon observations
THC	thermohaline circulation
TOC	total ozone content
TOMS	Total Ozone Mapping Spectrometer
TPP	total pure primary production
TPS	total aerosols
TsuBiMo Model	Tsukaba Biosphere Model
UCAR	University Corporation for Atmospheric Research
UNDP	UN Development Programme

UNEP	UN Environmental Programme
UNEP/GDP	UNEP/Gross Domestic Product
UNESCO	UN Educational, Scientific and Cultural Organization
UNFCCC	UN Framework Convention on Climate Change
UQL	upper quasi-homogeneous layer
WBGU	World Business Global Union
WCRP	World Climate Research Programme
WHO	World Health Organization
WSSD	World Summit on Sustainable Development

Figures

Tables

About the authors

Kirill Ya. Kondratyev received his PhD in 1956 at the University of Leningrad where he also became Professor in 1956 and was appointed Head of the Department of Atmospheric Physics in 1958. From 1978–1982 Kirill was Head of the Department of the main Geophysical Observatory, Leningrad, and from 1982–1991 Head of the Institute for Lake Research, Leningrad. Since 1992 he has been Councillor of the Russian Academy of Sciences. Kirill has published more than 100 books in the fields of atmospheric physics and chemistry, remote sensing, planetary atmospheres, and global change.

Vladimir F. Krapivin received his PhD in geophysics from the Moscow Institute of Oceanology in 1973. He became Professor of Radiophysics in 1987 and Head of the Applied Mathematics Department at the Moscow Institute of Radioengineering and Electronics in 1972. Since 1996 he has also been Director of the Institute of Ecoinformatics. Vladimir is a full member of the Russian Academy of Natural Sciences. He has specialized in investigating global environmental change by the application of modelling technology. Vladimir has published 12 books in the fields of ecoinformatics, theory of games, and global modelling.

Costas A. Varotsos received his BSc in Physics at Athens University in 1980 and his PhD in Atmospheric Physics in 1984. He was appointed Assistant Professor in 1989 at the Laboratory of Meteorology in the Physics Department of Athens University, where he also established the Laboratory of the Middle and Upper Atmosphere. In 1999 he became Associate Professor of the Department of Applied Physics at Athens University. Costas has published more than 200 papers and 15 books in the fields of atmospheric physics, atmospheric chemistry, and environmental change.

1

Global climate and carbon cycle

1.1 INTERACTIVITY OF GLOBAL CLIMATE CHANGE AND CARBON CYCLE DYNAMICS

The problem of the carbon cycle and its climatic implications (and respective feedbacks) has attracted attention from the days of S. Arrhenius (Kondratyev, 1992). However, during the last 2–3 decades the interest in this problem has grown substantially, much due to speculative statements with regard to anthropogenically induced global climate warming and apocalyptic predictions of catastrophic global warming with resulting harmful consequences for man's life and activity (Fleming, 1998; Connors *et al.*, 1999; Borisenkov and Pichugin, 2001; Grigoryev and Kondratyev, 2001a; Wofsy, 2001; Albrecht, 2002; Alverson *et al.*, 2002; Demirchian *et al.*, 2002). Cihlar *et al.* (2002) have justly emphasized that problems connected with the carbon flux variability turned out to be in the focus of not only present science but also politics.

1.1.1 Global warming: facts and illusions

Since the reality of global climate warming and the related exaggerations have recently been analysed in detail (Kondratyev, 2002), we shall only consider the related conceptual circumstances.

The failure of the International COP-6 Conference (Nilsson *et al.*, 2002) held in The Hague in November 2000, at which representatives of the states signed the International Framework Convention on Climate Change (FCCC), cannot be considered unexpected since the participants of COP-6 had not been adequately prepared and were lacking the necessary scientific information. It turned out that the most important problem of uncertain assessments of the reservoirs and fluxes of greenhouse gases (GHG) had been neglected.

After COP-6 the problem has become more acute because of President Bush's refusal to support the Kyoto Protocol (KP). On the other hand, the Third IPCC Report published in 2001 (Cihlar *et al.*, 2002; IPCC, 2001a–c) contains the conclusion that the available information enables one to convincingly speak about the GHG contribution to the formation of climate change and predict climate changes more reliably.

Of key importance are assessments of uncertainties of the estimates, which serve as the basis for conclusions about climate changes and required preventative measures (Kheshgi *et al.*, 1999a,b; Allen *et al.*, 2001; Mejer, 2001; Soon *et al.*, 2001; Kerr, 2002; Newell and Pizer, 2002; Reilly *et al.*, 2001; Varotsos, 2002a). Of special importance is the problem of evaluation of the level of GHG emissions into the atmosphere connected, first of all, with the solution of the problem of the global carbon cycle. Clearly, without a reliable verification of the available estimates of emissions, all discussions with respect to ecological profits of various measures and related expenditures are of abstract character (Nilsson *et al.*, 2002). How, for instance, can imposed fines for not following KP recommendations for GHG emissions reduction be justified if it is impossible to prove that emissions in 2012 will differ from those in 1990? So far, discussions on the KP have ignored, in particular, the quantitative estimates of uncertainties of the levels of GHG sinks (this specifically concerns the biosphere). However, uncertainties in assessments of CO_2 fluxes are rather substantial (exceeding 100% in conditions in Russia). Calculations of errors in estimation of total GHG fluxes vary by $\sim\pm 5$–25%, whereas the KP levels of GHG emissions reduction average is about 5%. A mean global situation can be illustrated, for example, by those uncertainties in evaluation of GHG emissions due to energy production systems which are approximately equal to those of the estimates of CO_2 assimilation by the biosphere and land (Kondratyev and Varotsos, 1995).

In this situation, solution of the problems of uncertain estimates (first of all it concerns reliable information on carbon cycle) and verification become of vital importance. A solution to the verification problem requires an agreement on its mechanisms, which is also very important from the financial point of view. Results of simulation modelling show, for instance, that if a 5.2% confidence level for GHG reduction is raised from 50 to 95%, it entails an increase of expenditure on measures to reduce GHG emissions by a factor of 3–4. The principal conclusion is that science should recommend measures in the field of ecological policy. In this context the COP-6 failure can be a healing shock (Nilsson *et al.*, 2002).

A serious uncertainty of the estimates of possible anthropogenic climate changes and their determining factors has brought forth heated discussions in scientific literature and the mass media (Baliunas, 2002; Burroughs, 2001; Ellsaesser, 2002; Gildor and Tziperman, 2001; Hartmann, 2002; Kondratyev and Grigoryev, 2002; Manabe, 1983; Moe and Tanger, 2000; Nordhaus, 2001; Philander, 2001; Pielke, 2001a, b, 2002; Pittock, 2002; Rosa and Ribeiro, 2001; Schönwiese, 2002; Wofsy *et al.*, 2001; Wuebbles, 2002). So, for example, in the interview with a journalist from *Scientific American* R. Lindzen (Professor of the Massachusetts Polytechnical Institute, USA) said that he started worrying about speculative discussions on climate change in the summer of 1988, when J. Hansen (Director of the Goddard

Institute for Space Studies, New York) stated that the global climate warming resulted from growing CO_2 concentration due to emissions to the atmosphere of products of fossil fuel burning. This statement made Lindzen clarify that climatology is at its initial stage of development and, in particular, there is no consensus on the causes of climate change. In the beginning of 2001 he reported on the problem of climate change at a meeting of the US Government. The fact that the global surface air temperature (SAT) has increased by about 0.5°C and the atmospheric CO_2 concentration by about 30% during the last 100 years is far from reflecting any cause and effect relation between increasing CO_2 concentrations and growing temperatures. Lindzen believes that the most reliable estimate of climatic sensitivity (SAT increase with a doubling CO_2 concentration) is 0.4°C, which means there is no basis for worrying about catastrophic changes in the future global climate.

According to the IPCC mandate, its task is to prepare an overview of "any climate changes in time, both natural and anthropogenic". Pielke (2001b, 2002) observed, however, that at least two climate-forming factors turned out to be either unreliable or not considered at all: (1) impact on global climate of anthropogenically induced changes in the characteristics of land surface; and (2) biological impacts of the growth of CO_2 concentration in the atmosphere (including the "fertilization effects"). If it is justified that both these factors exist, the conclusion suggests that an agreement of the results of the global climate numerical modelling (in the case of mean-global mean-annual SAT) is accidental.

To this end, Pielke (2002) discussed information that confirms the importance of these two climate-forming factors, and spoke about the possibility of checking the validity of such a conclusion. For this purpose the data can be used on the impact of anthropogenic changes in land surface characteristics on local, regional, and global climate. These changes illustrate the fact that this impact is as important as that of a doubled CO_2 concentration in the atmosphere (as well as the growth of other GHG concentrations). No less important is the fact that the atmosphere–surface interaction is characterized by various non-linear feedbacks, and therefore it may be impossible to predict climate changes for longer than one season.

As for possible biological impacts of CO_2 concentration growth, they are manifested through short-term (biophysical), middle-term (biogeochemical), and long-term (biogeographical) impacts of landscape-forming processes on weather and climate. The biophysical impact includes, for instance, the effect of transpiration on the relationship between latent and sensible heat fluxes as surface heat flux components. The biogeochemical impact incorporates the effect of vegetation growth ("fertilization effect") on evaporating leaf area, on surface albedo, and carbon storage. One of the manifestations of biogeographical impacts is a temporal change in the species composition of vegetation communities. Results of numerical modelling testify to the fact that without consideration of biophysical/biogeochemical impacts, the assessments of climate change cannot be considered reliable (Betts, 2000; Bounoua *et al.*, 2002; Bowling *et al.*, 1999).

Further development of climate models should consider, in particular, the following aspects of climate formation: (1) direct and indirect impacts of landscape dynamics through biophysical, biogeochemical, and biogeographical processes; (2) consideration of anthropogenic changes in land use on various (local, regional, and global) spatial–temporal scales (Houghton, 1999; Houghton *et al.*, 2000; Indermühle *et al.*, 1999; Jepma *et al.*, 2001); and (3) assessments of possibilities of climate prediction for a longer term than one season bearing in mind functioning of numerous non-linear feedbacks that determine the atmosphere–surface interaction (Hughes *et al.*, 1999; Joos *et al.*, 1999). With these and other problems unsolved, the importance of the IPCC-2001 Report and US National Report containing only assessments of global climate sensitivity to changes in some climate-forming factors, becomes rather unsignificant.

In the course of time-consuming preparations for the three published IPCC Reports (1990, 1996, 2001), analysis of climate-forming factors and various feedbacks has become more intense with each report (i.e., apart from GHGs, atmospheric aerosols have also been considered). However, the available complex numerical climate models still cannot be considered adequate from the viewpoint of taking into account every important climate-forming factor. A step forward in the IPCC-2001 Report was an account of forcings (F) on the climate by anthropogenic changes in land use, limited, however, by consideration of only the effect of land use dynamics starting from 1750, and on surface albedo. The estimates obtained gave an average F value of $-0.2\,\mathrm{W/m^2}$ with an uncertainty interval between 0 and $-0.4\,\mathrm{W/m^2}$. Not only are these estimates uncertain, there is also uncertainty in the sign of F (a more complete consideration of biophysical, biogeochemical, and biogeographical impacts on natural evolution and on climate suggests that in this case $F > 0$).

Great importance is given to the fact that the IPCC-2001 Report considers a new range of possible global warming values by the year 2100 (1.4–5.8°C) based only on numerical modelling data (which will inevitably change in the future). In addition, the problem is that the new range cannot be directly compared with similar estimates obtained earlier. From the viewpoint of reliability of estimates of future climate changes, of great importance is the use of the term "projections" instead of "predictions" in the IPCC-2001 Report, since the latter means that the factors left out will not be of importance in the future. It is the unacceptability of the latter supposition that prevents experts in numerical modelling from stating that climate prediction for the next 100 years is possible. This conclusion confirms the results of numerical modelling accomplished by Andronova and Schlesinger (2001).

The complicated problem of prognostic assessments of climate, and especially selecting the contribution of an anthropogenic component, is illustrated by the remaining controversy of analysis of the climatic impact of clouds. Tsushima and Manabe (2001) have analysed the effect of cloud feedback (CF) on the formation of the annual change of mean global SAT with the use of the *ERB* satellite measurement data, whilst bearing in mind an assessment of an adequate consideration of CF in numerical climate models by comparing the calculated and observed variations of mean global SAT. It follows from observations that the mean global annual change of SAT is in phase with the annual change of SAT in the northern hemisphere, and its amplitude is 3.3 K (this phase-coherence is determined by concentration in the northern hemisphere of the continents, contributing significantly to the formation of the amplitude of the SAT annual course).

Analysis of the observational data of the Earth radiation budget (ERB) components for the period from February 1985 to February 1990 has shown that the mean global estimates of both shortwave and longwave radiative forcings (SWRF and LWRF) depend weakly on the annual course of mean global SAT (the ERB data considered refers only to the range 60° N–60° S). Thus, the cloud cover dynamics neither intensifies nor reduces the SAT annual change. The considered data on SWRF and optical properties of clouds shows that not only albedo but cloud volume and cloud top height depend weakly on SAT and, hence, radiative forcing (RF) does not substantially affect the annual course of mean global SAT.

Based on this conclusion, we can say that the RF negligibly affects the annual course of mean global SAT. However, this speculative opinion is dangerous due to the rather complicated spatial field of SAT. Calculations with the use of three interactive climate models considering the dynamics of cloud microphysical properties, suggested the conclusion that the cloud top height albedo grows substantially with increasing SAT, which disagrees with observation data. This situation reflects the prospects for comparing the assessments of the role of CF from the data of numerical modelling and observations from the viewpoint of model validation.

Since J. Hansen was one of the principal initiators of the apocalyptic prediction of global warming, it is of interest to mark an evolution in his opinion in his recent work (Hansen and Sato, 2001), where it was stated that an increase of mean annual mean global SAT for the last 100 years by more than 0.5°C has been, at least partially, due to anthropogenic forcing on the climate. The authors calculated the radiative forcing (RF) due to observed growth of GHG concentration. The estimates testify to a considerable level of RF due to CH_4, whose contribution reached about half the RF due to CO_2: $1.4 \pm 0.2\,W/m^2$ including $0.5\,W/m^2$ due to direct and $0.28\,W/m^2$ due to indirect forcing. These results show the necessity for special attention to be given to CH_4 as a GHG and therefore its RF capabilities. New estimates of RF due to tropospheric ozone gave values of $0.7\,W/m^2$ instead of 0.3–$0.4\,W/m^2$, as supposed in the IPCC-2001 Report. Probably, an interval of 0.4–$0.8\,W/m^2$ can be assumed for this RF.

Considerable uncertainty (due to uncertain input data) in the calculations of RF due to "black" carbon (soot) gave $0.8 \pm 0.4\,W/m^2$. Importance should be given to the RF due to reflecting aerosols with values of $-1.4 \pm 0.5\,W/m^2$ with an additional contribution given by soil aerosols equal to $-0.1 \pm 0.2\,W/m^2$. Considerable contributions to total aerosol RF are made by organic aerosol ($-0.3\,W/m^2$ with an accuracy of a factor of 2) and ammonium sulphate ($-0.2\,W/m^2$). The sum of all positive components reaches $4.3 \pm 0.6\,W/m^2$, that is, it exceeds by three times the RF due to CO_2, whereas the sum of the negative components is $-2.7 \pm 0.9\,W/m^2$, therefore, the resulting RF is $1.6 \pm 1.1\,W/m^2$, which is close to that of CO_2 only.

Hansen and Sato (2001) performed an analysis of the trends of various GHG, from which it followed that the rate of CO_2 concentration increase grew rapidly during the period from the end of World War II to the mid-1970s, reaching a maximum of about 4% per year. It then decreased to a comparatively stable level of concentration growth 1.5 ppm/yr. During the same period the rate of CH_4 concentration increase grew rapidly (from 5 to 15 ppb/year), slowing down markedly after 1980 (for reasons unknown). The total trend of the rate of RF growth due to 13 chlorofluorocarbons (CFCs) was characterized by a rapid increase until the mid-1980s and a subsequent decrease due to reduced CFC emissions in the process of realization of measures stipulated by the Montreal Protocol. Apparently, before 2010 the RF due to CFCs should start decreasing. The rate of total RF growth (for a century) reached a maximum of about 5 W/m^2. At this rate of growth, the level of equivalent RF corresponding to a doubled CO_2 concentration should be reached by 2050. However, during the last two decades the rate of growth of total RF decreased to 3 W/m^2. With an account of possible reduction of GHG emissions, the global warming during the forthcoming several decades should constitute $+0.15 \pm 0.05$°C per decade, under the conditions that the level of CO_2 emissions will be "frozen". The remaining considerable uncertainty of the estimates considered is connected, first of all, with the difficulty of the aerosol-induced RF calculations.

1.1.2 Carbon cycle

Based on the data of the Third IPCC Report (IPCC, 2001a), let us consider the dynamics of the global carbon cycle and climate during the last century (some of the conclusions will be specified later on):

- Atmospheric CO_2 concentration in the pre-industrial epoch constituted 280 ± 10 ppm which during the past several thousand years had been gradually increasing (by 1999 it reached 367 ppm).
- The present CO_2 concentration exceeds that observed for the last 420,000 yr and possibly the last 14,000,000 yr. The rate of concentration growth is unprecedented, at least for the last 20,000 yr.
- The principal cause of the CO_2 concentration growth is fossil fuel combustion. Global CO_2 emissions into the atmosphere during 1980–1989 constituted 5.4 ± 0.3 Pg(petagram) C/yr, and during 1990–1999 reached 6.3 ± 0.4 Pg C/yr (the remaining part of the emissions dynamics is determined by the impact of land use and by CO_2 assimilation by the World Ocean).
- The annual growth of total atmospheric CO_2 content was 3.3 ± 0.1 Pg C/yr during 1980–1989 and decreased slightly to 3.2 ± 0.1 Pg C/yr during 1990–1999. The difference between the annual level of CO_2 emissions and its content in the atmosphere characterizes the scale of CO_2 assimilation by the World Ocean and land ecosystems. The dynamics of this assimilation determines the presence of substantial interannual variations in the growth of atmospheric CO_2 content from 1.0 Pg C/yr (1992) to 6.0 Pg C/yr (1998).

Between 1991 and 1997 about 6.2 Pg C/yr CO_2 were emitted to the atmosphere due to fossil fuel combustion. During this period the atmospheric CO_2 content increased only by 2.8 Pg C/yr. Using the observational data on atmospheric CO_2 concentration, Battle *et al.* (2000) have shown that during this period the biosphere of land and of the World Ocean assimilated, respectively, 1.4 ± 0.8 and 2.0 ± 0.6 Pg C/yr. A rapid assimilation of carbon by the biosphere during this period contrasts with conditions in the 1980s, when it was practically neutral. As for the World Ocean, it had been a powerful sink for carbon until 1995, but between 1995–1996 the assimilation of carbon decreased. Apparently, the storage of carbon on land is more variable than in the oceans.

The key role of the World Ocean in the formation of the carbon cycle is determined by photosynthesis of phytoplankton, and as a result, CO_2 transforms into organic carbon. This "biological pump" provides the means of removal of CO_2 from the atmosphere and facilitates carbon storage in the oceanic depths. Since, as local field experiments have shown, an addition of iron ("iron carbon") to sea waters stimulates the process of photosynthesis, an idea has been supposed to regulate the CO_2 concentration in the atmosphere through an intensification of the "biological pump" by "iron fertilization". Chisholm *et al.* (2001) have convincingly proved, however, that this approach is unacceptable. Real possibilities for atmospheric CO_2 concentration reduction are connected only with a reduction of CO_2 emissions into the atmosphere and with regulating the carbon stored in the land vegetation cover and in soil (the soil is known to contain about two thirds of the carbon reservoir on land) (Kondratyev, 2001b; Kondratyev and Isidorov, 2001; Kondratyev and Krapivin, 2001a,b, 2003a,b).

A characteristic feature of the ratio of concentrations of stable isotopes of light-weight elements such as hydrogen, carbon, and oxygen, is their weak variability in the substances of the Earth's crust. This variability is determined by differences in the chemical and physical "behaviour" of atoms of the same element, but with different masses, and can be an important source of information on geological, biological, climatological, and hydrological processes in the past, including an intensity of photosynthesis in sea waters, information about which is important for a number of reasons, but especially in the context of studies of the global carbon cycle (Hemming *et al.*, 1998). Bender (2000) performed an analysis of data on the concentration of oxygen isotopes in the euphotic zone of the oceans (near the Bermudas, where measurements were made, the depth reaches 100 m), and biological oxygen, with account of information about seawater mixing and gas exchange. This analysis gave estimates of photosynthetic productivity of seawaters. Analysis of measurement results has shown that in these conditions the photosynthetic productivity of the upper layer at about 10 m depth turns out to be sufficiently intense all over the year, decreasing with depth in the euphotic zone. As a rule, an intensified CO_2 emission into the atmosphere was observed during the years of El Niño events, although during a long El Niño event between 1991 and 1994 the rate of emissions was low. Apparently, an intensified rate of emissions can be explained by the impact of reduced CO_2 assimilation on land resulting from temperature rise, drought, and fires in the tropics. With this in mind, some conclusions can be drawn:

1. Since fossil fuel combustion is followed by consumption of atmospheric oxygen, this causes a slow reduction in O_2 content in the atmosphere, which turned out, however, to be slower than that determined by the consumption of O_2 in the

processes of burning alone – because part of the CO_2 emissions are assimilated by the land biosphere with the subsequent emission of oxygen into the atmosphere. With data on concentrations of CO_2, O_2, and $\delta^{13}C$ in the atmosphere, taken into account, the ocean–atmosphere CO_2 exchange turns out to be -1.9 ± 0.6 Pg C/yr (1980–1989) and -1.7 ± 0.5 Pg C/yr (1990–1999) (the minus sign reflects the loss of CO_2 by the atmosphere (i.e., CO_2 assimilation by the ocean). In the case of the land–atmosphere system, the respective fluxes constituted: -0.2 ± 0.7 Pg C/yr (1980–1989) and -1.4 ± 0.7 Pg C/yr (1990–1999). These estimates characterize the level of CO_2 assimilation by the land biosphere, which exceeded the loss due to changes in land use.

2. Estimates of contributions to the formation of the carbon cycle due to fossil fuel combustion, obtained from data on atmospheric CO_2 concentration, measurements, and on the ratio of $\delta^{13}C$ isotope concentrations, agree with results of observations of CO_2 and O_2 concentrations with rational suppositions with respect to global distributions of vegetation of the type C_2 and C_4, as well as the timescale of carbon cycle in the ocean and on land (Allison and Francey, 1998; Bergamaschi *et al.*, 2000a; Bilek *et al.*, 2001; Boutton *et al.*, 1998; Shimji *et al.*, 2000).
3. The CO_2 emissions to the atmosphere due to changes in land use in 1980 averaged, with various estimates taken into account, 1.7 Pg C/yr, with a possible variability of 0.6–2.5 Pg C/yr. These emissions were determined mainly by deforestation in the tropics (so far, there is no global data on land use changes in 1990).
4. The land–atmosphere CO_2 exchange is governed by the balance between CO_2 emissions due to changes in land use and CO_2 assimilation due to various factors ("residual sink on land"). The residual sink constituted -1.9 Pg C/yr (ranging from -3.8 to $+0.3$ Pg C/yr).
5. According to numerical modelling results the latitudinal distribution of CO_2 assimilation by the ocean is mainly determined by variations in natural CO_2 fluxes characterized by tropical degassing and assimilation in high latitudes. Estimates of the land–ocean CO_2 exchange north of 30° N between 1980 and 1989 ranged from -2.3 to -0.6 Pg C/yr, and in the tropics (30° N–30° S) from -0.1 to $+1.5$ Pg C/yr. Considering these estimates, the conclusion can be drawn that, it is necessary to distinguish between substantial "residual" sinks of carbon on land both in extratropical latitudes of the northern hemisphere and in the tropics. In both regions the spatial distributions of the CO_2 budget in 1980 and 1990 were similar, with a weak growing trend of CO_2 assimilation in 1990–1999 as compared with 1980–1989.
6. The longitudinal gradients of the atmospheric CO_2 concentration were much below the latitudinal ones. Attempts to estimate the CO_2 budget on a continental scale have given controversial results because of insufficient observational data.

The theoretical analysis of the dynamics of the functioning of the global carbon cycle has demonstrated that it is governed by complicated feedback mechanisms.

Below are enumerated the most important factors of the formation of the carbon cycle on time scales from decades to centuries.

1.1.2.1 Reaction of the carbon cycle to changing CO_2 concentrations

Formation of the global carbon cycle is determined by numerous feedback mechanisms and, in particular, feedbacks that govern the variability on a scale of decades and centuries. Here are some results:

1. An assimilation of anthropogenic CO_2 by the ocean is regulated by the general circulation of the ocean and its carbonate chemistry (Macdonald *et al.*, 2002a,b). In conditions of growing CO_2 concentration in the atmosphere, an assimilation of carbon by the ocean takes place due to the gradient of partial pressure of CO_2 between the atmosphere and the ocean. The efficiency of this assimilation decreases with growing CO_2 concentrations due to a lowering "buffer capacity" of the carbonate system. The efficiency of the oceanic carbon sink depends partly on the rate of the atmospheric CO_2 concentration growth, since an assimilation of CO_2 is limited to the rate of surface and deep water mixing.
2. An increase of atmospheric CO_2 concentration does not cause a substantial fertilization of marine bioproductivity but reduces pH. The 100 yr variations in marine bioproductivity due to intensified "calcification" with a low pH promotes an assimilation of CO_2 by the ocean intensified by several.
3. Due to bioproductivity processes, the transport of carbon takes place from surface to deep layers of the ocean, where its re-mineralization occurs. This process maintains the gradient of inorganic carbon concentration and preserves the atmospheric CO_2 concentration at a level 100–200 ppm below that in the absence of bioproductivity.
4. An assimilation of CO_2 on land is determined by net biome production (NBP), which is a balance between net primary production (NPP) and carbon loss due to heterotrophic breathing and processes of burning. According to observational data for many ecosystems, when CO_2 concentrations exceed the present level, an increase of NPP should take place (the degree of fertilization being rather variable), determined by reactions of photosynthesis and stomatal resistance to increasing CO_2 concentrations. With a high level of CO_2 concentration (800–1000 ppm) the effect of fertilization may be negligible.

The efficiency of carbon sinks on land depends on the transformation of carbon compounds into the forms characterized by long lifetimes (timber or modified soil organic matter). An intensified carbon sink due to increasing NPP can be manifested only as a result of inertia of these "slow" carbon reservoirs in the process of agricultural development of soils.

1.1.2.2 Feedbacks with climate dynamics that determine the carbon cycle

The climate warming reduces the CO_2 solubility and therefore weakens an assimilation of CO_2 by the ocean.

- Global warming can be followed by an intensified vertical stratification of the ocean, the probable consequences of which are: a reduction of CO_2 emissions due to upwelling and transport of excess carbon to deeper layers of the ocean as well as changes in bioproductivity.
- In the case of shorter timescales, the warming results in an increased rate of heterotrophic breathing on land, but the extent of the impact of this process on the long-term processes of CO_2 exchange remains unclear. Both the warming and the regional changes in spatial distributions of precipitation and clouds can lead to changes in the structures of land ecosystems, their geographical distribution, and primary production. The total impact of such forcings should depend on special features of the regional structures of climate change.

1.1.2.3 Impact of land use and other global changes on the carbon cycle

- Changes in land use and agriculture can substantially affect the carbon cycle on land. Apart from such large-scale measures as deforestation and afforestation, more local, but also substantial, events are possible. For example, fire extinction, which reduces both CO_2 emissions into the atmosphere due to biomass burning and the formation of inert (black) carbon, and also favours an invasion of forests (e.g., in savannahs). Supplies of soil carbon decrease with use of virgin soils, but can be restored in conditions of agricultural practice at a low level of soil cultivation.
- In some regions a deposition of anthropogenic nitrogen causes an increase of NPP on land, whereas the excess tropospheric ozone leads to an NPP reduction.
- Anthropogenic input of nutrients into the ocean through river sinks and deposition of atmospheric aerosols (especially nitrate as well as iron in the composition of atmospheric aerosols) can affect the marine bioproductivity, but qualitative estimates of this impact are still insufficiently reliable.

1.1.2.4 Theoretical assessments of the global carbon cycle components

The available developments in the field of the carbon cycle numerical modelling have given the following estimates:

- Between the period 1980–1989 the ocean–atmosphere carbon exchange varied between -1.9 and 2.9 Pg C/yr, which agrees with both the earlier model estimates and the data on the atmospheric carbon budget.
- Between 1980 and 1989 the land–atmosphere carbon exchange varied from -0.3 to -1.5 Pg C/yr, which agrees with the data on the atmospheric carbon budget. Contributions of fertilization due to CO_2 and depositions of anthropogenic N turned out to be substantial, reaching in total from -1.5 to -3.1 Pg C/yr. The effect of climate on the global carbon cycle formation in 1980 was weak and uncertain even in sign.
- With the IS92 scenario of anthropogenic CO_2 emissions to the atmosphere (IPCC, 2001a), by 2100 the carbon flux between the atmosphere and the

ocean should reach -5.5 ± 6.0 Pg C/yr. If the effect of possible climate changes is taken into account, this interval becomes 3.9 ± 6.0 Pg C/yr, that is, the assimilation of CO_2 by the ocean reduces.

- According to similar estimates for land, by 2050 the land–atmosphere carbon exchange should increase due to fertilization within -3.6 ± 8.7 Pg C/yr: an assimilation of CO_2 by land decreases when the effect of climate on this process is taken into account. It follows from some models that after 2050 an assimilation of carbon should rapidly decrease. Estimates for 2100 vary widely between +0.4 and −6.7 Pg C/yr.
- First experiments with the use of interactive global 3-D models "climate-carbon cycle" have led to the conclusion that a great uncertainty of prognostic estimates of CO_2 concentrations are connected mainly with an inadequate account of the impact of climate change on the processes taking place on the land surface.

The principal conclusion from these developments is that the formation of the level of CO_2 concentration in the 21st century will be mainly affected by CO_2 emissions into the atmosphere due to fossil fuel combustion. An important contribution is also made by changes in land use. According to available estimates, the global-scale deforestation will cause CO_2 emissions into the atmosphere to reach quantities 2–4 greater than the CO_2 amount assimilated as a result of afforestation. The World Ocean assimilates a great deal (up to 70–80% of emissions), but since it takes much time for the whole oceanic thickness to mix, the process of CO_2 assimilation will last for centuries. Lavorel *et al.* (2002) have shown that forest fires contribute significantly to the atmospheric global annual carbon budget (between 1.7–4.1 Pg C/yr).

After a detailed analysis of the present studies of the global carbon cycle Falkowski *et al.* (2000) emphasized that the following two, most important, questions demand special attention: Can we distinguish between contributions of anthropogenic and natural factors to the variability of biogeochemical cycles and climate? What is the sensitivity of the global climate to changes in atmospheric CO_2 concentrations? Answers to these questions can only be obtained within a systematic approach with consideration being given to the interactivity of the processes responsible for the formation of biogeochemical cycles and climate (Andronova and Schlesinger, 2001; Arneth *et al.*, 1998; Asner *et al.*, 1998; Sarmiento and Wofsy, 1999; Baldocchi *et al.*, 2001).

Large supplies of carbon (about 400–500 GtC) in the northern soils and in permafrost determine an urgency of analysis of the dynamics of the carbon budget in these conditions in the context of the problem of global climate changes (Zavarzin, 1999; Krapivin and Kondratyev, 2002; Romankevich and Vetrov, 2001; Brombach *et al.*, 2001; Cannel *et al.*, 1999). The process of carbon storage by the ecosystems of the Alaskan tundra that had functioned earlier as a sink for carbon (due to a low temperature and a high moisture content of soils, which reduced the rate of the organic matter decomposition), changed its direction as a result of climate warming and drying in the early 1980s, leading to a considerable loss of carbon.

Recent observations have shown, however, that the persistent climate warming and drying have reduced the carbon dioxide flux to the atmosphere, and in some cases the tundra functioned in summer as a sink for CO_2.

On this subject, Oechel *et al.* (2000) discussed the results of observations of the CO_2 exchange between the tundra ecosystems ("soil–plants") and the atmosphere, carried out between 1960 and late 1980 over two sites of the tundra. The results obtained show that the summertime tundra functioned as a sink for CO_2 in the periods of the warmest and driest climate during the last four decades and demonstrated the ability of the tundra ecosystems to adapt themselves metabolically to long-term (on a scale of decades and longer) climate changes.

Apparently, the mechanisms governing these processes are connected with the dynamics of the nutrient cycles, physiological acclimatization, and with the reorganization of populations and communities of vegetation cover. However, despite the observed acclimatization, the Arctic ecosystems turned out to be a source of CO_2 for the atmosphere, with an average annual amount of not less than 40 g $C/m^2/yr$ determined by wintertime emissions of CO_2 to the atmosphere. On this basis, further climate warming may intensify the CO_2 emissions to the atmosphere by the Arctic ecosystems.

The remaining deficit of observational data that would enable one to analyse the laws of formation of the carbon cycle has stimulated a number of field observations aimed at solving several related problems.

Large-scale efforts have been connected with an accomplishment within the Integrated Global Observing System (IGOS-P) programme of the partnership strategy in the field of global observations under the Terrestrial Carbon Observations (TCO) sub-programme of the use of various observational means and numerical modelling to monitor the carbon exchange between land ecosystems and the atmosphere or the oceans, and to estimate their contribution to the formation of the global carbon cycle (Braswell *et al.*, 1997). The related problem is that the lack of an adequate understanding of the laws of the global carbon cycle formation still does not allow one to "close" the annual budget of the global cycle, for which purpose reliable information is needed on the spatial–temporal variability of both the sources and sinks for carbon on scales of individual regions and ecosystems.

The data to be obtained within the framework of the TCO should give answers, in particular, to the following questions: what is the spatial–temporal variability of the carbon sources and sinks on land on continental and global scales, and what are the causes of this variability? In this regard it is important to take into account two types of processes determining the carbon cycle formation: (1) a relatively slow large-scale accumulation of carbon by vegetation cover and soils; and (2) relatively rapid emissions of carbon over small regions under the influence of various "disturbances". To obtain the observational data that would characterize the formation of the carbon cycle, various conventional and satellite observational means should be used, planned in the context of a global observational programme (Gildor and Tziperman, 2001). The programme includes the following sections:

Satellite observations

- Land surface and its changes, including seasonal variations and the effects of fires, with the use of various means of remote sensing in a broad wavelength interval from the visible to microwave.
- Biomass dynamics including the structure of vegetation cover.

In some cases (in particular, when it is necessary to calibrate the space-borne instruments for remote sensing) the satellite- and ground-based observations should be made simultaneously.

Observations on-site (together with satellite data)

- Carbon reservoirs (vegetation cover, soils, the atmosphere).
- The surface–atmosphere gas exchange (carbon fluxes), provided the needed spatial–temporal representativeness is available.
- Atmospheric chemical composition (including the data from aircraft observations).

This schematic programme does not contain any observational sections that should provide an accomplishment of other programmes. According to the adopted plans, by 2005 it should be demonstrated that use of the TCO data will allow an estimate of the surface–atmosphere carbon exchange on a subcontinental scale ($\sim 10^7\,km^2$) with an error of $\pm 30\%$. By 2008 the error should be reduced to $\pm 20\%$ for the global database.

Thematic developments in the context of studies of the global carbon cycle have been planned within the Global Carbon Project (GCP) (Young and Wefering, 2002; Crutzen and Ramanathan, 2000) including the following four principal directions:

1. Specifying the laws of the spatial–temporal variability of the present carbon cycle (Dolman *et al.*, 2002; Eastman *et al.*, 2001; Elderfield, 2002; Fan *et al.*, 1992; Fang *et al.*, 2001; Faure, 1990).
2. Analysis of anthropogenic impacts on the carbon cycle (Zavarzin, 1999; Borisenkov and Pichugin, 2001).
3. Simulation modelling of the interactive system "carbon cycle–climate–human activity" to reveal instabilities, threshold levels, and "surprises" (Danilov-Danilyan and Losev, 2000; Cox *et al.*, 2002; Craig *et al.*, 1998; Faluvegi *et al.*, 1999; Fan *et al.*, 1999).
4. Prognostic assessments of the evolution of carbon sources and sinks in the 21st century (De Lucia *et al.*, 1999; Elliott *et al.*, 2001).

Of course, the satellite-borne observations in the interests of TCO should be part of a broad programme of the global monitoring of the environment (Ervin *et al.*, 2002; Engelen *et al.*, 2001). In this regard Habib (2000) described the prospects for NASAs scientific instruments developed for the remote sensing of the Earth. In connection with limited financing, new developments are aimed at substantiation of small (and more economic) scientific complexes to solve specific problems instead of developing heavy satellites carrying huge multi-purpose observational complexes.

Many programmes for these developments are being completed or at a stage of preparation for launch. In the interests of the 2005–2010 perspective the strategy and specific satellite programmes are being worked out. These developments should improve the available methods for measuring the parameters (the content of ozone and minor gaseous component (MGC), temperature, humidity, precipitation, colour of the ocean, etc.), which are part of the programme of the existing observations, as well as a search for possibilities to retrieve the parameters inaccessible for remote sensing (ocean water salinity, atmospheric CO_2 content (carbon cycle), atmospheric properties at the tropopause, etc.) but can be accomplished with a breakthrough in the field of the instrument-making technologies. New technologies should provide answers to the following general questions: How is the Earth system changing? What are the principal causes of these changes? How does the Earth system react to natural and anthropogenic forcings? What are the possible consequences of changes in the Earth system for civilization development? Can we predict changes in the Earth system?

Key directions of prospective developments include the following problems (ACECHEM, 2001): (1) Seasonal and inter-annual climate changes; (2) global carbon cycle (including biology and biogeochemistry of ecosystems); (3) chemistry and dynamics of the atmosphere (including problems of atmospheric ozone); (4) global hydrological cycle; and (5) natural disasters.

Habib (2000) believes that it is possible to use remote sensing to retrieve soil moisture (microwave remote sensing), atmospheric CO_2 concentration (laser sensing), ocean water salinity (passive and active microwave sensing), precipitation rate (microwave sensing), oceanic mixing layer thickness (laser sensing), and the wind field in the troposphere (Doppler lidar sensing). Concrete objectives are: climate predictions (month to ten year scales), studies of changes in the biosphere and on the land surface, precise meteorological predictions (for up to two weeks), 12-hour advanced predictions of dangerous weather events, assessments of air and natural water quality, predictions of disasters, and assessments of anthropogenic forcings.

We shall mention some concrete developments of use in studying the carbon cycle.

Within the framework of the March–April 1999 Pacific Exploratory Mission in the tropics (PEM-Tropics B) as a part of the NASA Global Tropospheric Experiment (GTE) to study further transport of polluting minor gas components in the atmosphere of a remote (background) region of the Pacific (35° N–35° S), Staudt *et al.* (2001) performed aircraft measurements of the tropospheric chemical composition at altitudes up to 12 km. Data from these measurements have been analysed together with 3-D numerical modelling results (with the fields of meteorological parameters prescribed from the observational data) to study the laws of the polluting MGC transport from the northern hemisphere. They considered the results of this analysis for carbon monoxide (the CO lifetime is several months, which makes it possible to use CO as a tracer of the long-range transport).

Calculations of the 3-D field of CO concentration over the Pacific showed that in the northern hemisphere tropical Pacific the CO concentration in a free troposphere increased mainly due to biomass burning in south-eastern Asia, particularly intensive in the spring of 1999. But there was practically no CO increase in the atmospheric boundary layer, with the dominating pollution due to fossil fuel combustion in west Europe and in Asia. North of 35° N an input of pollutants from west Europe prevailed compared with long-range transport from Asia and was particularly great in spring due to CO accumulation at high latitudes in the winter. The CO sources located in North America played a secondary role in the formation of the CO concentration field in the atmosphere over the Pacific Ocean.

The transport of the Eurasian industrial pollutants from the region of large-scale circulation around the Pacific maximum to the zone of trade winds caused the formation of a "pollutants river" in the lower troposphere stretching from the north-eastern to western sectors of the Equatorial Pacific and reaching the south Pacific intertropical convergence zone (ITCZ). However, this process of long distance transport contributed little to the interhemispherical exchange.

A clear indicator of an intensive inter-hemispherical exchange through a narrow region of the west–eastern transport in the upper troposphere over the eastern sector of the equatorial Pacific ("western passage") is an increase of concentrations of CO and other MGC entering the upper troposphere over the south-eastern Pacific from the northern hemisphere. This transport turned out to be the most important means for global interhemispherical exchange during PEM-Tropics B, especially in the period of La Niña cooling.

Almost a quarter of moist tropical forests are located in Brazil, where the vegetation contains five times the amount of carbon than do the forests of Zaire, which are second to the Brazilian forests for carbon content. The Brazil–America basin covers a territory of 3,940,000 km^2, including 3,648,000 km^2 of forested and 292,000 km^2 of non-forested lands. Estimates of total surface biomass for dense forests are 151 ± 39 t/ha (Kabat and Nobre, 2000).

Carvalho *et al.* (2001) calculated the biomass consumption in the process of forest burning (to transform them into cultivated lands) in six test sites. The results obtained refer to Canabi Farm (near Alto Floresta) and the Brazilian state Mato Grosso, where five further sites were chosen (A, B, C, D, E) each with an area of 1 ha, where the forests were cut and burnt at different times. Site A bordered on the pasture (its three sides being fenced off by the forest). The forest was cut and burnt in 1997. The other four sites were located within the forests. On site B the forest was cut and burnt in 1997, and on site C (located on a deforested territory of 9 ha) in 1998. Site D was located within a deforested area of 4 ha, where the forest was cut in 1998 and burnt in 1999. The relative share of biomass in the process of the cut forest burning constituted 22.7%, 19.5%, 47.5%, 61.5%, and 41.8%, respectively. For sites A, B, C, D, and E the surface biomass and the amount of carbon for each site were, respectively, 496 t/ha and 138 t/ha. Taking this data into account, the amount of carbon emitted to the atmosphere in the form of various gases reached 69 t/ha and the amount of CO_2 and CO emitted during the process of burning was 228 t/ha and 15.9 t/ha, respectively.

From the data for Kenya and Brazil, estimates were obtained for emissions to the atmosphere of GHG and other MGC resulting from charcoal production widely used as a fuel not only in Kenya and Brazil but also in many other countries of Africa and the south-eastern Asia. The emission coefficients (EC), defined as an amount of pollutants in grams per kilogram of produced charcoal, were estimated for carbon dioxide, methane, nitrogen monoxide, carbon oxide, non-methane hydrocarbons (NMHC), nitrogen oxides (NO_x), and total aerosol (TPS). Depending on the type of furnaces used in charcoal production the share of wood carbon transformed to charcoal carbon varied from 37% to 69%. With the data for eight types of furnaces the EC varied (in g/kg) within 543–3027 (CO_2), 32–26 (CH_4), 143–373 (CO), 24–124 (NMHC), 0.011–0.30 (N_2O), 0.0054–0.13 (NO_x), and 13–14 (TPS). The typical relative distribution of wood carbon among various components was characterized by the following values: 51% (charcoal), 27% (CO_2), and 13% (products of incomplete combustion, PIC). Since the global warming potential (GWP) estimates for PIC exceed substantially the GWP estimates for CO_2 (calculated per carbon atom), in the process of charcoal production the atmosphere obtains a considerable amount of GHG. Calculations based on the use of the published GWP data for CO_2, CH_4, and N_2O showed that the atmosphere obtains from 0.77 to 1.63 kg C-CO_2 (carbon per equivalent amount of CO_2) per one kilogram of produced charcoal. The total contribution to the formation of GWP due to charcoal production in Kenya and Brazil reaches, respectively, about 2.7 and 7.5 Mt C-CO_2. These values are negligible, however, compared with, for instance, the GWP level (almost 1700 Mt C-CO_2) due to fossil fuel combustion in the USA.

According to the results of some recent studies, to balance the global sources and sinks for carbon dioxide, it is necessary to assume the presence of a CO_2 sink in the northern hemisphere. A considerable part of this sink should be connected with the biosphere–atmosphere interaction in the mid-latitudes (Barford *et al.*, 2001). To check grounds for this supposition, Haszpra *et al.* (2001) carried out regular measurements of CO_2 concentrations at various latitudes using instruments mounted at the heights 10, 48, 82, and 115 m on the radio-TV tower in western Hungary (46°57′N, 16°39′E). The land use around the tower was mainly agricultural fields and small forests with very small (100–400 people) settlements, without industrial enterprises.

Measurements of the vertical profiles of the CO_2 mixing ratio, temperature, humidity, and wind were started in September 1994, and direct (correlative) measurements of the CO_2 flux (at a height of 82 m) in 1997. Processing of weekly air samples for the content of CO_2, CH_4, CO, H_2N_2O, and ^{13}C and ^{18}O isotopes was carried out in two American laboratories. Analysis of the measurement results revealed that in the boundary layer of the Karpaty basin occasional accumulations of CO_2 occurred in winter and a diurnal course in the vertical profile of the CO_2 mixing ratio occurred in summer.

Sorokhtin and Ushakov (1999) substantiated an approximate technique to calculate the CO_2 flux from the data of gradient measurements that can be used to fill in the gaps in the results of the direct measurements.

An important source of information on the carbon dioxide fluxes (and MGC in general) between the surface (especially in the presence of forests) and the atmosphere are widely practiced observations with tower-mounted instruments. In these cases, however, of great importance is the assessment of the representativeness of these observations, which can be obtained, based on the conception of "imprints" (photoprint-F). The F is defined as a conditional probability $f(r - r_{\text{tower}})$ determining the contribution to the observed short-term changes in CO_2 concentrations at the site of the tower r, with CO_2 flux equal to $\varphi(r)$.

Variations are called short-term when they occur independent of the annual course or inter-annual trends. They are connected, as a rule, either with the synoptic-scale processes or with near-surface CO_2 flux changes. In this context the observed changes in the mixing ratio are determined by co-evolution of the imprints near the surface:

$$\chi(r_{\text{tower}}) = \int f(r - r_{\text{tower}})\varphi(r)\,dx\,dy + \chi_{\text{trend}}$$

where x and y are the horizontal coordinates. Using the processed data of measurements made at the TV tower in Wisconsin (USA), Gloor *et al.* (2001) showed that changes in the near-tower CO_2 mixing ratio used to calculate the CO_2 flux are affected by the input of air particles with a typical time of transport of up to 1.5 day. With this typical time and special trajectory of transport taken into account, the area of an "imprint" may be assumed in this case to reach 10^5 km^2, or one tenth of the USA territory. This areal representativeness determines the high information content of the data of the tower-based observations used to estimate the CO_2 fluxes over large territories. To obtain reliable information for the whole USA territory, data of observations on about ten towers is sufficient.

One of the principal factors of the formation of global carbon cycle is an atmosphere–ocean CO_2 exchange. Averaged over the time interval t, the CO_2 flux is defined as:

$$F_t = (ks(\Delta \mathrm{pCO_2}))$$

Where k is the exchange coefficient, s is the rate of CO_2 solution in the sea water, ΔpCO_2 is the difference of the CO_2 partial pressures between the sea water and the atmosphere. The coefficient k depends on wind speed, intensity of air bubble formation, air surface layer stability, and wind shear. The short-term variability of the CO_2 exchange remains almost unknown so far, although Bates and Mervilat (2001) showed that this variability is very important, however, being determined by two circumstances: (1) non-linear dependence of k on wind speed; and (2) importance of consideration of the averaged short-term wind speed variability. The respective quantitative estimates have been obtained from the data of frequented observations of wind speed and pCO_2 with the use of an autonomous sea buoy CARIOCA. A comparison of the results of hourly and daily observations showed that differences in CO_2 fluxes with these observations reach a factor of 3. Such differences are determined by a non-linear relationship between wind speed and the exchange coefficient k. This result is of great importance for correct calculations of the averaged regional and global estimates of the atmosphere–ocean CO_2 exchange.

One of the principal problems of the present dynamics and forecast of climate system evolution is a pre-calculation of the radiative forcing determined by various factors, including changes in GHG concentrations, which require, in particular, a reliable account of their exchange between the atmosphere and the surface. Possible solutions for this problem include a retrieval of spatially averaged GHG emissions to the atmosphere from the data from local observations. This approach is usually called a synthesizing reconstruction (SR) and consists of calculations of fluxes as a linear combination of local values with their spatial structure taken into account.

Kaminski *et al.* (2001) discussed the solution of such a problem for carbon dioxide to assess the effect of the spatial resolution of fluxes on the results of calculations of their spatially averaged values (in SR, the Earth's surface is divided into several vast regions for which averaged CO_2 fluxes are calculated). Calculations showed that the results of aggregation of local flux values with their averaging for large regions depend substantially on the level of spatial resolution. The related errors may be the same as the calculated fluxes. An algorithm has been suggested that would reduce such errors by a factor of 2. The remaining considerable uncertainty in the assessments of the global carbon cycle dynamics necessitates further intensive studies. Some topical ideas have been expressed by Ciais (1999a,b, 2002).

1.1.3 Carbon cycle and climate

We have already mentioned an interactivity of the carbon cycle and climate variability. Therefore serious efforts have recently been undertaken to carry out numerical interactive modelling of the global carbon cycle and climate change. Since this problem is very difficult to solve, it is necessary to search for and substantiate adequate schemes of parameterization of the processes that determine the dynamics of the carbon cycle and climate, firstly, within the framework of reproducing the processes in the atmosphere–ocean system.

In the earlier widely used interactive 1-D models of the atmosphere–ocean system one of the most important parameters is the coefficient of the vertical diffusion k_v in the ocean, characterizing the intensity of the heat- and mass-exchange between the ocean and the atmosphere. In 3-D models as in 1-D models, this parameter is prescribed a priori using an "adjustment" by comparing the results of numerical modelling with the observational data. However, during this adjustment there were substantial differences in the k_v values. In this regard, Harvey and Huang (2001) and Harvey (2001) developed a relatively simple interactive model of the atmosphere–ocean carbon cycle system, which can be used to numerically model the processes of assimilation of heat and CO_2 by the ocean and to assess changes in the level of the oceans due to thermal expansion.

The development of the model was mainly motivated by the desire to study the atmosphere–ocean interaction in the context of the formation of the carbon cycle and to further apply the model to analyse the climatic consequences with various scenarios of anthropogenic emissions of GHGs and aerosols. On the other hand, an important problem of the use of the model consists in retrieving the natural climate change and carbon cycle on geological timescales.

The discussed model is considered diagnostic but not prognostic, with emphasis on the global-scale processes. The latter determines a selection of three global regions (two polar regions and one non-polar) with special vertical resolution. The model makes it possible to reproduce the vertical distributions of oceanic parameters, such as concentration of dissolved inorganic carbon, phosphates, dissolved oxygen, alkalinity, temperature, and to describe the functioning of the "biological pump" that causes the formation of organic matter, calcite, and argonite. The formation of bottom waters takes place in one of the polar regions due to the interaction with the atmosphere and convective mixing in the ocean with subsequent transport of waters to the lower part of the non-polar region. The bottom waters form in the region of polar downwelling then suffer upwelling in the non-polar region, though parts of them get from intermediate depths to the second polar region.

The work discusses in detail specific features of the climate in the conditions of the considered interactive model, and in this connection a definition of the relative efficient coefficient of diffusion k_v has been substantiated for various tracers. Calculations showed that the k_v values are at a minimum for temperature, at an intermediate level for carbon, and at a maximum level for dissolved oxygen. A much lower value of k_v was obtained in the upper layer of the ocean compared to that used in earlier 1-D models ($\sim$0.2 cm^2/s instead of 0.6–1.0 cm^2/s) as well as a lower maximum velocity of upwelling (2 m/yr instead of 4 m/yr). Consideration of an apparent convective mixing resulted in a considerable effect of changes in the thermihaline circulation intensity on the time-dependent reaction of surface temperature and the rising level of the ocean. For this reason the reaction of surface temperature and ocean level to decreasing intensity of thermohaline circulation turned out to be quite different (faster) compared to that obtained earlier using a classical 1-D model. It means, in particular, that climatic sensitivity to external disturbing forcings, agreeing with the observational data for the last 140 years, is about 10% lower than in the case of the classical model.

A comparison of the results of the numerical modelling of the vertical profiles of various tracers (including the ratios of isotope concentration) with the use of an interactive 1-D model of the atmosphere–ocean carbon cycle system and observational data revealed a necessity to reproduce the following characteristics: (1) the rate of upwelling in a non-polar region reaching a maximum of about 2 m/yr at a depth of 1 km with constantly decreasing rate below and above this level of depth; (2) different values of the coefficient of the vertical diffusion for temperature and various tracers in the upper 0.5 km layer of the ocean; and (3) the ratio of the rates of formation of carbonate carbon and organic carbon equal to only 0.9. From the data of observations (together with information on mixing along isopycnic surfaces and the results of numerical modelling) it follows, in particular, that the effective coefficient of the vertical mixing in the upper layer of the ocean should be at a minimum for temperature and at a maximum for dissolved oxygen, with the values of alkalinity and concentration of phosphate somewhat lower than for dissolved inorganic carbon (DIC). The model parameters "adjusted" by the data of pre-industrial observations of the tracers distributions provided the best agreement with the results of observations of changes in the isotopes concentration.

The interaction between alkalinity (not earlier considered, as a rule, in such models) and DIC smoothes down the impact of changes in the model parameters on the stationary level of the pCO_2 in the atmosphere. Calculations showed that calculated estimates of the anthropogenic CO_2 assimilation by the ocean and estimates of the evolution of the CO_2 content in the atmosphere for the period up to 2200 are insensitive to the mixing parameters prescribed for the model (with these parameters being constant during this period). Estimates of sensitivity of the calculated values of pCO_2 to changes in the temperature of the warm surface of the ocean agree well with similar results obtained with the use of a 3-D model of the carbon cycle in the ocean.

Numerical modelling of the interactive dynamics of the atmosphere–ocean carbon cycle system is of special interest in the context of studies of the laws of palaeoclimate formation (Barker *et al.*, 2001; Beck *et al.*, 2001; Kukla, 2000). The data for a relatively warm climate on the ancient Earth have always been an enigma, since the extra-atmospheric insolation was then much lower than at present. According to geological data (which cannot be considered reliable enough), at the initial stage of its evolution the Earth's surface had been warmer than at present. An explanation of this paradox consists of supposition of a very high CO_2 concentration in the atmosphere at that time and, respectively, an exclusively intensive greenhouse effect by the atmosphere. Sleep and Zahnle (2001) have analysed a possible global CO_2 cycle on the ancient Earth and justified this supposition of a high CO_2 concentration.

In this regard, estimates have been obtained for a gigantic content of CO_2 in the Earth's crust. According to approximate estimates, about $6 \cdot 10^{21}$ mol CO_2 are contained in the Earth's crust and a comparable or greater amount in its mantle. The present-day estimation of CO_2 degassing by the mantle gives values of about $2.5 \cdot 10^{12}$ mol/yr, from which it follows that during a period of about 2.4 Gt/yr the content of carbonates in the surface layer of the Earth's crust could have doubled. In conditions of the early Earth, in the presence of geothermal sources which had been 2–3 times more powerful than today, the formation of the ocean's crust and its subduction could have taken place 4–9 times faster than now. Respectively, the degassing had been more intensive, and the time for doubling of the carbonates content in the Earth's crust could have decreased to 300–700 Mt/yr.

Apparently, when considering the long-term processes one should not leave out the effect of the mantle in the formation of the CO_2 cycle. The related possible processes include: subduction of carbonates directly deposited onto the sea bottom (pelagial carbonates); and subduction of continental carbonates and subduction of carbonates resulting from changes in the oceanic basalt (the latter process depends on the concentration of free CO_2 in the ocean and therefore can function as a buffer regulating the level of CO_2 concentration in the atmosphere). Sleep and Zahnle (2001) analysed the data on CO_2 content in the mantle and on CO_2 exchange between various reservoirs and the mantle within the conception of tectonic plates.

In this regard, the present global CO_2 cycle has been considered first, since even now the CO_2 content in the ocean and in the atmosphere depends on the processes connected with the mantle. There appears to be a feedback determined by the dependence of the level of carbonization of the ocean crust on the partial pressure of pCO_2.

At an early stage of the Earth's evolution powerful processes could have dominated in the mantle, and an intensive greenhouse effect due to CO_2 would have been impossible (it is the product of other GHG). Sleep and Zahnle (2001) suggested a model of the CO_2 cycle in which the ocean crust functions as a carbonate trap subject to subduction, which closes the mantle-connected CO_2 cycle. The factors that confine the subduction of carbonates could promote a concentration of CO_2 in the near-surface reservoirs. The weakest point of the model is, apparently, that although the capacity of the oceanic basalts or CO_2 reservoirs is high ($\sim 3 \cdot 10^{21}$ mol), it is only a part of the global CO_2 reservoir ($25 \cdot 10^{21}$ mol). Therefore it is possible that at an early stage of the Earth's evolution the effect of the buffer determined by carbonization of the ocean crust could have been dominated by the contribution of degassing of the Earth's crust subducting to the surface on the island arcs and, according to the model considered, it could have taken place, if the share of subducted carbonate that reached the mantle was $C_{deep} < 20\%$ (the present $C_{deep} = 64$–84%). In the case of $C_{deep} < 0.2\%$, a massive CO_2 atmosphere could have formed, and then the surface temperature could have risen to 200°C, making the Earth uninhabitable. The Earth could not become inhabited until C_{deep} became so great that CO_2 fluxes from the mantle to the surface became balanced by the subduction flux to deep layers.

Goddéris *et al.* (2001) studied the carbon cycle in the period of the Early Palaeozoic era using data on the isotopic composition of sea water, which is one of the most reliable sources of information about the Earth's evolution on the geological timescale. Diverse information about the evolution and interaction of lithosphere, biosphere, atmosphere, and hydrosphere is contained in the data reflecting changes in tectonics ($^{87}Sr/^{85}Sr$, $^{187}Os/^{186}Os$, and $^{44}Ca/^{40}Ca$) and (or) the redox state ($\delta^{34}S$, $\delta^{13}C$) and physical properties of the ocean ($\delta^{18}O$). The earlier review of data on the trends of concentrations of isotopes O, C, Sr, and S during the whole Phanerozoic period showed that the present values of isotopic "signals" are similar to those observed in the late Proterozoic period. The Neoproterozoic period was characterized by $\delta^{18}O \sim 26‰$, after which (during the transition from Proterozoic to Cambrian period) there was a decrease to $\sim 22.5‰$ with a subsequent, approximately linear, increase to present values of about 30‰. In a first approximation, oscillations with respect to this rising trend correlate strongly with the Earth's evolution, but the interpretation of this rising itself raises doubts.

A comparison with other isotope data suggested that changes in $\delta^{18}O$ are controlled by tectonic processes, especially by a relative intensity of relationships between the hydrosphere and oceanic and continental silicate formations. Therefore large-scale tectonic events could have broken the isotopic balance in the oceans. This scenario has been suggested, for example, as the cause of the high concentration ratio $^{87}Sr/^{86}Sr$ about 65 Myr ago, which was ascribed to the effect of the lifting and erosion of the Himalayan and Tibetan plateaus, determining the modulation of the global carbon cycle and climate (Kump *et al.*, 2000). During the Neoproterozoic, similar processes had taken place in Africa, and similarity of the present and Neoproterozoic isotopic characteristics of sea waters reflected a similarity of tectonic processes responsible for their formation.

In this regard Goddéris *et al.* (2001) discussed the results of the numerical modelling of interactive cycles C–alkalinity–S–Sr, from which it follows that during the early Palaeozoic period (from early Cambrian to the late Devonian period) the evolution of the isotopic composition of sea waters (Sr, O, C, and S) could have been determined by gradual oxidation of large reservoirs of the restored carbon released during orogenic processes in Africa. Based on a supposition that any changes in the isotopic composition of oxygen in sea water results from breaking an equilibrium of the organic carbon cycle due to the interaction of the cycles of the oxygen and carbon isotopes, changes of $\delta^{18}O$ in the brachiopods' shells are assumed to be responsible for the disturbing forcings. Calculated values of isotopic ratios of $\delta^{18}C$, $^{87}Sr/^{86}Sr$, and $\delta^{34}S$ agree well with the observational data, like the predicted evolution of pCO_2 in the atmosphere and its connection with the global climate. An important feature of the suggested model is that it admits a possibility of δ^{18} fluctuations in seawater within several ppm in the Palaeozoic period, which also agrees with observational results. Apparently, the suggested scenario is useless in conditions of the post-Devonian evolution of the planet, when the organic carbon cycle had been strongly affected by the land biosphere.

To solve the problem of the interactive palaeodynamics of the carbon cycle and climate, of great importance are estimates of CO_2 concentrations in the ancient atmosphere. Using a reverse relationship between indices of leaves' particles and CO_2 partial pressure for present plants *Ginkgo biloba* and *Metasequoia glyptostroboides*, Royer *et al.* (2001) retrieved the CO_2 concentration in the atmosphere from middle Palaeocene to early Eocene and during the period of middle Miocene from the data for the fossil remains of the same plants. The results obtained showed that, as a rule, the CO_2 concentration had not passed the limits of 300–400 ppm (by volume); that is, it had been close to the present concentration (with the exception of some high concentrations near the border of the Palaeocene (Eocene)). Thus it follows from these data that an increasing CO_2 concentration could not have led to an intensification of the atmospheric greenhouse effect that had compensated for the low level of the extra-atmospheric insolation.

1.1.4 Conclusion

The principal conclusion to be drawn from this information is that studies of such a complicated problem as the global carbon dynamics are still at an early stage of development, and the main difficulty is the lack of adequate observational data that characterizes the formation of the carbon cycle. One possible way to remove these contradictions and to overcome the available information uncertainties is to adopt the adaptive evolutionary technology described by Kondratyev *et al.* (2002a,b). This technology consists of complex usage of satellite and surface observations of the environment together with the global model of the nature–society system. It is necessary to improve the available versions of the global model, providing an agreement of the model with accumulated databases. A regular assessment of the differences between the results of modelling and the observational data will enable us to determine an efficient regime of the global geoinformation monitoring and to reliably assess the role of various regions and water basins in the formation of CO_2 fluxes in the "atmosphere–vegetation–soil–ocean" system.

1.2 PRESENT-DAY TRENDS OF CIVILIZATION DEVELOPMENT

1.2.1 Introduction

The globalization of man's impact on the environment with continuing population growth and the great uncertainty of the prospects of civilization development have put forward numerous principal problems of economic, political, and social character. One of them is an assessment and prediction of the state of the global carbon cycle, emphasizing the role of anthropogenic processes. Clearly, this problem cannot be selected and discussed separately from other problems inherent with the society–nature interaction. Nevertheless, to substantiate a set of relevant problems, it is necessary from the broad spectrum of natural–anthropogenic processes to select the key aspects, knowledge of which will enable one to answer the principal questions of sustainable development of the global system of nature–society. Although the notion of "sustainable development" has not been clearly defined, the term "sustainable development" is widely used in scientific literature. Various discussions on the origin and meaning of this term have long taken place among specialists in various fields. They include the widely known decisions of international conferences in Stockholm (1972), in Rio de Janeiro (1992), and in Johannesburg (2002), as well as an accomplishment of several international programmes, such as the International Geosphere–Biosphere Programme (IGBP), International Human Dimensions Programme of Global Changes (IHDP), the World Climate Research Programme (WCRP), and International Biodiversity Programme, DIVERSITAS (Braswell *et al.*, 1996; Kondratyev *et al.*, 2003a).

The principal conclusion of the July 2001 Amsterdam Conference "Challenges to Changing Earth" is that "the Earth System behaves as a single, self-regulating system of physical, chemical, biological, and human components. The interactions and feedbacks between the component parts are complex and exhibit multi-scale temporal and spatial variability." In reality, diverse important impacts of human activities on a global scale can be traced both on land and in the World Ocean. In some respects, the anthropogenic changes in the environment have become comparable or even exceeded the limits of the natural environmental changes. "Human-driven changes cause multiple effects that cascade through the Earth System in complex ways. These effects interact with each other and with local- and regional-scale changes in multidimensional patterns that are difficult to understand and even more difficult to predict":

> "In terms of some key environmental parameters, the Earth System has moved well outside the range of natural variability exhibited over the last half million years at least. The nature of changes now occurring simultaneously in the Earth System, their magnitudes and rates of change are unprecedented. The Earth is currently operating in a non-analogue state".

These fragments from the decisions of the Amsterdam Conference suggest that "an ethical framework for global stewardship and strategies for Earth System management are urgently needed. The accelerating human transformation of the Earth's environment is not sustainable." This problem can be solved by developing new scientific directions in global ecology (e.g., ecoinformatics) (Arskiy *et al.*, 1992; Krapivin and Kondratyev, 2002; Krapivin and Potapov, 2002). No doubt, this scientific direction should comprise a wide spectrum of natural and social sciences, creating an efficient mechanism for using available methods and technologies and ensuring their goal-directed development. A certain basis for this unification has already been created (Gorshkov, 1990; Kondratyev, 1990, 1998a; Gorshkov *et al.*, 2000; Yanovsky, 1999; Gerstengarbe, 2002; Beerling and Woodward, 1996; Bergamaschi *et al.*, 2000a, b; Berner and Hollerbach, 2001).

Following the work of Steffen and Tyson (2001), we shall analyse some formulations aimed at defining the Earth System as a complex system of interacting physical, chemical, biological, and human activity components, which determine the processes of transportation and transformation of substances and energy, and thus keep the planet habitable:

> "The Earth is a system that life itself helps to control. Biological processes interact strongly with physical and chemical processes to create the planetary environment, but biology plays a much stronger role than previously thought in keeping Earth's environment within habitable limits".

It should be mentioned that this statement has undoubtedly a fundamental meaning, but in this context it would be rather appropriate to refer to the conception of the biotically regulated environment extensively elaborated by Gorshkov (1990) and Gorshkov *et al.* (2000):

> "Global change is much more than climate change. It is real, it is happening now and it is accelerating. Human activities are significantly influencing the functioning of the Earth System in many ways; anthropogenic changes are clearly identifiable beyond natural variability and are equal to some of the great forces of nature in their extent and impact".

We have to emphasize that in the contemporary world, when the scientific discussion on the extent of anthropogenic pressure on climate has got transformed into a political discussion and reached the level of heads of state, it should be pointed out that global climate change deserves much keener attention in the context of the Kyoto Protocol. This has been done in a number of publications by Russian scientists (Kondratyev and Demirchian, 2001a; Kondratyev *et al.*, 2001a, b; Kondratyev, 2002). However we agree with Steffen and Tyson (2001) in the sense that climate is not the first priority in the totality of global change problems.

Over the past few generations humankind is in the process of exhausting fossil fuel reserves that were generated over several hundred million years. Nearly 50% of the land surface has been transformed by direct human action, with significant consequences for biodiversity, nutrient cycling, soil structure and biology, and climate. More nitrogen is now fixed synthetically and applied as fertilizers in agriculture than is fixed naturally in all terrestrial ecosystems. More than half of all accessible freshwater is used directly or indirectly by humankind, and underground water resources are being depleted rapidly in many areas. The concentrations of several climatically important greenhouse gases, in addition to CO_2 and CH_4, have substantially increased in the atmosphere. Coastal and marine habitats are being dramatically altered; 50% of mangroves have been removed and wetlands have almost halved. About 22% of recognized marine fisheries are over-exploited or already depleted, and 44% more are at their limit of exploitation. Both on land and in the World Ocean intensive processes of biodiversity reduction are taking place:

> "The human enterprise drives multiple interacting effects that cascade through the Earth System in complex ways. Global change cannot be understood in terms of a simple cause and effect paradigm. Cascading effects of human activities interact with each other and with local- and regional-scale changes in multidimensional ways".

> "The Earth's dynamics are characterized by critical thresholds and abrupt changes with catastrophic consequences for the Earth System. Indeed, it appears that such a change was narrowly avoided in the case of depletion of the stratospheric ozone layer. The Earth System has operated in different quasi-stable states, with abrupt changes occurring between them over the last half million years. Human activities clearly have the potential to switch the Earth System to alternative modes of operation that may prove irreversible".

Although the last two quotations are very important, we must keep in mind that until now there are only preliminary quantitative assessments and fragmentary data, there is not enough information for concrete judgement about the potential changes. This is true, for instance, of the case of the impact on the global ozone content in the stratosphere (Kondratyev and Varotsos, 2000). In this regard, the problem of the anthropogenic impact on the global biogeochemical cycles of carbon, sulphur, nitrogen, and other chemical compounds deserves serious attention (Kondratyev, 2000a; Krapivin and Kondratyev, 2002; Kondratyev *et al.*, 2002a; Ledley *et al.*, 1999; Zavarzin and Kolotilova, 2001; Kashapov, 2002; Irion, 2001):

> "In terms of key environmental parameters, the Earth System has recently moved well outside the range of the natural variability exhibited over at least

> the last half million years. The nature of changes now occurring simultaneously in the Earth System, their magnitudes and rates of change are unprecedented".

Such general assessments are important for developing an efficient technology to assess the global environmental changes (Zerchaninova and Potapov, 2001; Kontorovich *et al.*, 1999). In the context of perspectives of the contemporary civilization development a key factor is undoubtedly the continuing global population increase: during the second half of the 20th century the population doubled, the grain crops production tripled, the energy consumption increased fourfold, and economic activity increased fivefold. Gorshkov (1990) has clearly demonstrated that *homo sapiens* has long left his ecological niche, bringing forth the first signs of a developing ecological catastrophe.

Taking into account these conceptual circumstances, Steffen and Tyson (2001) have formulated two important conclusions, that society should respond to the emerging ecological threat and develop respective scientific ideas aimed at an adequate understanding of the processes that govern the variability of the Earth System:

> "Ethics of global stewardship and strategies for Earth System management are urgently needed. The inadvertent anthropogenic transformation of the planetary environment is, in effect, already a form of management, or rather mismanagement. It is not sustainable. Therefore the business as usual way of dealing with the Earth has to be replaced as soon as possible by planned strategies of good management".

> "A novel system of global environmental science is emerging. The largely independent efforts of various international research programmes and numerous national projects create the basis for an Earth System science that is capable of tackling the cognitive tasks suggested by the research findings above. This new science will employ innovative integration methodologies, organize itself into a global system with transnational infrastructures, and embark on a continuing dialogue with stakeholders around the world".

Analysis of the published suggestions, opinions, and decisions of international non-governmental organizations concerning the strategy of studies and development of environmental sciences in conditions of increasing anthropogenic forcings suggests the conclusion that there is no complete and objective list of priorities for solving the problems of global change. It means that humankind has no scientifically grounded strategy of interaction with the environment.

1.2.2 Contemporary global ecodynamics

The accumulated data on global environmental changes enable us to formulate a general pattern of these changes and characterize possible trends for the decades to come (Brown *et al.*, 2001; Losev, 2001). Here are some facts:

- The area of natural land ecosystems shrinks at a rate of 0.5–1.0%/yr. By the beginning of the 1990s about 37% of these ecosystems remained. This trend will lead to total abolition of natural land ecosystems.
- From the estimates for 1985, the rate of human consumption of pure primary biological production constituted 40% on land and 25% in the whole biosphere. In the future these estimates will reach 80–85% and 50–65%, respectively.
- Between 1972 and 1995 the GHG concentrations grew from a rate of tenths of % to several % annually. Concentrations of CO_2 and CH_4 are expected to grow at the expense of an increasing rate of biota destruction.
- An annual 1–2% depletion of the ozone layer, the growth of the ozone hole in the Antarctic, and growth in the areas of other ozone holes. This trend will remain level with decreasing emissions of chlorofluorocarbons (CFCs) (Varotsos, 2002b; Varotsos, 2003a,b; Varotsos and Kondratyev, 2003).
- Shrinking forest areas, especially tropical forests. During the period 1990–1995 the forest areas shrunk at a rate of 13 Mha/yr, and the ratio "afforestation/forest cutting" was 1:10. By 2030 the area of the moderate-belt forests will decrease, and that of the tropical forests will have halved.
- There will be an increase in arid land areas of 40%. This growth of technogenic desertification will be caused by the broken water cycle, forest cutting and soil contamination. There will be a reduction of agricultural lands per capita, fertility reduction, growing acidity, and salinity of soil.
- During the last decades of the 20th century the level of the World Ocean was rising by 1–2 mm/yr. This rate may increase to 7 mm/yr.
- Natural disasters and technogenic accidents in the late 20th century were characterized by an annual increase of 5–7% with increasing damage of 5–10% and increase in the amount of victims by 6–12%. Preservation and intensification of these trends is a natural result of global processes.
- Biological species in the second half of the 20th century disappeared 100–1000 times faster than at any other time in the past.
- There was a qualitative depletion of land water caused by increasing volumes of sewage, a growing number of point and areal sources of pollution, and a broadening spectrum of pollutants and pollutant concentrations. With this trend preserved, by 2030 more than 60% of the population will suffer water shortage.
- There is a persistent growth in the mass and number of pollutants accumulated in media and organisms. A migration of pollutants, especially radionuclides in food chains, is intensifying.
- Worsening quality of life, growing number of diseases connected with the ruining of man's ecological niche, and environmental pollution. Poverty, food shortage, high infant mortality, high level of morbidity, inadequate provision of pure water in developing countries, high accident rate, growing consumption of medicine, an increase of allergic diseases in the developed countries, a decrease of immune status. In the 21st century all these trends will remain and increase with areal extent.

Table 1.1. World population (1950–2000).

Year	G	ΔG	Year	G	ΔG
1950	2.555	38	1984	4.774	81
1955	2.780	53	1985	4.855	83
1960	3.039	41	1986	4.938	86
1965	3.346	70	1987	5.024	87
1970	3.708	78	1988	5.110	86
1971	3.785	77	1989	5.196	87
1972	3.862	77	1990	5.284	83
1973	3.939	76	1991	5.367	83
1974	4.015	74	1992	5.450	81
1975	4.088	72	1993	5.531	80
1976	4.160	73	1994	5.611	80
1977	4.233	72	1995	5.691	78
1978	4.305	75	1996	5.769	78
1979	4.381	76	1997	5.847	78
1980	4.457	76	1998	5.925	78
1981	4.533	80	1999	6.003	78
1982	4.613	81	2000	~6.080	77
1983	4.694	80			

Notation: G – size of population (billions); ΔG – population increment (M/yr).

- The global distribution of supertoxicants through the trophic chains (including the human chain). A damaged endocrine system affects the reproduction system, the work of the brain and other vital organs. All this will result in a growing number of childless couples.
- Artificial introduction and accidental invasion of alien species into the ecosystems. Transport of pests and diseases of plants, animals, and humans, leading to a reduction of biodiversity. These trends will remain in the 21st century.
- The changes observed in the World Ocean: destruction of reefs, reduction of the mangrove ecosystems, depletion of fish supplies as a result of intensive fishing, reduction of whale populations and pollution of inland and coastal waters ("red tides").

These trends suggest two principal conclusions: (1) in the second half of the 20th century there was a continued worsening of environmental characteristics; and (2) relevant expected trends will not only remain negative in the future but may even worsen. Thus the trends of unsustainable development will intensify. To confirm this, we shall give some quantitative estimates.

Table 1.1 illustrates the dynamics of the global population, the principal features of which are the continuing growth in the absolute size of the population, with its growth rate decreasing. Of course, the mean tendencies mask the strongest regional differences. Practically in all industrial countries the size of population is either slowly increasing or decreasing. The USA is an exception, where the increase in

Table 1.2. Gross world product (GWP), 1950–2000.

Year	U	U_G	Year	U	U_G
1950	6.4	2502	1984	26.2	5485
1955	8.1	2921	1985	27.1	5582
1960	10.0	3306	1986	28.0	5673
1965	12.8	3822	1987	29.0	5778
1970	16.3	4407	1988	30.3	5938
1971	17.1	4505	1989	31.2	5996
1972	17.8	4599	1990	31.9	6031
1973	19.0	4810	1991	32.0	5957
1974	19.4	4829	1992	32.4	5941
1975	19.7	4816	1993	33.2	6000
1976	20.7	4977	1994	34.5	6150
1977	21.5	5083	1995	35.8	6295
1978	22.4	5210	1996	37.3	6475
1979	23.1	5282	1997	39.0	6666
1980	23.6	5306	1998	39.9	6732
1981	24.2	5329	1999	41.2	6871
1982	24.4	5280	2000	~43.2	~7102
1983	25.1	5341			

Notation: U – net GWP (10^9 US$ in 1999); UG – per capita GWP (US$ in 1999).

the population is 1%/yr partly due to immigration, which constitutes one third of this. 95% of the global increase of the population in 2000 is attributed to the developing countries including Asia (57% with an absolute increase of 45 mln), Africa (23%), Latin America (9%), and the Middle East (5%). About half of the global increase is attributed to India, China, Pakistan, Nigeria, Bangladesh, and Indonesia. A number of factors favour a reduction of birth rate and the rate of population growth, including: improvement of the economic situation and the public health system in many countries; progress in women's education alongside their increasing status. So, for instance, in Iran the rate of population growth decreased from 3.2% in 1986 to 0.8% in 2000. The decrease of birth rate has been caused by negative factors, such as AIDS (about 3 million people died in 2000 of AIDS, and the overall total reached 22 million people).

Naturally, during the second half of the 20th century the global world domestic product (GWP) continued growing (though slowing down) (Table 1.2). Relative growth of the scale of the global economy in 2000 constituted 4.7% and exceeded the 1999 level (3.4%). The gross world product (GWP) reached $43 billion, with the USA providing the average GWP per capita of $7102. This progress has been achieved due to a successful development of the economy in the USA and Western Europe, recovery of the economy in Asia after the 1997 financial crisis, recovery of the economy in Latin America after the 1998 crisis, as well as a marked improvement in the situation in countries with transitional economies. Note that the growth of the economy in China in 2000 constituted 7.5% and

Table 1.3. Global fossil fuel consumption (1950–2000) (10^6 t of oil equivalent).

Year	C	O	G	Year	C	O	G
1950	1043	436	187	1983	1918	2632	1463
1955	1234	753	290	1984	2001	2670	1577
1960	1500	1020	444	1985	2100	2654	1640
1965	1533	1485	661	1986	2135	2743	1653
1970	1635	2189	1022	1987	2197	2789	1739
1971	1632	2313	1097	1988	2242	2872	1928
1972	1629	2487	1150	1989	2272	2921	1904
1973	1668	2690	1184	1990	2244	2968	1938
1974	1691	2650	1212	1991	2189	2967	1970
1975	1709	2616	1199	1992	2179	2998	1972
1976	1787	2781	1261	1993	2171	2969	2012
1977	1835	2870	1283	1994	2186	3027	2019
1978	1870	2962	1334	1995	2218	3069	2075
1979	1991	2998	1381	1996	2298	3150	2170
1980	2021	2873	1406	1997	2285	3224	2155
1981	1816	2781	1448	1998	2243	3241	2181
1982	1878	2656	1448	1999	2034	3332	2277
				2000	~2034	~3332	~2277

Notation: C – coal, O – oil, G – gas.

was the most significant in Asia. Figures for India, Pakistan, and Bangladesh are, respectively, 6.7, 5.6, and 5%, which is especially important for the region with a population exceeding 1.2 billion people, where the per capita income is about $1 per day or less.

Fossil fuel consumption is a significant indicator of the development of economies and technologies (Table 1.3). As shown in Table 1.3, the year 2000 was the second consecutive year to show a decrease in the global level of total fossil fuel consumption (after a long period of the growth in this level by a factor of 3.5 starting from 1950). Now the contribution of the fossil fuel consumption in the commercial production of energy is about 90%, the share of coal being 25%. This mean-global share is gradually decreasing, but in the USA the consumption of coal (~25% of the global use) increased in 2000 by 1.6% due to the development of electric power stations burning coal. Contrary to this, the coal consumption in China (which is also 25% of the global use) decreased by 3.5% (from 1996 the decrease has reached 27%). In India, the third largest consumer of coal (7%), the level of consumption in 2000 increased by 5.4%.

Global oil consumption (its share in energy production is 41%) has increased by 1.1%, and in the USA, the leading petroleum user (26%), by only 0.1%, whereas in the Asian countries of the Pacific region (27% of the global oil consumption) the use of oil has increased by 2.6%. The respective figures for Western Europe are 22% and 0.2%. As for natural gas (24% of the commercial energy production), the level of its consumption has increased by 2.1%. In the USA (where 27% of the global natural

Table 1.4. Global CO_2 emissions due to fossil fuel combustion (1950–2000) and its concentration in the atmosphere (1960–2000).

Year	H_1	C_A	Year	H_1	C_A
1950	1612	–	1984	5098	344.2
1955	2013	–	1985	5271	345.7
1960	2535	316.7	1986	5453	34.0
1965	3087	319.9	1987	5574	348.7
1970	3997	325.5	1988	5789	351.3
1971	4143	326.2	1989	5892	352.7
1972	4305	327.3	1990	5931	354.0
1973	4538	329.5	1991	6020	355.5
1974	4545	330.1	1992	5879	356.4
1975	4518	331.0	1993	5861	357.0
1976	4776	332.0	1994	6013	358.9
1977	4910	333.7	1995	6190	360.9
1978	4950	335.3	1996	6315	362.6
1979	5229	336.7	1997	6395	363.8
1980	5155	338.5	1998	6381	366.6
1981	4984	339.8	1999	6340	368.3
1982	4947	341.0	2000	~6299	~369.4
1983	4933	342.6			

Notation: H_1 – CO_2 emissions into the atmosphere (10^6 tC); C_A – CO_2 concentration in the atmosphere (ppm).

gas is consumed), the scale of consumption has increased by 2.4%. The consumption of natural gas in the Baltic countries has also strongly increased: Lithuania, 29%; Estonia, 30%; and Latvia, 45%. South Korea and Spain are the leaders of a 16% growth of natural gas consumption in Asia and West Europe. Recent years have been characterized by high oil prices, reaching a maximum ($34.5 a barrel) in 1985, which has drawn attention to the possibility of oil and gas extraction on the Arctic shelf of Russia and Alaska, although the Alaskan resources are comparatively small, and huge investments would be needed for Russia.

According to the International Energy Agency, the total fossil fuel consumption can increase by 57% (2%/yr) by the year 2020, preserving a 90% share of consumption in energy production. The supposed global trend of coal consumption increase will constitute 1.7%/yr, with two thirds of this increase due to consumption by India and China. Petroleum will remain the main source of energy (increasing by 1.9%/year) and its share in primary energy production will reach 40%. The highest rate of consumption increase (2.7%/yr) is characteristic of natural gas consumption (mainly due to gas-operating electric power stations).

Data in Table 1.4 reflect the dynamics of CO_2 emissions into the atmosphere and the increase of its concentration in the atmosphere. Starting from 1950, the atmosphere has gained an amount of CO_2 equivalent to 217 Gt C with an annual fourfold increase of emissions. An important fact is that the fossil fuel use per unit produc-

Table 1.5. Global grain production, 1950–2000.

Year	P	P_G	Year	P	P_G
1950	631	247	1984	1632	342
1955	759	273	1985	1647	339
1960	824	271	1986	1665	337
1965	905	270	1987	1598	318
1970	1079	291	1988	1549	303
1971	1177	311	1989	1671	322
1972	1141	295	1990	1769	335
1973	1253	318	1991	1708	318
1974	1204	300	1992	1790	328
1975	1237	303	1993	1713	310
1976	1342	323	1994	1760	314
1977	1319	312	1995	1713	301
1978	1445	336	1996	1872	324
1979	1411	322	1997	1881	322
1980	1430	321	1998	1872	316
1981	1482	327	1999	1869	311
1982	1533	332	2000	~1840	~303
1983	1469	313			

Notation: P – net grain production (10^6 t); P_G – per capita consumption (kg).

tion, calculated with CO_2 emissions taken into account, decreases: in 2000 this decrease constituted 3.6% (this figure is equivalent to 148 t C per $1 million GWP), and over 50 years was 41%. Contrary to the Kyoto Protocol, CO_2 emissions in the western industrial countries increased by 92% during the period from 1990, and in the USA, by 13%, on average. The respective contribution of developing countries has reached 22.8%. According to the IPCC, by 2020 the annual global CO_2 emissions may reach 9–12.1 Gt C and by 2050 they may be within 11.2–23.1 Gt C. Therefore the problem of global climate change due to CO_2 emissions is discussed in detail and studied in this book. As far as possible, a complete analysis of the biogeochemical CO_2 cycle is given in the following chapters.

The most important component of economic development is grain production (Table 1.5), which has recently started decreasing from 1869 Mt in 1999 to 1840 Mt in 2000. The corn crop in 2000 turned out to be 2% below the maximum grain production reached in 1997 (1881 Mt). The main reason for this decrease was a reduced corn crop in China from 391 Mt in 1998 to 353 Mt in 2000 (approximately 10%). This resulted from low grain prices which discouraged the farmers from increasing grain production, as well as drought and a water deficit in the northern region of the country. Grain production in the USA, second to China, increased during the same period from 332 to 343 Mt, mainly due to the increased yield of maize.

The mean global grain yield decreased slightly in 2000 (2.75 t/ha) compared to 1999 (2.77 t/ha), when it reached its multi-year maximum. During the past years

Table 1.6. Foreign debt of the former USSR republics and Eastern European countries (1970–1999).

Year	V	Year	V
1970	0.26	1985	1.47
1971	0.29	1986	2.57
1972	0.32	1987	2.73
1973	0.37	1988	1.68
1974	0.42	1989	1.70
1975	0.51	1990	1.77
1976	0.59	1991	1.80
1977	0.74	1992	1.85
1978	0.86	1993	1.97
1979	0.98	1994	2.15
1980	1.07	1995	2.30
1981	1.18	1996	2.35
1982	1.23	1997	2.40
1983	1.33	1998	2.61
1984	1.35	1999	2.57

Notation: V – sum of debt (10^9 US$, 1999).

there have been only slight fluctuations about an average of 2.75 t/ha, but in 2000 the consumption per capita decreased to 303 kg, which was 13% below the maximum level of grain consumption per capita in 1984. This decrease was mainly due to Western European nations, the former USSR, and Africa. In 2000 and during the previous two years the maize production (588 Mt) for the first time exceeded the level of wheat production (580 Mt), whereas the production of rice constituted 401 Mt. About 43% of the global maize yield have been spent by the USA on stock-breeding. For wheat production, the league table is as follows: China, India, and the USA. China also dominates the rice production. The USA exports more than 75% of its maize production to other countries. The main exporter of wheat is the USA, then France, Canada, and Australia. The leading exporters of rice are China, Thailand, USA, and Vietnam. Until recently, Japan had been the main importer of wheat, but during the last years of the 20th century the main importers have been Brazil, Iran, and Egypt.

One of the key indicators of the level of socio-economic development is a country's amount of foreign debt. Table 1.6 totals the data for the developing and East European countries, as well as for the former USSR republics. As seen, a continuously growing debt, at a maximum in 1998, has stabilized, somewhat reducing in 1999 (US$ 2.57 billion) due to inflation. Serious changes in the financial trends in Brazil, Indonesia, Russia, and South Korea after 1996 are largely explained by the financial crises in 1997 and 1998.

To assist with understanding and predicting the global changes in the environment, of great importance is the data on human health (Brown *et al.*, 2001). In Table 1.7 is given a fragment of the most demonstrative trend of AIDS propagation.

Table 1.7. Global data on the number of HIV infected people and AIDS deaths (1980–2000).

Year	G_C	M_G	Year	G_C	M_G
1980	0.1	0.0	1990	10.0	1.7
1981	0.3	0.0	1991	12.8	2.4
1982	0.7	0.0	1992	16.1	3.3
1983	1.2	0.0	1993	20.1	4.7
1984	1.7	0.1	1994	24.5	6.2
1985	2.4	0.2	1995	29.8	8.2
1986	3.4	0.3	1996	35.3	10.6
1987	4.5	0.5	1997	40.9	13.2
1988	5.9	0.8	1998	46.6	15.9
1989	7.8	1.2	1999	52.6	18.8
			2000	~57.9	~21.8

Notation: G_C – HIV infected (10^6 people); M_G – AIDS deaths (10^6 people).

Only this disease is able to reduce the size of the able-bodied population of the planet, it concerns especially the nations of Africa and the Caribbean basin. Expenses on health services reaching about US$ 3 billion vividly demonstrate the urgency of the problem. However, these expenses have been distributed rather unevenly. The nations with a low and average income and with 84% of global population have got only 11% of global expenses on health services, despite the fact that 93% of the damage caused by diseases occur in these countries. If the expenses on health services in the world average 5% GWP, in the USA they reach 13.7%, in Somalia only 1.5%. Respectively, the annual expenses per capita range from $50 in poor countries to $4100 in the USA.

One important indicator of socio-economic trends is the level of militarization. Table 1.8 characterizes the dynamics of nuclear armaments as an indicator of technical progress. Here the cause–effect feedbacks of the world system "nature–society" are directly regulated by political aspects.

To summarize the results of global civilization development during the second half of the 20th century, it is necessary first of all to underline that this was a period of unprecedented fast changes in the global size of the population, in the biosphere, economy, and society as a whole. The world has become more rich economically but poorer from an ecological point of view. The following most important trends have been observed during this period:

- The global population has exceeded 6 billion, increasing by 3.5 billion during the last half century (i.e., more than doubling). Most of the growth has taken place in the developing countries, which are already overpopulated. The growth was especially rapid in the urban areas where the population increased almost fourfold. A remarkable feature of the demographic dynamics in industrial countries is an increasing share of old people.

Table 1.8. World nuclear arsenal (1945–2000).

Year	Y	Year	Y
1945	2	1983	66,979
1950	303	1984	67,585
1955	2490	1986	69,478
1960	20,368	1987	68,835
1965	39,047	1988	67,041
1970	39,691	1989	63,645
1971	41,365	1990	60,236
1972	44,020	1991	55,772
1973	47,741	1992	52,972
1974	50,840	1993	50,008
1975	52,323	1994	46,542
1976	53,252	1995	43,200
1977	54,978	1996	40,100
1978	56,805	1997	37,535
1979	59,120	1998	34,535
1980	61,480	1999	31,960
1981	63,054	2000	~31,535
1982	64,769		

Notation: Y – number of nuclear warheads.

- The growth of the world economy has exceeded (approximately sevenfold during a 50 yr period) the rate of population growth, which provided (on average) a significant increase of living standards. However, at the same time 1.2 billion people still live in severe poverty and 1.1 billion do not have access to pure drinking water.
- The global grain yield has nearly tripled since 1950, which allowed people to enrich their diet, but led to dangerous ecological consequences: lowering the level of subsoil waters, and intensifying natural water pollution due to massive over use of fertilizers and pesticides.
- Only a small part of primary forests (boreal and tropical) have been preserved on the planet, more than a half of all wetlands and over a quarter of coral reefs have been lost. This was accompanied by considerable damage to biodiversity and, more seriously, the mechanisms for biotic regulation of the environment have been broken (Gorshkov *et al.*, 2000; Gorshkov, 1995).
- The anthropogenic impacts on the global climate and ozone layer are of great concern, though there are more uncertainties than established trends.

It is especially important that socio-economic and ecological dynamics during the second half of the 20th century was that, as rule, the observed trends were unpredictable. As the world becomes more and more complicated and inhomogeneous, climate prediction becomes more difficult but necessary, since an adequate planning aimed at minimizing the risk and maximizing favourable conditions for life depends on its accuracy.

1.2.3 Sustainable development

The paradox of the contemporary situation is that despite the unprecedented scale of discussions on sustainable development, the world is continuing to follow the path of unsustainable development. The definition of the term "sustainable development" proposed by the International Commission on Environment and Sustainable Development ("Bruntland Commission") as "a development that provides the needs of the present generations of mankind, and at the same time, gives the opportunity to the coming generations to satisfy their needs as well" is too general and non-constructive. This term should, of course, reflect the key role of biospheric dynamics, which has been mentioned in the works by Losev (2001), Kondratyev (1997, 1999c), Gorshkov (1990, 1995), Gorshkov *et al.* (2000). In these works it has been shown that the communities of the organisms within the biocenosis and the whole biosphere regulate the chemistry of the most important media – atmosphere, hydrosphere, and soil. A necessity has also been emphasized to determine the limits to permissible disturbances in these media. Losev (2001) points to the following four problems to be solved to achieve sustainable development.

1. *Clash of interests of our civilization with nature.* The modern civilization is the result of 10,000 yrs of spontaneous human development, when people could build their history without any restrictions by the biosphere. But by the beginning of the 20th century, as a result of continuing expansion, humanity had gone beyond the limits, with subsequent global environmental changes. As a result, the 20th century has become the century of severe ecological crisis, a fact that has still not been realized by people and politicians, though its denial or reduction does not explain the global, nature protection infrastructure created during the last 25 years, environmental protection expenses, and application of resource-preserving technologies.

   The ecological crisis is a signal that humanity cannot continue with its random development but must coordinate its history with the laws of the biosphere from which man is inseparable. First of all, development should be coordinated with the law of energy flow distribution in the biosphere, that determines the limits of the corridor of ecologically safe civilization development.

   The signals from the anthropologically destroyed biosphere either do not yet affect the majority of the people enough or people do not yet connect these signals with the rapidly developing environmental crisis. At the same time, as long as the virgin ecosystems exist on our planet, there is hope that irreversible changes have not yet occurred, and that these damaging processes can be stopped and reversed. It is important that these signals are adequately perceived as a guide for action. None of the national programmes or strategies on sustainable development contain an adequate response to such signals because no theoretical scientific basis has been developed to perceive signals. Some ideas on the development of such a basis and technology to perceive signals from the biosphere have been discussed by Kondratyev *et al.* (2000):

2. *Economical.* It is the development of economies that has led to the destruction of natural mechanisms of environmental stabilization and to the development of these ecological crises. The ecological costs of civilization have begun to convert into economical, social, and demographic costs. They have reduced the world economy efficiency, the investing activity, slowed down the rate of growth, stimulated poverty, broken the genetic programmes, and undermined human health. In its interaction with nature, humankind still remains an appropriator, not only using natural resources free of charge but also thoughtlessly destroying the biosphere – which is the basis of life. In the 21st century mankind will have to radically change its attitude towards the organization of production cycles, exceeding sustainable limits for extracting resources free of charge, and producing unlimited levels of waste.
3. *The social sphere.* The existing social–economic system has not only led civilization to clash with nature but ironically neither has it solved social problems. The gap between the poor and rich is widening, and the difference between the living standards in poor and rich countries is growing. This means that the global socioeconomic system does not correspond to the requirements for the sustainable development of humankind.
4. *Demographic crisis.* This leads to a decrease in the areas of arable lands and of the volume of food production per capita. It intensifies the ecological crisis, which, in turn, deteriorates an important social indicator, human health. The social inequality of people destabilizes the situation in developing countries, causes conflicts and migration of population, gives rise to extremism and terrorism, and finally evokes a global scale conflict.

Thus, the modern civilization is in a state of system crisis, a fact which was repeatedly discussed at a number of international conferences. The way out of this crisis is in the transition towards sustainable development. This crisis is somehow connected with the problem of growing atmospheric CO_2 concentration, which worries both the economists and politicians. It is important to predict the consequences of global climate change to work out the strategies of regional and global economic and political initiatives in the future.

In this regard, the scientists of the US are accomplishing a complex programme to study the global carbon cycle. The main goals of this programme are as follows:

- To develop techniques to measure, monitor, and model the elements of the carbon cycle in nature in order to synthesize an integrated control system for these elements.
- To work out a scientific basis to assess the ability of land and oceanic ecosystems to extract and accumulate CO_2, with the existing anthropogenic activity taken into account.
- To develop techniques to identify and assess the sources of CO_2 and other GHGs on regional and global scales.
- To search for an algorithm to describe the functions of these sources in the future and to provide information on the prognostic estimates of climate trends.
- To assess the potential strategies of carbon extraction from the environment.

- To develop an efficient technology for the information fields for objective planning of climate studies in the future, and realization of the decisions when managing the land ecosystems.
- To create a unified information base for the available and planned environment research programmes.

1.3 THE GLOBAL CLIMATE CHANGE AND THE KYOTO PROTOCOL

There is no doubt that the success of both the Second UN Conference on Environment and Development in Rio de Janeiro in 1992 and the Special Session of the UN General Assembly "Rio+5" held in New York 5 years later consisted of drawing attention of the governments and the public to the problems of global changes and sustainable development. Unfortunately, neither off these forums could formulate and substantiate the priorities of global humankind development. One global problem mentioned earlier is the problem of climate change caused by breaking the biogeochemical cycles of chemical substances, carbon in particular. Unfortunately, the principal feedbacks in the environment that govern biotic regulation are poorly understood. The problem of "global warming" has become the problem of first priority, which resulted in adoption of the FCCC, which was poor, confusing, and unjust with respect to the developing nations. The FCCC was focused on the anthropogenic origin of the observed global climate change and on recommendations (to industrial countries) to reduce the GHG emissions to the atmosphere (in the first place, carbon dioxide).

In December 1997 more than 60 countries participated in the Third Conference of the FCCC signatory powers held in Kyoto (Japan). The long and sharp discussions were concentrated on whether it was possible to reduce CO_2 emissions by about 5% by 2008–2012 with respect to emissions of the 1990 level (Table 1.9). Naturally, the position held by the developing countries was that their first priority was to raise the living standard of people but not to curtail industry for the sake of CO_2 emissions reduction. Therefore it is practically impossible to arrive at a globally agreed-upon decision regarding the problem of CO_2 emissions reduction whilst taking into account the opinion of numerous experts who have proven the inadequacy of the Kyoto Protocol conclusions.

One may think that this situation stems from the poor development of scientific principles applied to the problems of global climate change. This conclusion is valid only in part since as far back as 1990 the basic concept of the biotic regulation of the environment was published (Gorshkov, 1990; Kondratyev, 1990, 1991, 1992, 1993, 1996) which proved the unfounded character of the "greenhouse" hypothesis of global warming and drew attention to the necessity of studying the climate system "atmosphere–ocean–land–cryosphere–biosphere" taking into account the complexity of its interactive component feedbacks.

The most important facts testifying to the delusion of many about the problems of global climate change are as follows:

1. Observational data (so far, inadequate from the viewpoint of its completeness and reliability) does not confirm the fact of "global warming" (this especially concerns the data of surface observations in the US (Harris *et al.*, 1992), the Arctic (Sturm *et al.*, 2001), as well as results of satellite remote sounding.
2. If the atmospheric greenhouse effect enhancement due to the supposed doubling of CO_2 concentration in the atmosphere constitutes about $4\,W/m^2$, then the uncertainties connected with an account of the climate-forming role of atmospheric aerosols and clouds as well as a number of other factors, amount to 10 and perhaps $100\,W/m^2$.
3. The results of numerical modelling of the climate, that substantiate the hypothesis of the "greenhouse global warming" and ostensibly agree with the observational data, are nothing more than models forced to fit the observational data.
4. Recommendations of the levels of the GHG emissions reduction based on these results are senseless (their realization can have far-reaching negative socio-economic consequences, however).

In the context of global changes, the problems connected with the assessments of the present and future global climate changes are, of course, of primary importance. The concept of "global warming" still dominates such assessments, as can be seen from the Third IPCC report published in 2001 (Kondratyev and Demirchian, 2001a,b). These assessments should be treated only as the inertia of the earlier,

Table 1.9. Recommendations of the Kyoto Protocol for CO_2 emissions reduction by various countries by 2008–2012 compared to the 1990 level.

L	Country	*Y*	*D*	*Q*	*G*
108	Australia	288,965	16,516	16.91	0.98
94	Canada	462,643	19,705	17.44	0.81
92	Czech Republic/Slovakia	165,792	3049	16.00	5.25
92	France	366,536	20,966	6.49	0.31
92	Germany	1,014,155	21,861	12.76	0.59
92	Ireland	30,719	11,349	8.77	0.68
92	Japan	1,155,000	24,205	9.35	0.39
92	Luxemburg	11,343	34,614	30.41	1.10
92	The Netherlands	174,000	18,939	11.22	0.61
100	New Zealand	25,476	12,192	7.61	0.59
101	Norway	35,514	26,387	8.37	0.31
94	Poland	414,930	1459	10.87	7.04
92	Portugal	42,148	6774	4.00	0.63
100	Russia	2,388,720	3897	16.11	4.13
92	England	577,012	16,723	10.08	0.59
93	USA	4,957,022	22,046	19.83	0.90

Notation: *L* – CO_2 emissions (%); *Y* – amount of CO_2 emissions (10^3 t C); *D* – per capita GNP ($); *Q* – per capita emissions of CO_2 (t C); *G* – CO_2 emissions per unit GNP (t C/$\$10^3$) (Kok and Verweij, 1999; Watson *et al.*, 2000; Kondratyev and Demirchian, 2001b).

far from scientific, speculative opinions, which have been convincingly demonstrated by Boehmer-Christiansen (2000). Clearly, for an objective analysis of all the cause–effect feedbacks and processes in the Earth's climatic system, a constructive mechanism is needed to collect and analyse the data. Here we shall briefly discuss the well-known data (Kondratyev and Demirchian, 2001a).

1.3.1 Atmospheric temperature

According to SAT observations for the period from 1860, the mean annual mean global value has increased by 0.6 ± 0.2°C. This exceeds the value given in the IPCC 1996 report (Kondratyev, 1999b) by about 0.15°C, which is due to a relatively high SAT between 1995 and 2000. The observational data revealed a very strong spatial–temporal variability in the mean annual SAT on a global scale. For example, the climate warming in the 20th century took place during two periods: 1910–1945 and from 1976 to the present. It follows from new data on the global climate diagnostics that the climate warming in the northern hemisphere in the 20th century was, strongest during the last 1000 years; the 1990s were the warmest decade, with 1998 being the warmest year. An important feature of the climate dynamics was that, on average, the rate of the nocturnal (minimum) SAT increase on land nearly twice exceeded that of the diurnal (maximum) SAT increase, starting from 1950 (0.2°C against 0.1°C/10 yr). This favoured the long duration of the frostless periods in many regions at moderate and high latitudes.

The IPCC 2001 report does not mention an enhanced climate warming in the northern hemisphere high-latitudes, which was emphasized earlier as a characteristic indicator of anthropogenically induced global warming. An analysis of the data of direct SAT measurements made at North Pole stations over the last 30 years and dendroclimatic indirect data for the last 2–3 centuries, performed by Adamenko and Kondratyev (1999), have not revealed any enhancement of climate warming; the climate changes observed during the last century and the last decade were characterized by a strong spatial–temporal variability – in the Arctic, regions of both warming and cooling of climate formed simultaneously.

From the data of the aerological observations network (starting from the 1950s, when this network became more or less adequate), the trends of the mean global SAT and of the lower troposphere temperature were almost identical (about 0.1°C/10 yr). According to the data of satellite microwave remote sounding (starting from 1979), the mean global lower troposphere temperature has been increasing at a rate of 0.06°C/10 yr, significantly below the SAT increase (~0.15°C/10 yr). This difference in climate warming manifested itself mainly in the regions of the oceans in the tropics and subtropics, the reasons for these differences being unknown (Christy *et al.*, 1998).

1.3.2 Snow and ice cover

With the exception of Australia, other continents have as regional indicators of climate change – glaciers and snow cover extent. Starting from the late 1960s, the extent of snow cover has decreased by about 10% globally, and the annual duration of the ice cover on lakes and rivers has become two weeks less at middle- and high-latitudes of the northern hemisphere. In non-polar regions mountain glaciers have retreated. The extent of the sea ice cover in spring and summer in the northern hemisphere has decreased by 10–15%, starting from the 1950s. Probably, during the last decades of the 20th century (from late summer to early autumn) there has been a 40% decrease in the Arctic Sea ice-cover thickness, with a much less significant thinning in the winter period. During the period of regular satellite observations (starting from the 1970s) no marked trend in the ice cover extent has been observed in the Antarctic.

One of the indicators of the GHG climatic effect is a change in the World Ocean level. Unfortunately, so far there are no completed studies of the balance of the glacier masses in the Antarctic and Greenland, and therefore there is no reliable data on changes in sea level, although many authors believe that an atmospheric temperature increase should inevitably lead to melting of the planetary ice sheet. The use of the airborne laser altimeter during the period 1993–1998 over the southern edge of the Greenland ice sheet showed that the ice thickness at an altitude of about 2 km decreased at a rate of 0.5 ± 0.7 cm/yr, at lower altitudes this rate being 1 m/yr.

1.3.3 Surface level and heat content of the upper layer of the ocean

During the 20th century the World Ocean level rose between 10 to 25 cm (18 cm on average), probably partly due to thermal expansion of sea waters and ice melting on land caused by global warming. In the 20th century the rate of the World Ocean level rise exceeded tenfold the rise observed during the last 300 years. Starting from the late 1950s (when the SAT measurements were regularly taken globally), the heat content of the upper layer of the ocean has increased.

1.3.4 Other climatic parameters

As seen from the observational data, during the 20th century a 0.5–1.0%/10 yr precipitation increase was observed over most of the terrestrial regions in the northern hemisphere mid- and high-latitudes and a 0.3%/10 yr decrease over most of the subtropical terrestrial regions, which has weakened, however, over the last years. As for the World Ocean, the lack of adequate observational data prevented scientists from producing reliable precipitation trends. During the last decades, precipitation in the northern hemisphere mid- and high latitudes was more frequent, intensive, and extreme. Since the mid-1970s, the El Niño Southern Oscillation (ENSO) events have become more frequent, persistent, and intensive. Such ENSO dynamics were reflected in specific features of regional precipitation and SAT variations in most of the tropics and subtropics (Varotsos and Deligiorgi, 1991). Isolated and inadequate observational data on the intensity and frequency of occurrence of tropical and extratropical cyclones as well as local storms does not allow one to reveal any clear trends (Kondratyev and Grigoryev, 2000).

1.3.5 Concentration of GHGs and anthropogenic aerosols in the atmosphere

During the period 1850 to present day the atmospheric CO_2 concentration has increased by one-third, reaching a maximum over the last 420,000 yrs (possibly over the last 20 Myrs) as judged from paleoclimatic data. About two thirds of the CO_2 concentration increase for the last 20 years are determined by emissions to the atmosphere due to fossil fuel combustion (the rest of the share is determined by deforestation and, to a lesser extent, the cement industry). It is of interest that by the end of 1999 the CO_2 emissions in the USA exceeded the level of 1990 by 20%, and by the year 2008 their expected increase should exceed this by a further 10%. Nevertheless, according to the Kyoto Protocol, the emissions should be reduced by 7% by 2008 compared to the level of 1990, which requires a total reduction of emissions of about 25%.

According to the available observational data both the World Ocean and land surface are now global sinks for CO_2. In the ocean it is governed by chemical and biological processes, and on land it is connected with an intensified "fertilization" of vegetation due to increasing concentrations of CO_2 and nitrogen and with changes in land use. More details have been given in other sections of this book and in numerous publications of other authors (Braunlich *et al.*, 2001; Chase *et al.*, 2000; Chen *et al.*, 1999; Cihlar *et al.*, 2002; Cox *et al.*, 2000).

There is no doubt that fossil fuel combustion will remain the principal factor of the CO_2 concentration growth in the 21st century, with the biosphere of both the ocean and land acting as a barrier to the concentration growth diminishing with time. According to the IPCC 2001 report, by the end of this century CO_2 concentrations may range between 540 and 970 ppm (pre-industrial and present values are, respectively, 280 and 367 ppm). An important factor in the global carbon cycle is changes in land use, but even if total carbon emitted to the atmosphere due to land use is assimilated by land biota, this will only reduce CO_2 concentrations by 40–70 ppm. As for the prognostic estimates of the concentrations of other GHGs, they vary widely. So, for instance, it follows from some estimates that the role of tropospheric ozone as a greenhouse gas can be equal to the contribution of methane, being the factor of deterioration of air quality over most of the northern hemisphere.

Since 1850 atmospheric methane concentration has increased by a factor of 2.5, compared to the observed one (from indirect data), and continues to grow. The annual rate of CH_4 concentration growth slowed down and became more variable in the 1990s compared to the 1980s. Starting from 1850 the concentration of nitrous oxide has increased by 16%. As a result of implementation of the recommendations of the Montreal Protocol and its subsequent supplements, the concentrations of several halocarbons acting as GHGs and ozone-destroying gases either increased more slowly than before or started to decrease. On the other hand, concentrations of their substitutes and some other synthetic compounds (e.g., perfluorocarbons (PFC) and sulphur hexafluoride (SF_6)) started rapidly increasing.

Estimates of changes of the RF that characterizes an enhancement of the atmospheric greenhouse effect and is determined by the growing concentration of MGCs in the atmosphere totalled 2.42 W/m^2, with the following contributions of various

MGCs: CO_2 (1.46 W/m^2), CH_4 (0.48 W/m^2), halocarbons (0.33 W/m^2), and N_2O (0.15 W/m^2). A decrease in the total ozone content observed during the last two decades could lead to a negative RF constituting −0.15 W/m^2, which may drop to zero in this century if the measurements taken to protect the ozone layer are successful. The tropospheric ozone content increase since 1850 (approximately by one-third) could have led to a positive RF of about 0.35 W/m^2.

Since the publication of the IPCC 1996 report, substantial changes have taken place in assessments of RF determined by not only purely scattering sulphate aerosol but also by other types of aerosol, especially carbon (soot) (characterized by a considerable absorption of solar radiation) as well as organic sea salt and mineral aerosols. A strong spatial–temporal variability of the atmospheric aerosol content and its properties seriously complicates the assessments of the climatic implications of aerosols (Kondratyev, 1999a). The results of new calculations made by Hansen *et al.* (2000) have radically changed the ideas regarding the roles of various factors of RF formation. According to these calculations, there exists an approximately mutual compensation of climate warming due to the growing CO_2 concentration and cooling due to anthropogenic sulphate aerosol. In these conditions the anthropogenically induced emissions of methane (mainly due to rice paddies) and carbon (absorbing) aerosols should play a more important role.

The climate-forming factors to be taken into account are changes in the extra-atmospheric solar radiation. Contributions of these changes into RF since 1850 could have reached ~20%, compared to the CO_2 contribution mainly due to an intensified insolation during the second half of the 20th century (of importance is an account of an 11-year cycle of insolation). However, possible mechanisms for enhancing the climatic implications of solar activity are far from being understood.

All this confirms the conclusions drawn by many authors that recommendations like those of the Kyoto Protocol might have some sense provided they were made based on a complex analysis of every climate-forming process, which is impossible using the present-day numerical climate modelling techniques. The recent considerable progress in the development of more complete numerical climate models with an account (interactive, as a rule) of principal components of the climate system "atmosphere–hydrosphere–cryosphere–biosphere" raises no doubts, however. An interactive simulation of the global carbon cycle within the framework of climate theory has started, at last. Extreme complexity of climate models and the large number of applied empirical parameterization schemes for various (sub-grid, in particular) processes hinders analysis of adequate models, especially from the viewpoint of their use in the future climate predictions. Therefore, so far, the attempts undertaken to compare the results of numerical climate modelling with the observational data have been rather schematic, contradictory, and unconvincing.

Also unconvincing were, for example, the conclusions concerning the secular trend of the mean annual mean global SAT for the 150 years. If according to the IPCC 1996 report, the observed and calculated (with the growth of CO_2 and sulphate aerosol concentrations taken into account) changes of SAT agree well, then, following Hansen *et al.* (2000), an account of methane and carbon aerosols is more important. Unfortunately, in both these cases the conclusions are based on arbitrary opinions, and an agreement with observations is nothing else than forced fitting. It is also clear that practically any meaningful comparison of the theory with observations should contain a consideration of regional climate changes (not only SAT), not only average values of climate parameters, but also their variability characterized by higher-order moments.

The IPCC 1996 report contains a disputable conclusion:

> "The balance of evidence suggests a discernable human influence on global climate" (Kondratyev, 1998b).

It also states that the "anthropogenic signal" has already manifested itself against the background of natural climate changes. According to the IPCC 2001 report, "the detection and attribution studies regularly indicate the presence of the anthropogenic signal in the climate observation data for the last 35–55 years.... The nature-induced impacts could play a role in the warming observed in the first half of the 20th century, but they cannot explain the warming in the second half of the century". This report also states, however, that "the reconstruction of the climate for the last 1000 years, as well as model estimates demonstrate a low probability that the observed climate changes in the second half of the 20th century could be of completely natural origin". Also, the report emphasizes a high uncertainty in the quantitative estimates for the anthropogenic warming, especially from the viewpoint of contributions from various factors of warming (primarily, atmospheric aerosols). The contradictory and unconvincing character of the cited opinions is apparent and does not require any comment. Of course, the leading role in the substantiation of future climate predictions should belong to integral simulation models describing the dynamics of interactions between the maximum possible number of factors of the global natural–anthropogenic system (Tol, 2000). It is unclear, however, how real climate predictions can be made in the case of inadequate input information. In the near future the integral models can only give rather conditional scenarios. The development of detailed models of the climate systems individual components, such as biogeochemical cycles of greenhouse gases, will make it possible to reduce the existing uncertainties in assessing their role in the dynamics of climatic characteristics.

From new data for various scenarios of growth of the concentrations of GHGs and aerosols, the mean-global mean-annual SAT should rise by 1.4–5.8°C between 1990 and 2100, whereas according to the IPCC 1996 report, this rise would be 1.0–3.5°C. In this regard it is symptomatic that both an improvement in, and an increasing number of, the models has brought forth not a narrowing but a broadening of the divergence of this process. It is important that the divergence of calculated SAT values using various models with one scenario of GHG emissions and one with different scenarios of emissions are nearly similar. As for the regional climate, the models are still statistically unreliable. However, the following conclusion may be considered reliable: the climate in almost all terrestrial regions will warm up faster compared to mean global warming, especially in high-latitudes in the cold half of the year. The climate warming estimated for the northern regions of North America and for northern and central Asia, where it exceeds the mean-global value by about 40%, is especially marked. Contrary to this, in the summertime in southern and south-eastern Asia and in the wintertime in southern South America the warming should be weaker than the mean-global value. The numerical modelling testifies to a forthcoming increase of atmospheric moisture content and enhanced precipitation. In particular, there may be a precipitation enhancement in the northern hemisphere mid- and high-latitudes as well as in the winter months in Antarctica (this conclusion is of special importance in the context of the problem of glacier dynamics). In low-latitudes there may be regions with both intensified and reduced precipitation (depending on the choice of GHG emissions scenarios).

Calculations of anthropogenic ("greenhouse") climate changes reveal the possibility of a future weakening of the thermohaline circulation (THC) in the northern hemisphere oceans. However, even the models showing such attenuation still reflect the preserved "greenhouse" warming in Europe. But it is not clear whether the THC collapse would be irreversible and what threshold conditions would correspond to such a collapse. None of the available models predict a complete THC collapse over the next 60 years.

According to the results of the numerical modelling of global warming processes, there will be a further reduction in the snow and sea ice cover extent in the northern hemisphere. Further retreat of glaciers is expected in the 21st century (except the ice sheets of Greenland and the Antarctic). With the chosen scenarios of the GHG concentration increasing between 1990 and 2100 the World Ocean level may rise by 0.14–0.8 m (~0.47 m, on average), which exceeds by two to fourfold the rate of ocean level rise in the 20th century.

The anthropogenic impacts on the global climate will remain for a long time. This determines the following specific features of the respective processes:

- The supposed stabilization of the level of atmospheric CO_2 concentration requires a considerable reduction of CO_2 emissions into the atmosphere as well as a large decrease of other GHGs emissions.
- The effect of CO_2 emissions on the CO_2 concentration in the atmosphere is likely to be of long duration. Even several centuries after cessation of emissions the share of CO_2 remaining in the atmosphere may reach 20–30% of the whole volume of emissions.
- The rise of the mean-global SAT and the World Ocean level (due to thermal expansion) may also continue for several centuries after the stabilization of the CO_2 concentration, due to the gigantic thermal inertia of the oceans.
- The response of ice sheets to earlier climate changes may continue for several thousand years after the climate stabilizes. According to model calculations, maintenance of the local mean-global warming by more than 3°C during thousands of years may lead to the complete melting of the Greenland ice sheet. With a 5.5°C warming, the World Ocean level should rise by 3 m during 1000 years (due to the Greenland ice sheet melting). The present models of the ice sheet dynamics in western Antarctica show that in the process of its melting the ocean level may rise by not more than 3 m during

1000 years, but one should take into account that the possible long-term dynamics of the western Antarctic cryosphere has not been properly studied.

Of course, one of the important components of the climatic system is the global biogeochemical CO_2 cycle, specified characteristics of which could remove many uncertainties and reduce the contradiction in the estimates of the greenhouse effect.

1.4 THE ATMOSPHERIC GREENHOUSE EFFECT AND GLOBAL CARBON CYCLE

During the last decade the term "greenhouse effect" has been used in numerous publications regarding the problems of global climate change on the Earth (Danilov-Danilian and Losev, 2000; Demirchian and Kondratyev, 1999c; Kondratyev, 2000b; Ledley *et al.*, 1999; Fearnside, 1997; Field and Fung, 1999; Friedlingstein *et al.*, 2001; Holtmann and Morais, 2000; Kheshgi *et al.*, 1999a; Sorokhtin and Ushakov, 1996, 1998). This term implies a totality of results of simulating the effects caused in the climate system and associated with a number of natural and anthropogenic processes. On the whole, the term "greenhouse effect" refers to an explanation of changes in the atmospheric thermal regime caused by the impacts of some gases on the process of radiation transport. Many gases are characterized by a high stability and long residence time in the atmosphere (see Table 1.10). Carbon dioxide is one of them. Other greenhouse gases are CH_4, N_2O, CFC-11 (CCl_3F), and CFC-112 (CCl_2F_2). Their observed concentration growth in the atmosphere is characterized by the following values. Since the pre-industrial period to 1997 the CO_2 concentration has grown, on average, from 280 ppm to 364 ppm. From estimates by Ledley *et al.* (1999) in 1996 the atmosphere gained 6.5 GtC due to fuel combustion and by-products of the cement industry. From the data in Fan *et al.* (1998) an additional 1.6 ± 1.0 GtC were emitted to the atmosphere due to changes in land use. Such a high uncertainty in this estimate is explained by the poorly studied spatial pattern of land use. Atmospheric CH_4 concentrations increased from

Table 1.10. The characteristics of the most important greenhouse gases (Mintzer, 1987; Ledley *et al.*, 1999).

Greenhouse gas	Lifetime in the atmosphere (years)	Average concentration	Rate of concentration increase (%)
Carbon dioxide	2–10	362 ppmv	0.5
Nitrogen oxide	100–150	308 ppbv	0.25
Methane	10–11	1815 ppbv	1.0
Chlorofluorocarbons:			
CFC-11, $CFCl_3$	50–75	0.34 ppbv	7.0
CFC-12, CF_2Cl_2	102–111	0.54 ppbv	7.0

700 ppb in the pre-industrial period to 1721 ppb in 1994 (Houghton *et al.*, 1996). The sources of anthropogenic CH_4 connected with fuel combustion give approximately 70–120 Mt CH_4/yr. At the expense of rice paddies, biomass burning, stockbreeding, and dust-heaps, the atmosphere obtains a further 200–350 Mt CH_4/yr. Among the non-anthropogenic sources of CO_2 and CH_4 of special importance are marsh biogeocoenoses. According to the estimates by Dementyeva (2000), in 1998 and 1999 the rates of CO_2 emissions from marshes during the vegetation period were 75.8–216.2 mg $CO_2/m^2/h$ and 94.4–104.5 mg $CO_2/m^2/h$, respectively. These figures demonstrate the wide range of seasonal variations of CO_2 fluxes from the soil to the atmosphere, which necessitates a detailed inventory of land biogeocoenoses.

Atmospheric N_2O concentrations increased from 275 ppb in 1900 to 312 ppb in 1994 (Houghton *et al.*, 1996). The anthropogenic N_2O emissions varied between 3–8 Mt N/yr. The principal sources of anthropogenic N_2O emissions to the atmosphere were agriculture and industrial production of acid.

Before 1950 there had been practically no chlorofluorocarbons (CFC-11 and CFC-12) in the atmosphere. With the use of these components in refrigerators and other devices the problem of ozone depletion appeared, which led to the Montreal Protocol that foresaw a substantial reduction of the CFC-11 and CFC-12 content in the atmosphere of the 21st century.

The stability of some GHGs is estimated from their average lifetime in the atmosphere, which is defined as the time needed to remove 63% of an anthropogenic emission from the atmosphere (Table 1.10). It is very difficult to estimate this for CO_2 because of the different timescales of its fluxes between biospheric reservoirs. The model estimates of the CO_2 lifetime show that 70–85% of anthropogenic CO_2 is removed from the atmosphere annually.

Numerous long-term observations in different latitudinal belts show a high level of correlation between temperature and CO_2 content. From the estimates by numerous authors the largest contribution to this dependence is made by the atmosphere–ocean interaction (Nefedova and Tarko, 1993). Although the atmosphere and the ocean are in a certain balance with respect to the CO_2 exchange, this balance gets regularly broken. The most serious reasons for this breaking if atmosphere–ocean balance are: (1) changing SAT; (2) changing volume of the ocean; and (3) changing regime of the oceanic vertical circulation. In general, these reasons can be characterized, from the viewpoint of their efficiency, by the ratio of their forcing on CO_2 concentration in the atmosphere. The changing SAT introduces a change of about 65% to the CO_2 partial pressure in the atmosphere (p_a). The remaining 35% can be attributed to the changing ocean volume and regime of the oceanic vertical circulation. Quantitatively, this ratio is characterized by a 6% increase in the CO_2 partial pressure per °C rise in the upper ocean layer temperature. A 1% decrease of the ocean volume raises p_a by 3%.

An estimation of the greenhouse effect requires complex consideration of interactions among processes of energy transformation on the Earth. However, there is a certain hierarchy in the significance of the diverse processes (from astronomical to biological) affecting the climatic system on various timescales. This hierarchy cannot be constant since the role of individual processes can vary within a wide range of significance for climate changes. Consideration of individual factors simplifies the analysis of their effect on climate. The level of the greenhouse effect is determined by an excess of surface temperature T_s over the effective temperature T_{eff}. The Earth surface temperature T_s depends on emissivity ε. The effective temperature T_{eff} is a function of the emissivity α of the atmosphere–ocean–land system. In general, the values ε and α depend on many factors, including the atmospheric CO_2 concentration. There are a lot of simple and complex mathematical models used for the parametrical description of these dependences. Unfortunately, so far, there is no model which would meet the requirements of adequacy and reliably simulate the prehistory of the climatic trends on the Earth. Nevertheless, one can state that the greenhouse effect depends non-linearly on the difference $T_s - T_{eff}$, that is, on atmospheric transparency, especially in the longwave region. The greater the CO_2 content in the atmosphere, the stronger atmospheric attenuation expected. In the longwave interval 12–18 μm the effect of CO_2 on atmospheric transparency is the strongest. Its weakest effect is in the intervals $\frac{7}{8}$, $\frac{9}{10}$, 2.0, 2.7, and 4.3 μm. It is clear that with an increasing partial pressure of CO_2 in the atmosphere the role of various CO_2 bands will increase, meaning that with an intensified CO_2 absorption the upward longwave radiation flux will decrease. At the same time, on the Earth's surface, the downward longwave flux will increase. From the available estimates a decrease of the upward and increase of the downward fluxes are characterized by the values 2.5 and 1.3 W/m^2, respectively (Ramanathan and Coakley, 1978).

To assess the level of the greenhouse effect due to CO_2, one should be able to predict its concentration with an account of all the feedbacks in the global biogeochemical carbon cycle as well as a correlation of this cycle with the cycles of other GHGs. From the estimates by Gale and Freund (2000), the contribution of various gases into the greenhouse effect are averaged at: CO_2 – 63.5%, CH_4 – 20.5%, nitrogen oxides – 4.5%, and CFCs – 11.5%. One of the most important problems of the global model is temporal variations of various GHGs.

A complete list of CO_2 fluxes is given in Table 1.11, with independent sub-cycles of CO_2 in the World Ocean and on land. The relationship between these components is realized through the atmosphere and is due to river run-off. From the available estimates, the intensity of the mean global CO_2 circulation constitutes about $6 \cdot 10^9$ tC/yr. Some quantitative characteristics of the CO_2 global biogeochemical cycle have been published by many authors (Bach *et al.*, 1987; Biutner, 1986; Eliasson *et al.*, 1999; Kelley, 1987). The present proportion of the carbon content in various global reservoirs demonstrates vividly the role of each of them in the global CO_2 cycle. The proportions are: atmosphere/land/ocean/geosphere = 1/3/50/10. It follows that the role of the World Ocean in the CO_2 cycle exceeds all the other reservoirs. The law of carbon flow in the ocean is connected with the carbonate system of the ocean, where carbon is divided into organic and inorganic constituents. Carbon dioxide dissolving in the ocean forms carbonic acid. An equilibrium condition of the reaction of carbonic acid formation depends on the partial pressure of gaseous CO_2, concentration of dissolved carbonic acid, temperature, pressure, and salinity of sea water. Carbonic acid gives bicarbonate and carbonate ions. The latter is the final product of the reaction of dissolved CO_2. The concentra-

Table 1.11. Description of the global CO_2 cycle.

CO_2 flux	Processes responsible for the CO_2 flux formation
H_1	Fuel burning
H_2	Desorption
H_3	Sorption
H_4	Erosion of land surface
H_5	Volcanic eruptions
H_6	Photosynthesis in the World Ocean
H_7	Plant respiration
H_8	Vegetation burning
H_9	Humus decomposition
H_{10}	Human activity
H_{11}	Biotic activity in the World Ocean
H_{12}	Human vital functions
H_{13}	Animal vital functions
H_{14}	Plant decay
H_{15}	Animal decay
H_{16}	Bottom deposits
H_{17}	Solution of bottom deposits in the World Ocean
H_{18}	Decay of dead organic matter
H_{19}	Water rise
H_{20}	Water lowering and gravitational sedimentation
H_{21}	Photosynthesis of land vegetation
H_{22}	River run-off into the World Ocean

tion of carbonate ions in the ocean increases until the limit of the calcium carbonate solubility is exceeded.

The process of photosynthesis is a source of organic carbon in the ocean. In the reaction of photosynthesis 106 mol CO_2, 16 mol nitrate, and 1 mol phosphate are used. Under the influence of the trophic pyramid of the ocean, the photosynthesis production changes substantially, and as a result, the concentration of inorganic matter is regulated through deposits of calcium carbonate and organic matter oxidation. In this regulation, the process of the vertical mixing of the ocean contributes significantly. The result of all these processes is a CO_2 exchange on the *ocean–atmosphere* border. The CO_2 cycle in the *atmosphere–vegetation–soil* system has been described in detail by Krapivin *et al.* (1982), who give the base fluxes of carbon in the land ecosystems. Their photosynthesizing elements assimilate CO_2 from the atmosphere and transform it into substratum and humus; 0.5% being washed into underground water and further into the oceans. The spatial structure of these fluxes is determined by the global distribution of different types of soil–plant formations. Table 1.12 characterizes the role of various types of biocoenoses in the global cycle of CO_2. It is clear that an accuracy of the estimates of land vegetation productivity and the reliability and details of its structural classification are important parameters in specifying the scheme and parameters of the global CO_2 cycle.

Table 1.12. Biomass and productivity (dry matter) of some biocoenoses (Watson *et al.*, 2000).

Biocoenoses	P	B	τ
Tropical humid forests	2200	45,000	70
Tropical seasonal forests	1600	35,000	80
Moderate-zone forests:			
evergreen	1300	35,000	40
broad-leaved	1200	30,000	40
Arctic forests	800	20,000	90
Bushes	700	6000	50
Savannah	900	4000	50
Moderate-zone meadows	600	1600	60
Tundra and alpine vegetation	140	600	100
Desert and semi-desert vegetation	90	700	100
Complete deserts (sand, ice, stones)	3	20	40
Cultivated land	650	1000	80
Marshes and bogs	3000	15,000	40

Notation: P – primary productivity ($g/m^2/yr$); B – biomass (g/m^2), τ – carbon cycle time (years).

The global CO_2 balance is defined by numerous functional dependencies applied by various investigators in their models of this balance. Since, on the whole, globally there is no single available database, including adequate descriptions of all processes of CO_2 transformation in the ocean and on land, at present all conclusions about catastrophic consequences of the expected warming due to increasing partial pressure of atmospheric CO_2 should be assessed as hypothetical. Of course, not all fluxes given in Table 1.11 are equal in their influence on changes in atmospheric CO_2 concentrations. But such fluxes as CO_2 exchange on the border of the atmosphere with the ocean and land prevail when estimating the atmospheric CO_2 gradients. However, it is these fluxes that are still poorly assessed and practically lacking a adequate parameterizations. Most of the models are based on the fact that the CO_2 flux at the ocean–atmosphere border is in proportion to the difference of its partial pressures in these domains. But as shown in Alekseev *et al.* (1992), this flux is described more reliably by the model in which the flux is considered to be in proportion to square root of this difference.

1.5 MODELLING THE ANTHROPOGENIC PROCESSES

1.5.1 Anthropogenic factors in the global environmental dynamics

The human–nature (N) interaction is a function of a broad and complex group of factors functioning both in human society and in the environment (Kondratyev,

1999a; Demirchian and Kondratyev, 1998; Kondratyev *et al.*, 1997a; Marchuk and Kondratyev, 1992; Gorshkov *et al.*, 2002; Jenkins *et al.*, 2002). The principal problem of this interaction is the globalization of the human effect on the environment due to a rapid development of megalopolises concentrating the sources of anthropogenic impacts on the local environment. Almost half the global population live in cities occupying only about 3% of the land area. This trend of concentrated populations in cities and the transformation of the latter into megalopolises will intensify, judging from numerous expert estimates. The number of large cities (with populations of 3–8 million) and megalopolises (>8 million) in the 21st century will increase in the developing countries, with limited resources for environmental safety. Therefore the ecological conditions of the territories containing 80% of the global population creates concern and demand for additional investment (Kondratyev, 1996).

Many cities are contaminated by waste, huge territories have been transformed into waste heaps, the discharge of contaminated sewage into the hydrosphere and the emission of polluting matter and gases to the atmosphere have increased. Processes of transformation and even vegetation cover destruction have intensified. Of course, human economic activity has always taken place during historical development. Since the pre-historical epochs people have gradually increased their influence on nature, intensified the use of mineral resources, contaminated the environment, and broken the connections between various natural events established in the process of evolution, which have led, in particular, to a biodiversity reduction, deforestation of huge territories, soil salination, depletion of biological resources of the seas and oceans, and, finally, threatened the existence of life itself on the Earth. In the period from the Palaeolithic epoch to the epoch of feudalism these forcings had been first local and then regional. The technical progress in the 18th and 19th centuries had contributed to broadening the scale of human impacts on nature, causing an extermination of some species of animals, wholesale destruction of forests, and rapacious exploitation of other natural resources. In the 20th century these anthropogenic forcings have become global in scale and dangerous for the biosphere. The search for a strategy of an optimal interaction of the human–nature system began. In this regard, due to accomplishment of numerous international and national programmes of environmental studies, data have been compiled that make it possible to assess the level and direction of anthropogenic processes. The data in Tables 1.13 and 1.14 are indicative of this.

The principal regularity noted by many authors (Kondratyev, 1998a,b; Watson *et al.*, 2000) is that the rate of impact of human civilization on the biosphere depends directly on population growth. Such categories as energy production, amount of consumed mineral resources, and investments depend directly on the size of population (Demirchian and Kondratyev, 1998). The result is that the planetary forcing of society on natural processes has reached a level where the future survival of humankind as a biological species is in doubt. To survive, the human race should learn to co-evolve with nature, and to do that, it is necessary to thoroughly study and understand it (Essex and McKitrick, 2002; Kondratyev, 1992, 1993; Kondratyev *et al.*, 1992; Nakicenovic *et al.*, 1998).

Table 1.13. Dynamics of the global water consumption (km^3/yr).

	Water consumption W_G and of irretrievable losses P_G specific consumers						Total water consumption and irretrievable losses	
	Municipal services		Industry		Agriculture			
Year	W_G	P_G	W_G	P_G	W_G	P_G	W_G	P_G
1900	20	25.0	30	6.7	350	74.3	430	67.5
1940	40	20.0	120	5.0	660	72.7	870	61.0
1950	60	18.3	190	4.7	860	73.3	1190	59.1
1960	80	17.5	310	4.8	1510	76.7	1990	63.2
1970	120	16.7	510	3.9	1930	78.9	2630	61.5
1975	150	16.7	630	4.0	2100	76.2	3080	60.0
1985	250	15.2	1100	4.1	2400	79.2	3970	56.4
1995	320	15.1	1560	3.9	2760	78.4	4750	54.9
2000	440	14.8	1900	3.7	3400	76.5	6000	50.0

Table 1.14. Characteristic of regional water consumption in the 20th century.

	Population		Industry		Agriculture			Total
Region	*a*	*b*	*a*	*b*	*a*	*b*	Σ	*b*
Europe	29	11	160	5	125	67	314	24
Asia	40	15	60	6	1400	79	1500	74
Africa	4	18	3	7	110	82	117	78
North America	41	21	270	30	210	62	521	28
South America	4	15	8	10	55	82	67	74
Australia	1	10	8	8	13	77	22	49

Notation: *a* – net water consumption (km^3); *b* – % of irretrievable losses; Σ – total amount of water consumption (km^3).

According to the general opinion of experts with various scientific backgrounds, the basic trends in human economic activity are to regulate the global processes in the biosphere, such as production of energy, industrial materials, and food. On average, the efficiency of these processes calculated per capita tends to increase. For instance, in the USA the per capita amount of energy increases annually by ~2.5%. The mean global energy increase exceeds that of population by 2%. So, for the territory of the former Soviet Union the average population increase constituted 1.4%/yr, and the annual growth of energy production is estimated at 7%. On the whole, the energy potential has been, so far, one of the characteristics of the level of human society development. Other parameters determining the state of production and the economy of the region, territory, country, and world, depend directly on the energy potential.

The atmosphere, hydrosphere, and soil are becoming more and more contaminated. Practically, every principal geosphere within which there is life, suffers from growing anthropogenic forcing. For instance, during the last years of the 20th century more than 4 Mt of oil have been discharged into the World Ocean. The concentration of mercury in the oceans has almost doubled.

Since all the domains of the geosphere are interconnected, a destructive forcing on one of them threatens all the others. Therefore, for example, burning of the final product of sewage purification cannot be considered a solution to the problem of the pollutants control, since this technology only transfers the load from the hydrosphere to atmosphere. Also inefficient is the cleaning of soil by pumping organic pollutants into sub-soil waters. In short, transporting the pollutants into any domain of the geosphere damages all spheres of life. Of course, all spheres react differently to contamination, and a redistribution of anthropogenic forcings among various geospheres can be an element governing the environment. Therefore whilst analysing the processes of pollution generation, it is necessary to determine the complex characteristics of the anthropogenic processes. For this purpose we shall consider each geosphere and formally describe the processes of their pollution.

One aspect of human impact on the Earth's climate is atmospheric pollution. Along with the GHG biogeochemical cycles that break the natural heat balance of the Earth, a relevant global model should take into account other pollutants that change the optical properties of the atmosphere, such as industrial dust, combustion gases, radioactive substances, smoke, molecular admixtures, and vapours of water and other liquids. For instance, molecular admixtures within the atmosphere can markedly change atmospheric temperatures; the most dangerous admixtures being those whose absorption spectrum is in the atmospheric "transparency windows". Examples of atmospheric molecular pollutants are methane, water vapour, sulphur dioxide, hydrogen sulphide, ozone, carbon monoxide, and nitrogen oxides. These components are divided into stable and short-lived (Table 1.15). The short-lived pollutants, such as SO_2 or mercury vapour are unstable and cannot propagate over long distances, thus cannot transform local anomalies into global ones. Contrary to this, the long-lived pollutants like CO_2 tend to accumulate in the atmosphere and propagate over large territories. Therefore their concentration over long time periods becomes uniform in the global atmosphere.

Dust and aerosols get into the atmosphere from many sources. Small solid particles are emitted to the atmosphere together with gases from internal combustion engines and industrial smoke from metallurgical works and power stations. About half the solid particles ejected with combustion gases contain lead that tends to propagate for long distances. For example, anthropogenic lead was discovered in the Greenland mainland ice. A dangerous component of atmospheric pollution is mercury whose dispersed liquid particles are harmful to all life. From the available estimates the total amount of mercury emitted annually into the atmosphere reaches 1100 t. Mercury sulphate with the concentration 10 ml/l kills living organisms in several days, and with a lower concentration it substantially violates life cycles. For example, tests on *Monodonta articulata* using mercury salts with a concentration of (sulphate, acelate or chloride) 0.01 mg/l were found to reduce its O_2 consumption.

Table 1.15. Atmospheric residence times for some atmospheric components.

Component	Atmospheric residence time
Carbon dioxide	5 years
Carbon oxide	0.1–3 years
Water vapour	10 days
Sulphur dioxide	3 days
Ozone	10 days
Hydrogen chloride	3–5 days
Nitrogen oxide	5 days
Nitrogen dioxide	5 days
Nitrous oxide	120 years
Ammonia	2–5 days
Methane	3 years
Freons	50–70 years

A discharge of mercury and copper into the sea off California killed gigantic focus algae after only four days of exposure to negligible concentrations of these metals (0.05 g mercury and 0.1 g copper per 10^6 g [or metric tonne] of water). Dangerous concentrations of mercury (for humans) is found in fish and mollusks in more and more water bodies of the World Ocean (coastal waters of the USA, Sweden, Norway, Canada, and the Mediterranean) (McIntyre, 1999).

Radioactive substances are also dangerous – some of them can remain in the atmosphere for a long time. Of significant danger are accidents at atomic reactors and atomic bomb explosions – most intensive sources of radioactive elements.

Both technogenic and natural processes affect changes in the chemical and optical properties of the atmosphere. Table 1.16 lists some sources of atmospheric pollution. Each source is special. This list is constantly renewed due to the appearance of new spheres of human activity.

Atmospheric pollution changes human and animal habitats and causes climatic variations. The spectrum of human activity broadens towards the higher standard of living, and this fuels an increase of energy consumption and productivity of natural and man-made ecosystems (Table 1.17). Feedbacks appear that require changes in technologies of natural resources exploitation. On the other hand, the scales of atmospheric pollution grow linearly with the growing size of population.

The next important sphere of the anthropogenic activity manifestation is the pollution of the World Ocean with oil products, organic and inorganic compounds, and various chemicals. From the available estimates, the total amount of oil products getting into the oceans annually varies between 32–60 Mt (Klubov *et al.*, 2000). Oil pollution is not uniform in the oceans, in many water bodies it exceeds the admissible normal levels by thousands of times (0.5 mg/l). From the estimates of many authors, the concentrations of oil hydrocarbons on the surface microlayer of

Table 1.16. Sources of atmospheric pollution (Brenninkmeijer and Rockmann, 1998, 1999; Demirchian *et al.*, 2002; Grigoryev and Kondratyev, 2001a).

Sources of pollution	Pollutant
Natural	
Volcanoes, fumaroles, and solfataras	Gases, volcanic dust, and mercury vapour
Natural emissions of natural gas and oil	Hydrocarbons
Deposits	
mercury	Mercury vapour
sulphides	Sulphur dioxide
radioactive ore	Radon
Blowing away from the sea and ocean surfaces	Chlorides, oil, and sulphates
Underground coal fires	CO_2, CO, SO_2, and hydrocarbons
Natural forest and steppe fires	Smoke
Plant transpiration	Water vapour, aromatic and other volatiles
Anthropogenic	
Burning of solid and liquid organic matter	CO_2, CO, SO_2, lead, hydrocarbons, mercury vapour, cadmium, and nitrogen oxides
Metallurgy of black, non-ferrous, and rare metals	Dust, SO_2, mercury vapour, and metals
Atomic industry	Radioactive substances
Nuclear explosions	Radioactive isotopes
Cement industry	Dust
Construction explosions	Dust
Human-caused forest and steppe fires	Smoke
Oil and gas fields	Hydrocarbons
Auto mobile transport	CO, soot, and nitrogen oxides

the World Ocean exceeds maximum permissible norms by a factor of 19.8 in the Atlantic Ocean, 3.6 in the Pacific Ocean, 4.4 in the Indian Ocean, 29.5 in the North Sea, 25 in the Norwegian Sea, and 4.8 in the Caribbean Sea. The property of oil is such that after a leakage it spreads over the water surface forming a film $1.02 \cdot 10^{-3}$ mm thick. The areal spread of an oil spot of 1 t reaches 12 km^2. At present about 30% of the World Ocean's surface is covered with an oil film (Kondratyev *et al.*, 2003). This film hinders the atmosphere–ocean gas exchange as well as decreasing the coefficient of water transparency. Settled oil accumulates in bottom sediments, which then violates the natural process of the vital functions of benthos animals and bacteria. Important trophic and energetic feedbacks in the structure of the marine ecosystem get broken, and damage by oil pollution is transported through these trophic chains of the marine ecosystems (Stephens *et al.*, 2000; Patin, 1999).

Table 1.17. Some characteristics of the energy parameters of the biosphere.

Characteristic	Estimate
Energy of the Earth's mass (J)	10^{46}
Global consumption of all kinds of energy (kJ/yr)	$(2.1–3) \cdot 10^7$
Energy utilized by the biosphere (J/yr)	10^{20}
Energy used by humans (J/yr)	10^{21}
Energy of tides and ebbs (kWhr/yr)	$6.4 \cdot 10^{14}$
Energy of rivers (kWhr/yr)	$23 \cdot 10^{12}$
Energy of wind (kWhr/yr)	$15 \cdot 10^{23}$
Solar energy consumption (%)	
Falling on the Earth surface	100
Accumulated in photosynthesis	0.1
Used as food	0.001
Energy needs satisfied due to burning of the fossil products of photosynthesis (%)	95
Consumption of photosynthesis products as food (%)	1
Coefficient of performance (non-dimensional)	
Stockbreeding	0.2
Agriculture	
Extensive	20
Intensive	2
Greenhouse	0.02
Energy spent in agriculture on nitric fertilizer production by chemical fixation of atmospheric nitrogen (%)	30
Annual rate of the energy consumption increase (%)	4
Volume of burnt organic fuel (10^9 t/yr)	6–7
Annual expenditure of natural gas (10^{12} m^3)	1.4
Geological supplies of coal and brown coal (10^9 t)	$(14–15) \cdot 10^3$
Energy content in these supplies (10^9 J)	$201 \cdot 10^{12}$
Natural gas supplies (10^{12} m^3)	145–340
Explored oil deposits (10^9 t)	65
Extracted oil (10^9 t)	147
Global oil output (10^9 t/yr)	7.4
Share of the sea oil products in the global oil output (%)	19–20
Extraction of mineral fuel from the ocean bottom	
Oil (10^6 t/yr)	450
Gas (10^9 m^3/yr)	180
Known supplies of combustibles in units of conditional fuel (10^9 t)	$27 \cdot 10^3$
Volume of extracted oil equivalent (10^9 t/yr)	6
Total energy of fossil fuels (kJ)	10^{20}
Oil (% of total energy)	10
Gas (% of total energy)	10
Coal (% of total energy)	80
The ratio of the amount of explored deposits of fossil combustibles to their annual extraction	
Oil	37
Gas	41
Coal	720
Energy emitted by burning timber (kJoule/yr)	$2 \cdot 10^6$
Power of all powerstations over the globe (kW)	10^{10}

Characteristic	Estimate
Share of total anthropogenic energy emissions over the Earth in its radiation budget	0.02
Energy expenditure/consumption (%)	
Heat power stations and central heating	30–35
Industry	30–35
Transport	25–30
Household needs	5–10

Along with the oil pollution of the World Ocean in the late 20th century, dangerous toxic pollutants of the marine domain have become appreciable, such as lead, mercury, arsenic, and other heavy metals, which damage the marine organisms and accumulate within them. This makes these organisms unfit for consumption. Quantitatively, the natural background concentrations of heavy metals in numerous water bodies of the oceans exceed between five and tenfold the admissible levels and in some places even reach dangerous levels. So, with a total input of mercury into the oceans of 5000 t/yr due to its local use in chemical and the paper industry the highest concentrations of mercury are observed in the coastal waters of Japan, Scandinavia, the Netherlands, and Canada.

Approximate estimates put the annual input of lead into the World Ocean at 40,000 t, half of which settles out of the atmosphere. This half contains a substance especially dangerous for marine organisms – tetraethyl lead, usually added to petroleum. Lead, like mercury, can accumulate in marine organisms.

The World Ocean pollution spectrum is rather broad. Solid wastes and sewage play a marked role in the total pollution of the oceans. From rivers and ships a huge amount of rubbish gets into the oceans. All this leads to degeneration of the oceanic domain and worsening of the living conditions for marine animals.

Along with pollutants directly harmful for marine domains, huge amounts of nutrient elements are introduced into the oceans together with everyday sewage and fertilizers washed out from the fields (e.g., various compounds of nitrogen, phosphorus, and silicon). In some cases such inflows are harmless or even useful, but they can often lead to negative consequences causing eutrophication, or secondary biological pollution. In this case excessive reproduction and development of algae takes place which in turn consumes all oxygen dissolved in the water, killing large numbers of living organisms. The eutrophication processes can take place without anthropogenic interference (e.g., in the region of the Peru upwelling). These processes occur in a balanced regime developed in the process of evolution and, on the whole, are not dangerous for the ocean ecosystem dynamics.

Bearing in mind that the annual global input of sewage into the sea has exceeded 450 km^3 and in the future it may increase between 10–15 times due to population growth, the eutrophication processes will become all-embracing with unpredictable consequences. Therefore one of the goals of eco-informatics is to assess these processes and their consequences. Clearly, the technogenic pollution processes and natural self-cleaning ability of the oceanic waters are in a constant struggle. A complex network of physical, chemical, and biological processes of pollutant decomposition and transformation, resulting in non-toxic substances, provides the ocean with a powerful barrier to anthropogenic forcing. However, the rivers, for example, have already lost this ability since the water medium does not have time to clean itself completely, allowing the pollutants to accumulate. As a result, many rivers have been transformed into ditches (Vega *et al.*, 1998; Ragas and Leuven, 1999).

The World Ocean is capable of self-cleaning of such stable pollutants as oil. Due to the atmosphere–water surface interaction, oil gradually evaporates: during the first 24 hours 15%, during the next ten days ~20%. Storm winds forming foam crests on the water surface favour an input of water microdroplets into the atmosphere which also carry oil particles. Solar UV radiation, chemical oxidation, and bacterial decomposition also destroy the oil products. Micro-organisms use oil hydrocarbons as a source of carbon and energy and transform these hydrocarbons into bacterial cells, carbonic acid, and water. Besides, the micro-organisms and many zooplankton species promote the transport of oil into the ground.

Anthropogenic forcings on the environment should be assessed together with natural processes, in order to work out a technology to reliably predict the consequences of human activity. For this purpose it is necessary to analyse the sphere of anthropogenic activity – utilization of natural resources. On the whole, the following groups of resources are considered: fossil (geological and mineral), water, soil, vegetation, fauna, atomic, climatic, planetary, cosmic, chemical compounds of the Earth's crust deposits containing oil, gas, and coal, various salts, etc. Natural resources are divided into renewable and non-renewable, exhaustible and inexhaustible. The renewable resources are soil, vegetation, and fauna. The non-renewable resources are fossil fuels. Solar radiation and wind energy are inexhaustible resources. Water resources and atomic energy are practically globally inexhaustible but can become scarce in some regions. Depending on time and scientific and technical progress, the anthropogenic processes can change both their scale and direction.

Assessments of fossil fuel supplies make it possible to predict, to some extent, levels of their anthropogenic utilization. Numerous elements of industrial value are contained in the Earth's crust in relatively small amounts. For instance, the share of aluminum and iron is, respectively, $\frac{1}{24}$ and $\frac{1}{30}$. Almost half of all chemicals contain oxygen. Land contains the basic source of minerals and raw materials for humankind. Chemical elements from the World Ocean are only utilized in small amounts.

At the dawn of industrial activity humankind had used only 19 chemical elements and compounds: by the 18th century this had risen to 28, and in the early 20th century there were more than 100. With the development of science and technology the use of minerals and their extraction has intensified. So, during the last 100 years the annual consumption (and, hence, extraction) of coal, iron, manganese, and nickel has increased between 50–60 times; that of tungsten, aluminum, molybdenum, and potassium between 200–300 times. Annually about 100 Gt of ore and mineral fuel are extracted, as well as more that 300 Mt of mineral fertilizers.

The per capita amount of extracted fuel is the principal indicator of energy supply. In 1972 the per capita amount of oil equivalent fuel constituted about 2 t, but in the early 1970s the annual output of oil equivalent fuel constituted almost 6 Gt, accounted for as follows: transport – 25/30%, heat power stations – 30/35%, industry – 30%, and everyday needs – 5/10%.

The basic types of minerals are fuel resources, ores of black metals and ores of alloy metals, non-ferrous and noble metals, and non-metal raw materials. According to an assumed classification, the trends of extracting these resources are as follows. Fuel energy raw materials are oil, gas, and coal. In the 20th century oil and gas output has increased from 10 to 1800 Mt and from 5 to 1300 m^3, respectively. At the same time, the output of coal increased by only four to five times. The development trends of the output of other minerals such as iron and manganese ores, chromites, copper, lead, tin, molybdenum, etc., are growing. According to the available estimates, this pace of extraction will completely exhaust the acceptable supplies of aluminum in 570 years, iron in 250 years, zinc in 23 years, and lead in 19 years, and by the mid-21st century most of the metals will disappear completely. The dangers of the possible realization of this prediction makes one intensify the extraction and use of fossil non-metals, such as asbestos, black lead, mica, melted spar, etc.

Noble metals (platinum, gold, and silver) are special among minerals. Their volumetric production is small, but they are important not only in the development of industry and scientific instrument-making but also in the formation of the socio-economic relationships in human society.

Consumption of natural resources includes also the use of fresh water and forests. The global water balance has been discussed by Krapivin and Kondratyev (2002). It includes a component that affects the natural water cycle. In particular, although the water used in households returns to its source eventually, it does so in quite a different way than it would during natural exchange between water and individual biospheric elements. Water consumption for household needs is followed by irreversible losses, which, for instance, in the USA and Russia constitute 10–20%. Water supply in agriculture is characterized by 20–40% of irretrievable losses. On the whole, the volume of irretrievable water consumption by the population exceeds 20 km^3/yr or 17%. Industrial water consumption constitutes 500 km^3/yr globally.

Irrigation causes especially large fluctuations in the water cycle. The total amount of water used for irrigation is determined by physico-geographical conditions, agricultural crops, and technical equipment of irrigation systems and their application technology. The general characteristics of the water consumption dynamics is given in Tables 1.13 and 1.14.

Along with the quantitative change of the water balance in the 20th century there has developed a process of exhaustion of water resources connected with increasing pollution of natural waters. Most sewage containing harmful chemicals is discharged into the hydrophysical network without preliminary purification, causing a change in the water resources quality and making it unfit for further use without purification. This circumstance has led to a preferential use of underground, less polluted waters. As a result, conditions have been created for a sharp violation of the natural water cycle in the biosphere.

Forest is one of the catastrophically exhausted natural resources. Forests are in the process of being exterminated over the globe. For example, in the USA only 18 Mha of the once 365 Mha of forested areas remain (Watson *et al.*, 2000). The area of tropical forests has decreased by two-thirds. In many countries the forested areas occupy a small part of their territory: in India – 18%, Greece – 15%, Spain – 12.5%, China – 9%, and Cuba – 8%. On Madagascar the forests have been exterminated almost completely.

Soils as an independent natural body are one of the natural resources of great importance for the biosphere. Parent rock, living organisms (plants, animals, and micro-organisms), climate, and relief are factors of soil formation. The anthropogenic forcing on soil is manifested through its impoverishment. The impoverished ploughed soil is easily subject to erosion. Deep shaft and open cast mining are factors promoting soil destruction. Fertile soils become barren and deprived of vegetation producing so-called "industrial deserts".

Thus anthropogenic factors in these conditions are versatile in their functions and affect practically all the natural processes. Clearly, a complete account of the whole spectrum of anthropogenic forcings on the environment is impossible because of a lack of data. In modelling the global carbon cycle a unit is needed in the models responsible for the parameterization of anthropogenic processes and providing their consideration in simulation experiments.

1.5.2 Modelling the demographic processes

The demographic situation determines the dynamics of anthropogenic processes. Therefore attempts to develop models of demographic processes are justified. The existing predictions of changes of the size of population and variations in its spatial distribution enable us to synthesize the scenarios to be used in a global model as well as to try to solve the problem of its verification. An adequate model of the global demographic process requires an extensive database covering the characteristics of changes in the standards of demographic behaviour, detailed information on intensities of demographic processes in various regions of the planet, assessments and criteria of demographic policy, and the structural indicators of the human society, which are especially important (Demirchian and Kondratyev, 1999).

Many authors have considered the problem of parameterization of the demographic processes. The demographic processes are considered as a part of the planetary biospheric processes in which *Homo sapiens* plays the role of a user of biocoenoses products and a regulator of energy and matter fluxes among them.

1.5.2.1 Matrix model of the population size dynamics

In accordance with the possibility of and the needs of the global model, the unit of population dynamics includes the impact of the following factors (Krapivin and Marenkin, 1999; Logofet, 2002):

- per capita food provision F (calculated as the sum of several shares of vegetation and animal population within the region as well as fish catches);
- the share A of animal protein in the human diet (determined from the contribution to the F value from animals and fish);
- the level of public medical service M (in this model version it is in proportion to the per capita financing); and
- genetic load for human population G (grows slowly with population development and depends on the level of environmental pollution).

The sexual structure of the population and the processes of population migration between regions are ignored. The age structure contains three groups (0–14 years, 15–64 years, and 65 years and older). There is also a fourth group within the population (invalids and disabled persons outside of these three age groups). For terminology convenience the fourth group is considered as not affecting the age structure.

Let S_t be the age structure at a time moment t, then the population size dynamics can be described by the following matrix equation:

$$S_{t+1} = D \times S_t$$

where D is the demographic 4×4 matrix including the effect of the factors F, A, M, and G:

$$D = \begin{Vmatrix} d_{11} & d_{12} & 0 & 0 \\ d_{21} & d_{22} & 0 & 0 \\ 0 & d_{32} & d_{33} & 0 \\ d_{41} & d_{42} & d_{43} & d_{44} \end{Vmatrix}$$

Matrix D differs from Leslie's (1945) traditional matrix for the model of population with the age structure, firstly, by its diagonal elements inequality to zero, which is explained by overlapping of the next generations, and, secondly, by the last line that reflects the non-age character of the fourth group. The diagonal elements of matrix D are determined from an apparent balance of relationships:

$$d_{ii} = 1 - \mu_i - \sum_{j>i}^{4} d_{ji} \qquad (i = \tfrac{1}{4}) \tag{1.1}$$

which include the coefficients of mortality μ_i of the ith group, descending functions of the per capita food provision, and the medical service level $\mu_i = \mu_i(F, M)$ with:

$$\lim_{F,M\to\infty} \mu_i(F, M) = \mu_{i,\min} > 0$$

where $\mu_{i,\min}$ characterizes a minimum of physiological mortality with optimal food provision and medical service.

It is supposed that the reproductive potential of the human population is entirely concentrated in the second age group. The d_{12} coefficient value is considered to be a regional constant. Note that the birth rate coefficient is a complex and poorly studied function of many variables, the ethnic traditions being one of the important variables. The birth rate is affected by religion (i.e., the Moslems consider children to be their wealth, in south Asia the religious norms order each family to have at least one son). Catholicism also markedly affects the birth rate index. Therefore if the function μ_i is well parameterized by the statistical data, the coefficient d_{12} is still an unidentified parameter. Nevertheless, the d_{12} value for different global regions can be estimated from the birth rate data.

The coefficients of transition to the following age groups d_{21} and d_{32} are determined from the duration of the respective age group and the hypothesis of a uniform distribution of ages within the group, namely: $d_{21} = \frac{1}{15}$ and $d_{32} = \frac{1}{50}$.

Of all possible transitions to the invalid group the coefficient of transition from a younger group reflects the effects of genetic diseases and a deficit of proteins during early age. Thus $d_{41} = d_{41}(G, A)$ is the age function of its arguments. Coefficients d_{42} and d_{43} are small compared to d_{41} and determined by the sum of the vital parameters of the environment, the level of medical service, and other indicators of anthropogenic medium (investments). In a first approximation, suppose $d_{4k} = \Delta_k \mu_k (k = 23)$, where $\Delta_k \ll 1$.

Demographic matrix D has a certain set of properties which make it possible to reveal the typical trends of the demographic process. A maximum eigenvalue is:

$$\lambda_{max}(D) = \max\{\lambda_1; d_{33}; d_{44}\} > 0$$

where

$$\lambda_1 = 0.5(d_{11} + d_{22}) + 0.25[(d_{11} + d_{22})^2 + d_{12}d_{22} - d_{11}d_{22}]^{1/2}$$

It follows from (1.1) that if $\lambda_{max}(D) \geq 1$, then $\lambda_{max}(D) = \lambda 1$, that is, the rate of *Homo sapiens* population growth in the neighbourhood of the stationary age distribution is determined by λ_1 value. The stationary age structure is calculated as an eigenvector corresponding to $\lambda_{max}(D)$. The coefficients and modifying dependences of matrix D calculated, as a rule, from the data of Stempell (1985), give the dominating eigenvector $P_D = (0.369; 0.576; 0.05; 0.005)$. The first three components of vector P_D coincide, to within second sign, with the global data on the age complement of the global population.

The matrix version of the global model demographic unit has a time step of one year. An inclusion of this unit in the chain of other global model units with an arbitrary time step $\Delta t < 1$ yr is needed as an adjusting procedure. This is possible, for instance, if the demographic unit is included only at time moments of yearly multiples. In this version the spasmodic changes of the population size will cause some imbalance in continued trajectories of the global system. Another procedure free of this drawback consists in the use of a prognostic equation:

$$S_{t+\Delta t} = S_t + \Delta t(D - I)S_t \tag{1.2}$$

where I is an identical matrix. One can demonstrate that for the age structure vectors coinciding in direction with the eigenvector of matrix D, the main term of a relative error accumulated during one year with a time step $\Delta t = 1/n$ constitutes $(1 - 1/n)(\eta - 1)^2/2$. Let the right-hand part of (1.1) be denoted by $f(S, \Delta t)$. Then if e is the eigenvector of matrix D with an eigenvalue η, then with the time step $\Delta t = 1/n$:

$$f(e, 1/n) = e + (1/n)(D - I)e = (1 + [\eta - 1]/n)e$$

It is apparent that $S_{t+n\Delta t} = f^{(n)}(S_t, 1/n)$, and $S_t = ef^{(n)}(e, 1/n) = (1 + [\eta - 1]/n)^n e$. But if $\Delta t = 1$, then $S_{t+1} = f(e, 1) = \eta e$, so that a relative error resulting from a subdivision of the 1-year time interval is:

$$(1 + [\eta - 1]/n)^n - \eta = n(n-1)(\eta - 1)^2/(2n^2) + 0((\eta - 1)^3) \qquad \text{QED}$$

In other words, the accuracy of an approximation of (1.2) increases with Δt approaching unity.

1.5.2.2 Differential model of population dynamics

The matrix model considered above enables us to use the demographic statistics by age groups but requires knowledge of many parameters, which leads to uncertainties in the global model. Therefore a second version of the demographic unit is suggested which simulates the dynamics of the total size of the population only. This version assumes that an impact of numerous environmental factors and social aspects on the population size dynamics G_i in the ith region is manifested through birth rate R_{G_i} and mortality M_{G_i}:

$$dG_i/dt = (R_{G_i} - M_{G_i})G_i \qquad (I = 1, \ldots, m) \tag{1.3}$$

Birth rate and mortality depend on food provision and quality, as well as environmental pollution, living standards, energy supply, population density, religion, and other factors. Within the global model all these factors will be taken into account following the principle:

$$R_G = (1 - h_G)k_G G H_{GV} H_{GO} H_{GG} H_{GMB} H_{GC} H_{GZ}$$

$$M_G = \mu_G H_{\mu MB} H_{\mu G} H_{\mu FR} H_{\mu Z} H_{\mu C} H_{\mu O} G + \tau_{GO} \tau_{GC}^{\omega(G)}$$

where for simplicity purposes we shall omit the i index attributing the relationship to the ith region; h_G is the coefficient of food quality ($h_G = 0$ when the quality of food is perfect); coefficients k_G and μ_G point to levels of birth rate and mortality, respectively; indices τ_{GO} and τ_{GC} characterize the dependence of the regional population mortality on the atmosphere's oxygen and carbon dioxide concentrations – indicators of the environmental state (within the global model it is the content of O_2 and CO_2 in the atmosphere) manifested through human physiological functions; coefficient $\omega(G)$ characterizes an extent of the influence of population density on mortality (in the present conditions $\omega(G) \approx 0.6$); and functions $H_{GV}(H_{\mu FR})$, $H_{GO}(H_{\mu O})$, $H_{GC}(H_{\mu C})$, $H_{GMB}(H_{\mu MB})$, $H_{GG}(H_{\mu G})$, and $H_{GZ}(H_{\mu Z})$ describe, respectively, an impact on birthrate (mortality) of the environmental factors, such as food provision, atmospheric O_2 and CO_2 concentrations, living standards, population density, and environmental pollution. Functions $H_{\mu O}$ and $H_{\mu C}$ approximate the medico-biological dependences of mortality on the atmospheric gas composition. Consider all these functions in more detail. For this purpose we shall formulate a number of hypotheses concerning the forms of dependences of mortality and birth rate on various factors.

The results of numerous studies with national specific features taken into account enable us to assume the following dependence as an approximation of the function H_{GV}: $H_{GV} = 1 - \exp(-V_G)$, where V_G is an efficient amount of food determined as a weighted sum of the components of *Homo sapiens* food spectrum:

$$V_{G_i} = k_{G\Phi_i}\Phi + k_{GF_i}\left(F_i + \sum_{j\neq i} a_{F_{ji}} Fi\right) + kG_{Gr_i} I_i (1 - \theta_{Fr_i} - \theta_{ur_i})$$
$$+ k_{GL_i} L_i + k_{GX_i}\left[(1 - \theta_{FX_i})X_i + (1 - \nu_{FX_i})\sum_{j\neq i} a_{X_{ji}} X_j\right]$$

Here Φ is the volume of vegetable food obtained from the ocean; F is protein food; L is food provided by forest; X is vegetable food produced by agriculture; I is fishery products; coefficients $k_{G\Phi}$, k_{GF}, k_G, k_{GL}, and k_{GX} are determined using the technique described in Chapter 4; $a_{F_{ji}}$ and $a_{X_{ji}}$ are, respectively, shares of the protein and vegetable food in the jth region available for use by the population of the ith region; θ_{FX_i} and ν_{FX_i} are shares of vegetable food produced and imported by the ith region, respectively, to produce the protein food; θ_{Fr_i} and θ_{ur_i} are shares of fishery spent in the ith region on the protein food production and fertilizers, respectively.

With an increasing food provision the population mortality drops to some level determined by the constant $\rho_{1,\mu G}$ at a rate $\rho_{2,\mu G}$, so that $H_{\mu FR} = \rho_{1,\mu G} + \rho_{2,\mu G}/F_{RG}$, where the normalized food provision F_{RG} is described by the relationship $F_{RG} = F_{RG}(t) = V_G/G(F_{RG}(t_0) = F_{RGO} = V_G(t_\theta)/G(t_0))$. Similarly, we assume that the birth rate depending on the living standard M_{BG} of the population is described by the function with saturation so that a maximum of birth rate is observed at low values of M_{BG}, and at $M_{BG} \to \infty$ the birth rate drops to some level determined by the value of a_{*GMB}. The rate of transition from maximum to minimum birthrate with changing M_{BG} is set by the constants $a_{1,GMB}$ and $a_{2,GMB}$:

$$H_{GMB} = a_{*GMB} + a_{1,GMB} \exp(-a_{2,GMB} M_{BG})$$

where

$$M_{BG} = (V/G)\{[1 - B - U_{MG} - U_{ZG}]/[1 - B(t - o) - U_{MG}(t_o) - U_{ZG}(t_o)]\}$$
$$\times [E_{RG}(t)/E_{RG}(t_o)]$$
$$E_{RG}(t) = 1 - \exp[-k_{EG} M(t)/M(t_o)]$$

The dependence of mortality on the living standard is described by the decreasing function:

$$H_{\mu MB} = b_{1,\mu G} + b_{2,\mu G} \exp(-b_{*\mu G} M_{BG})$$

This function shows that population mortality with an increasing per capita share of capital drops with the rate coefficient $b_{*\mu G}$ to the level $b_{1,\mu G}$.

The birth rate and mortality in certain limits are, respectively, decreasing and increasing functions of population density:

$$H_{GG} = G_{1,G} + G_{*G} \exp(-G_{2,G}Z_{GG}) \qquad H_{\mu G} = \theta_{1,\mu G} + \theta_{2,\mu G}Z_{GG}^{\omega\mu G}$$

where $Z_{GG} = G(t)/G(t_o)$.

Finally, an important aspect of the *Homo sapiens* ecology is the environmental state. In this connection the problems of anthropogenic biocoenology have been widely discussed in scientific literature and many authors try to identify the required regularities. Without going into details of these studies, most of which cannot be used in the global model, we shall confine ourselves to the following dependences:

$$H_{GZ} = l_{1,G} \exp(-l_{*G}Z_{RG}) \qquad H_{\mu Z} = \eta_{1,\mu G} + n_{2,\mu G}Z_{RG}$$

$$\tau_{GC} = \begin{cases} \tau_{1,GC} + \tau_{2,GC}(C_a - C_{1,G}) & \text{for } C_a > C_{1,G} \\ \tau_{1,GC} & \text{for } 0 \le C_a \le C_{1,G} \end{cases}$$

$$H_{\mu O} = f_{1,\mu G} + f_{2,\mu G}/O(t) \qquad H_{\mu C} = \exp(k_{\mu G}C_a)$$

$$H_{GO} = 1 - \exp(-k_{GO}O) \qquad H_{GC} = \exp(-k_{GC}C_a) \qquad Z_{RG} = Z(t)/Z(t_0)$$

$$\tau_{GO} = \begin{cases} \tau_{1,GO} & \text{for } O > O_{1,G} \\ \tau_{2,GO} - (\tau_{2,GO} - \tau_{1,GO})O/O_{1,G} & \text{for } 0 \le O \le O_{1,G} \end{cases}$$

where C_a is the atmospheric CO_2 concentration, $C_{1,G}$ and $O_{1,G}$ are safe (for humankind) levels of the atmospheric CO_2 and O_2 content.

1.5.3 Parameterization of anthropogenic processes

The biosphere Ω as a complicated unique system is functioning following the laws of co-evolution of its sub-systems, the human society H and nature N being its basic ones. The impact of human activity on nature, being of comparatively small scale, can be apparently only assessed with a newly developed technology. Clearly, for this purpose a system approach is needed to formalize the ecological, technological, economic, and geopolitical interactions of the sub-systems H and N. In general, the system H has at its disposal technologies, science, economic structure, size of population, etc. The system N has a set of mutually dependent processes, such as climatic, biogeocoenotic, biogeochemical, geophysical, etc.

From the viewpoint of the theory of systems, H and N are open systems. Their division is a conditional procedure aimed at selecting controlled and non-controlled components of the environment. Without going into philosophical and methodical aspects of this procedure, we assume that both systems are symmetrical from the viewpoint of their simulation, that is, we assume that each system has a goal, structure, and behaviour of its own. Let $H = \{H_G, H_S, |H|\}$ and $N = \{N_G, N_S, |N|\}$, where H_G and N_G are the goals of the systems, H_S and N_S are strategies of the systems behaviour and $|H|$ and $|N|$ are the systems structure, respectively. Then, the H–N interaction can be described by the process of (V, W)-exchange, such that in each of the systems, to reach its goal, it spends the resources V and, in exchange, obtains a new resource of an amount W. Each system is aimed at an optimization of the (V, W)-exchange with another system, that is, to maximize W and minimize V.

Now write the equations for (V, W)-exchange:

$$W_H(H^*, N^*) = \max_{\{H_s,|H|\}} \min_{\{N_s,|N|\}} W(H, N) = \min_{\{N_s,|N|\}} \max_{\{H_s,|H|\}} W(H, N) \qquad (1.4)$$

$$W_H(H^*, N^*) = \min_{\{H_s,|H|\}} \max_{\{N_s,|N|\}} W(H, N) = \max_{\{N_s,|N|\}} \min_{\{H_s,|H|\}} W(H, N) \qquad (1.5)$$

where H^* and N^* are optimal systems. Here, in contrast to traditional game-theory models, there exists a power spectrum of the H–N interaction covering the final intervals of changes in the payoffs W_H and W_N depending on the aggressiveness of each of them. A concrete definition of the payoff function requires a certain systematization of the mechanisms of both the human and natural co-evolution. One of the widely used models of the balanced development of the world community and nature subjected to criterion (1.4) and (1.5) consists in identification of the system H with a totality of large cities with adjacent industrial and recreation zones. There are numerous versions for consideration and formal description of such structures. In particular, there exist a well-known method of logic-information modelling of the processes of rational nature use and a simulation method of controlling the ecological-economic systems (Ugolnitsky, 1999). According to these methods, to solve a concrete problem, it is necessary to conceptualize the information base of the model and to select the most general relationships between the elements of the interacting systems. This procedure is completed with enumeration of all functional elements of the systems and determination of the capacity loads on their elements. The whole procedure ends with a synthesis of the simulation model, which, within certain assumptions, is an instrument of investigation. In the case considered we assume that the structure of the system H includes population G, pollutants Z, and natural resources M, that is, $|H| = \{G, Z, M\}$. Similarly, the structure of the system N consists of elements, such as the climate parameters, temperature T, environmental quality Q, areas of forests σ_L, and agricultural lands σ_X, that is, $|N| = \{T, Q, \sigma_L, \sigma_X\}$.

The behavioural strategy of the system H is formed from the distribution of investments into the retrieval of resources U_{MG}, struggle with pollutions U_Z, and agricultural investments U_{BG}, that is, $H_S = \{U_{MG}, U_{ZG}, U_{BG}\}$. The behavioural strategy of the system N is identified with the rate of investments ageing T_V, population mortality μG, agricultural productivity H_X, the cost of resources retrieval G_{MG}, and time constant of biospheric self-cleaning of pollutants T_B, that is, $N_S = \{T_V, \mu_G, H_X, G_{MG}, T_B\}$.

Equations (1.4) and (1.5) are basic equations for the model of survival. In general, this model is formulated in terms of the theory of evolutionary technology of modelling. If all possible states of the biosphere with acceptable conditions for human life constitute the multitude $\Gamma = \{\Gamma_i\}$, then as a result of the effect of C_k on the biosphere, two outcomes are possible: $C_k(\Gamma_i) \rightarrow \Gamma_i \in \Gamma$ and $C_k(\Gamma_i) \rightarrow \Gamma_i \notin \Gamma$. When the sequence of biospheric states $\{C_k(\Gamma_i)\} \in \Gamma$, then we can speak about the persistent co-evolution of the system $H \cup N$. No doubt, there is a problem of an adequacy between real processes and their simplified presentation as a model. Nevertheless, despite the philosophical doubt in expedience of mathematical modelling for the perspective assessment of the kinetics of biospheric parameters, the model approach has proved to be profitable. The use of the biospheric model instead of the biosphere itself is convenient, firstly, because there is more information about the model than about the biosphere, secondly, because the model is easier to handle, and thirdly, because direct experiments with the biosphere are dangerous. All these aspects are the subject of studies for global eco-informatics aimed at achieving a sufficient similarity of the observed behaviour of the system $H \cup N$ and the model. This is possible with the constant renewal of databases and broadening of knowledge in accordance with the technology of the information multitude of the biospheric parameters suggested by Krapivin and Kondratyev (2002).

Trends in human activity are determined by a great number of factors. However, all of them are reduced to factors of economy and energetics, whose interconnections can be described by linear regression $V = k_{Ve}e + b_{Ve}$, where V is the regional budget and e is the energy produced in this region. The constant coefficients k_{Ve} and b_{Ve} reflect specific features of the economic activity and its efficiency. The general scheme of a possible model of the socio-economic processes includes V_i as the size of the funds in the ith region, which can be used when planning its anthropogenic activity:

$$\frac{dV}{dt} = G_{VG}G_{VMG} - \frac{V}{T_{VG}}$$

where T_{VG} is the time constant of basic funds deterioration and coefficients G_{VG} and V_M determine the rate of the funds generation. The function V_M defines the dependence of the rate of the funds generation on the living standard M_B of the population of the region. This dependence is described by the logarithmic function:

$$V_{MG} = k_{MGV} \ln (1 + k_{1,MGV} M_{BG})$$

where the coefficient k_{MGV} is chosen from the condition:

$$k_{MGV} \ln (1 + k_{1,MGV} M_{BG}(t_0) = 1$$

In this case the coefficient G_{VG} represents the volume of the funds generated per capita at the moment t_0 in the region. The time constant of the basic funds deterioration is a function of the scientific-technical progress and is considered in the model as a controlling parameter.

The basic funds determine the intensity and direction of anthropogenic activity. In particular, the generation of pollutants and their utilization are substantially regulated by this distribution.

Let Z be the concentration of an arbitrary pollutant and Z_{VG} the rate of the pollutant's assimilation due to realization of the technologies of environmental purification. Then:

$$\frac{dZ}{dt} = Z_{VG} - Z_{TV}$$

where $Z_{TV} = Z_T + Z_V, Z_T$ is the natural rate of the pollutant decomposition, and Z_V is the rate of the pollutant assimilation due to environmental purification.

Assume that the rate of pollution generation in the region is in proportion to population density with the coefficient k_Z and depends on the per capita volume of the funds $V_{RG} - V/G$:

$$Z_{VG} = k_Z G Z_{*VG}$$

where

$$Z_{*VG} = Z_{VG,max}[1 - \exp(-G_{1,ZG} V_{RG})].$$

The rate of an artificial assimilation of pollutants is determined by the capital share U_{ZG} to be used to intensify the cleaning processes in the environment: $Z_V = U_{ZG} V/G_{ZV}$ where G_{ZV} is the cost of the environmental cleaning of unit pollution.

The rate of natural removal of harmful wastes is in direct proportion to the pollutants concentration and in indirect proportion to the decomposition time $T_Z : Z_T = Z/T_Z$. Naturally, the processes of generation and utilization of pollutants are more complicated. Therefore the respective unit of the global model needs further development considering new knowledge in the field of pollution utilization. The generated pollutant is scattered throughout the environment getting partially into the atmosphere, soil, and the hydrosphere. Possible ways of pollutant propagation in these media are described in other chapters. However, for a complex analysis of the pollutant an equation for the global model is needed which would reflect the general state of the atmosphere. Here we introduce the atmospheric turbidity index B measured by the weight of admixtures in the atmosphere per unit area of the biospheric surface. A change in this index is determined by the share N_B in the general pollution Z generated in the region, by the amount of waste produced by the energy-generating enterprises (the coefficient N_A), by the water vapour content in the atmosphere W_A and the rate of its natural cleaning $1/T_B$ (assumed to be equal to inverse value of the time T_B of sedimentation of dust and smoke particles and water droplets). The effect of water vapour on atmospheric transparency is manifested through the derivative $dW_A/dt = \rho_B \, dT/dt$ where T is the atmospheric temperature and ρ_B is the coefficient of proportionality between the temperature variation and cloud amount. As a result, the dynamic equations for the system variable B will be:

$$\frac{dB}{dt} = N_B Z + N_A b_{GC} G + \rho_B \frac{dT}{dt} - \frac{B}{T_B} + B_n$$

where B_n is the rate of atmospheric pollution resulting from natural processes (rock weathering, erosion, volcanism, etc., $\sim$0.78 t/km^2 yr).

The horizontal transport of pollutants and the respective change in B are simulated with an account of wind speed and the spatial grid $\Delta\varphi r \Delta\lambda$ simulated with T_B using the formula $T_B = h_e/\nu_s$ where h_e is the efficient height of the pollutant scattering in the atmosphere, ν_s is the rate of particle sedimentation calculated with the Stokes formula: $\nu_s = 2g\rho_k r^2/(9\varsigma)$. Here g is the acceleration due to gravity, ρ_k is the particle density, r is the particle radius and ς is the coefficient of molecular viscosity. For instance, for particles with $r \in \lfloor 10^{-5}, 5 \times 10^{-3} \rfloor$ cm and

Table 1.18. Annual amount of particles with radius >20 μm emitted into the atmosphere.

Type of particles	Flow of particles (10^6 t/yr)
Natural particles of soil and stones	100–500
Particles of forest fires and timber industry wastes combustion	3–150
Sea spray	300
Volcanic dust	25–150
Particles generated by gas production:	
sulphates from H_2S	130–200
ammonium salts from HN_3	80–270
nitrates from NO_x	60–430
hydrocarbons from vegetable compounds	75–200
Synthetic particles	10–90

$\rho_k[1,4]\,\mathrm{g\,cm}^3$ we obtain $\nu_s = 10^2$ cm/s, which for calm atmospheric conditions gives $T_B = 7$ h.

Some characteristics of atmospheric aerosols are given in Tables 1.15, 1.16, and 1.18.

The scattering process of pollutants in the water domain and in the soil layer are the most complicated processes. These are simulated in the respective chapters in studies of the hydrospheric components and associated processes. The relationships between natural and anthropogenic formation of air or water quality deduced and applied in simulation experiments contain the parameters determined by the pace of expenditure and renewal of natural resources. The pace dynamics are determined by the rate of population growth and civilization development. On average, the annual global acceleration of natural resources consumption constitutes about 4% with a 2% mean annual population increment. Hence, the value of per capita natural resources consumption R_{MG} is an important indicator of the present civilization development, since the annual consumption of accessible resources constitutes $\Delta M = R_{MG}G$. The consumption of natural resources increases with increasing living standard but the extent of this growth decreases with the increasing investments into science. Clearly, this relationship is multi-faceted and has been poorly studied. Therefore to consider this problem, assume the following parameterization:

$$R_{MG} = m_G \ln(1 + M_{BG})$$

where the parameter m_G is the time function set in the simulation experiment as a scenario.

The protection and retrieval of natural resources requires certain material expenses. Assume that the effect of these measures is in proportion to the share U_{MG} of the respective investments and in inverse proportion to the cost G_{MG} of the retrieved conditional unit of resources:

$$\frac{dM}{dt} = \Delta M + U_{MG}V/G_{MG}$$

The investments spectrum also includes the sphere of agricultural production, including expenses on scientific investigations in the field of selection, plant physiology, agrotechnology and reclamation and solution of technical and social problems. Let q_V be the share of capital invested in agriculture. We obtain:

$$\frac{dq_V}{dt} = (U_{BG}B_{FG}B_{qG} - q_V)/T_B$$

where the controlling parameter $U_{BG} = U_{BG}(t)$ in the simulation experiments is described by the scenario and the remaining components are determined by the following dependences:

$$B_{FG} = \exp(-b_{BG}F_{RG}), \qquad B_{qG} = b_{1,BG} + b_{2,BG}(q_{MG}/q_{FG})^{\alpha SG}$$

$$q_{MG} = b_{1,qG} + b_{2,1G}(M_{BG})^{\alpha qG}, \qquad q_{FG} = a_{qG}(F_{RG})^{\beta qG}$$

1.5.4 Modelling natural and anthropogenic processes in a megalopolis

A megalopolis is an area of the Earth's surface exhibiting high urbanization, developed industry, and other human activity attributes concentrated in a limited territory. The number of such territories in the world is constantly growing and their total area increases with the growing size of the global population. Characteristic examples of large megalopolises are Moscow, Tokyo, New York, Ho Chi Minh City, etc. Some typical estimates of the environmental parameters in megalopolises are given in Table 1.19. For instance, the Moscow megalopolis is characterized by high concentrations of the sources of anthropogenic pollution over a limited territory

Table 1.19. Comparative characteristic of the developed and developing megalopolises.

Parameter	Estimate of the parameter	
	Moscow	Ho Chi Minh City
Area of the megalopolis (km^2)	100	31
Size of population of the megalopolis (10^3 people)	8894	8563
Planted area (km^2)	14	16
Number of enterprises in the megalopolis (10^3)	25	14
Number of pollution sources (10^3)	100	70
Emissions of harmful pollutants into the atmosphere (10^3 t/yr)	1153	248
Length of the open water bodies on the territory of the megalopolis (km)	398	101
Volume of sewage (10^6 m^3/year)	3036	226

(energy enterprises, chemical industry, and automobile transport). Their share in total emissions constitutes 90.4%. The polluted air plume from Moscow can be observed at a distance of about 100 km from the city. The state of the water bodies within the megalopolis are determined by the input of sewage and surface run-off as well as run-offs from industrial enterprises into the Moskwa river and Yausa river, as well as into 70 other rivers and springs within the territory of the megalopolis. The concentration of chemicals in the Moskwa river both within the city boundaries and downstream of the city varies widely depending on the season. For instance, the copper ion content varies during the year from 0.004 to 0.013 mg/l (4–13 is permissible concentration level (PCL)) with a maximum in the spring. The content of oil products varies within 0.25 to 0.6 mg/l (5–12 PCL). A similar situation exists in the smaller Yausa river where the concentration of copper varies from 0.007 to 0.12 mg/l and that of oil products, from 0.38 to 0.7 mg/l. The environmental service of the megalopolis monitors the local concentrations of pollutants forming the respective time series of data on the state of the environment. There is a global network of megalopolises, consideration of which in the global model results in an increase in its accuracy. Many megalopolises are in the early stages of formation. Ho Chi Minh City is an example of a young megalopolis as well as its adjacent provinces of South Vietnam – Dong Nai, Bin Zyong, and Baria-Vung Tau. This megalopolis occupies an important place in the economy of Vietnam, generating 75% of the industrial production of South Vietnam and 50% for the whole country (Bui, 1997). The development of the megalopolis infrastructure includes services of ecological and sanitary control of the environment at its initial stage, which makes possible the perspective planning of the environment.

An analysis of the data on the structure of the environment of the megalopolises of the world reaches the conclusion that for complex assessment it is possible to develop a sample system of the models simulating the transport and propagation of pollutants in the atmosphere and water bodies. The input information for this model is obtained both from the monitoring systems and the global model.

Characteristic linear dimensions of a megalopolis constitute tens of kilometres. This means, that to simulate the processes of atmospheric transport of pollutants, a model of the Gaussian type can be used. To assess the quality of the atmosphere in the megalopolis it is sufficient to form a composition of the Gaussian streams and at the points of their intersections to summarize the concentrations of the respective types of pollutants.

Assume that the source of pollutant of the type s is located at a point (0, 0) in the coordinate system (x, y, z) where the 0_x axis is in the wind direction and the 0_z axis is normal to the Earth's surface at a height h. Then the concentration C_s of the pollutant of type s at any point in (x, y, z) can be calculated with the formula:

$$C_s(x, y, z) = Q(2\pi U \sigma_y \sigma_z)^{-1} \times \exp -y^2/[2(\sigma_y)^2] - (z+H)^2/[2(\sigma_z)^2] - (z-H)^2/[2(\sigma_z)^2] \quad (1.6)$$

where $H = h + \Delta h$, $\sigma_y = (U/[2k_y(x)])^{1/2}$, $\sigma_z = (U/[2k_z(x)])^{1/2}$, k_y and k_z are the coefficients of eddy diffusion corresponding to axes y and z, U is the wind speed, and the parameter Δh is calculated from the Briggs formula: $\Delta h = \Delta_1$ for unstable and neutral conditions of the lower atmospheric layer and $\Delta h = \Delta_2$ for stable conditions of the lower atmospheric layer thus:

$$\Delta_1 = \begin{cases} 1,6F^{1/3}x^{2/3} & \text{for } x \leq x_p, \\ 1,6F^{1/3}(3,5x_p)^{2/3} & \text{for } x > 3,5x_p, \end{cases}$$

$$\Delta_2 = \begin{cases} \min\{2,6(FU^{-1}/S_n)^{1/3}, \quad 1,6F^{1/3}(3,5x_p)^{2/3}\} & \text{for } U > 1,4\,\text{m/s}, \\ 5,3F^{1/4}S_n^{-3/8} - D/2 & \text{for } U \leq 1,4\,\text{m/s}, \end{cases}$$

where $F = g(T_s - T_a)V/(\pi T_s)$ is the buoyancy coefficient, $S_n = (g/T_a) \cdot (\partial T_a/\partial z + 0.01)$ is the coefficient of stability, $\partial T_a/\partial z$ is the potential temperature gradient at the point of pollutant emission, T_s is the temperature of the emitted gas (°K), T_a is the ambient air temperature (°K), D is the diameter of the pollution source (chimneys, etc., m), V is the gas expenditure per unit time (m^3/s), g is gravitational acceleration, and x_p is the distance along the wind direction where the eddy diffusion dominates over the:

$$x_p = \begin{cases} 14F^{5/8} & \text{for } F \leq 55, \\ 34F^{2/5} & \text{for } F > 55, \end{cases}$$

To calculate the pollutant concentration at an arbitrary point within the megalopolis territory, it is necessary to take into account the impacts of all the sources. For this purpose we choose an arbitrary system of coordinates $x0y$. Assume that the wind direction vector is at an angle α to the 0_x axis. Denote the coordinates of the pollution sources $Q_i (i = 1, \ldots, n)$ by (a_i, b_i). Then the net concentration of the pollutant of the s type at an arbitrary point (u, v) at height z will be:

$$C_s(u, v, z) = \sum_{i=1}^{n} C_s(u, v, z, a_i, b_i)$$

where $C_s(u, v, z, a_i, b_i)$ is the concentration of the pollutant of the s type at the point (u, v) at height z due to the source Q_i. Knowledge of the angle α makes it possible to reach an agreement between the coordinates of the systems $u0v$ and $x0y$.

The hydrological cycle parameters for a megalopolis are calculated using the general model described by Krapivin and Kondratyev (2002). However, simplifications are possible here due to the absence of some elements of the standard scheme or changes in their weights. The water flow by canals (pipelines) is most characteristic of a megalopolis. In this case we have the following equation:

$$\sigma^{-1}\frac{\partial Q}{\partial t} - Q\sigma^{-2}\frac{\partial \sigma}{\partial t} + Q\sigma^{-1}\left[\sigma^{-1}\frac{\partial Q}{\partial x} + Q\sigma^{-2}\left(\frac{\partial \sigma}{\partial x} + \frac{\partial \sigma}{\partial y}\cdot\frac{\partial y}{\partial x}\right)\right] + g\left(\frac{\partial z}{\partial x} + \frac{\partial y}{\partial x} + p\right) + E_{IL} = 0 \quad (1.7)$$

where z is the level of the canal, y is the water depth in the canal measured from the level z, p is the canal bottom friction, E_{IL} is the impact of inflows and outflows, σ is the effective area of the canal cross section at x (x is the water flow direction), and Q is the water flow at x.

In addition to (1.7), an equation is needed for the water medium continuity: $\partial\sigma/\partial t + \partial Q/\partial x = q$, where q is the intermediate inflow per unit time per unit length of the canal section.

A change of temperature T_w of an element of the water volume V is described by the relationship:

$$\frac{V\partial T_w}{t}\partial t = (lc^{-1}r^{-1})\frac{\partial(\sigma_x\Delta_x\partial T_w/\partial x)}{\partial x} - l\frac{\partial(Q^* T_w)}{\partial x} + [SH_{al} + \theta]c^{-1}\rho^{-1} - T_w\frac{\partial V}{\partial t}$$

where l is the length of the canal section, S is the section surface area bordering with the atmosphere, ρ is the water density, c is the specific heat capacity of water, σ_x is the area of the cross section of the canal section at x, Q^* is an advective heat flux through the section boundaries, θ is the outer heat source, Δ_x is the coefficient of the eddy diffusion, and H_{al} is the heat flux velocity through a unit surface area.

The heat flux through the surface on the border with the atmosphere is the sum of the following components: $H_{al} = E_1 + E_2 - E_3 - E_4 - E_5$, where E_1 and E_2 are fluxes of the incoming shortwave and longwave radiation, respectively, E_3 is the outgoing longwave radiation flux, E_4 is the evaporation-driven outgoing heat flux, and E_5 is the convective heat exchange. The E_1 parameter is a complex function of cloudiness, dust volume, sun location, and other geophysical parameters. For the E_2 flux we use the dependence $E_2 = 1.23 \cdot 10^{-16}[1 + 0.17\chi^2](T_a + 273)^6$, where χ is the amount of cloud. The E_3 flux is estimated from the expression $E_3 = \varepsilon\sigma_1(T_a + 273)^4$, where ε is the water emissivity (~ 0.07), σ_1 is the Stephan-Boltzman constant ($1.357 \cdot 10^{-8}\,\mathrm{cal/m^2/s/^\circ C^4}$). For $T_a \in [0^\circ, 30^\circ]$C an approximation $E_3 = 7.36 \cdot 10^{-2} + 1.17 \cdot 10^{-3}T_a$ is valid.

Heat losses through evaporation are presented as the dependence $E_4 = \rho L e_a$, where ρ is the water density (kg/m^3), L is the latent heat of evaporation (kcal/kg), and E_a is the rate of evaporation (m/s). The latter parameter is approximated by the relationship $E_a = (a + bW)(ps - p_a)$, where W is the above-surface wind speed, p_s is the pressure of the saturated vapour over the surface (hPa), p_a is the vapour pressure in the atmosphere (hPa), and a and b are empirical coefficients. The p_s parameter is well approximated by the linear function $p_s = \alpha_p + \beta_p T_a$ whose coefficients are given in Table 1.20.

And, finally, the convective heat transport $E_5 = R_4 E_4$, where $R_4 = 6.1 \cdot 10^{-4}p_a(T_a - T_d)(p_s - p_a)$ is the atmospheric pressure (hPa) and T_d is the dry air temperature (°C). We now obtain $H_{al} = \mu_{al} - \lambda_{al}T_a$, where $\mu_{al} = E_1 + E_2 - 7.36 \cdot 10^{-2} - \rho L(a + bW)(\alpha_p - p_a - 6.1 \cdot 10^{-4}p_a T_d)$ and $\lambda_{al} = 1.17 \cdot 10^{-3} + \rho L(a + bW)(\beta_p + 6.1 \cdot 10^{-4}p_a)$.

The atmosphere–water gas exchange is described with the equation $V\partial C/\partial t = SK_C(C^* - C)$, where $K_C = K_2\theta^{(T_a-20)}$, C is the concentration of the water-dissolved gas, C^* is the concentration of saturated dissolved gas at local temperature, K_2 is the coefficient of gas re-airing at 20°C, and θ is the temperature coefficient. The K_2

Table 1.20. Coefficients of the temperature dependence of the atmospheric vapour pressure.

T_a (deg)	α_p	β_p
0–5	6.05	0.522
5–10	5.10	0.710
10–15	2.65	0.954
15–20	−2.04	1.265
20–25	−9.94	1.265
25–30	−22.29	2.151
30–35	−40.63	2.761
35–40	−66.90	3.511

coefficient at known wind speed W(m/s) and coefficient of molecular diffusion D_g is calculated with the formula:

$$K_2 = D_g \cdot 10^6[200 - 60W^{1/2}]^{-1}$$

The coefficient D_g is equal to $2.04 \cdot 10^{-9}$ for O_2 and $1.59 \cdot 10^{-11}$ for CO_2. The system of geo-information monitoring in the megalopolis is based on the data of measurements of the environmental parameters and on equations (1.4) and (1.5). It allows us to analyse and assess the most vulnerable areas of the megalopolis's structure to natural catastrophes as well as reactions of the system N to the behaviour of the system H in the realization of the projects of the natural complexes reconstruction. As a result, an interaction takes place between the external shell of the global model with the details of the description of the environment by the scale $\Delta\varphi \times \Delta\lambda$ and the model of the megalopolis where the spatial resolution is more detailed. The resulting diverse schemes of agreement between the global model and the megalopolis model are determined both by their specific features and by the formal system circumstances. Their study and development are the subject of future investigations on the realization of constructive ideas discussed in the monograph by Kondratyev (1999a).

1.6 SCENARIOS OF THE ENVIRONMENTAL DYNAMICS

As experience shows, the development of the strategy of global change monitoring with the present development of science on global processes is impossible using any models that simulate the functioning of the nature–society system. No doubt, the developed methods of modelling the type of evolutionary technology give hope to creating a global model which, based on accumulated knowledge, will make it possible to obtain reliable long-term predictions of global changes and to substantiate thereby conditions for sustainable development of the environment. Obviously, success in this direction depends on the complex use of modelling technologies and environmental observations. So far, the global models need the schemes, which would accomplish a parameterization of various processes in the nature–society system in the form of scenarios. The construction of the latter is the result of expert knowledge and a certain amount of imagination on the part of the investigator. This problem has been discussed in detail by Alcano *et al.* (2001), Gyalistras (2002), and in a special IPCC report (IPCC, 2001c). To some extent, the Kyoto Protocol can serve as one of the scenarios. Some countries of the European Economic Community (EEC), who signed this protocol, agreed to reduce the GHG emissions by 12.5% by 2008–2012, with CO_2 emissions decreasing (due to the reduction) by 20% by 2010 compared with the year 1990. In each country the diverse scenarios of this type bring forth a high level of global uncertainty, a solution of which is impossible without certain generalizations. Therefore we shall dwell upon some generalizing scenarios.

1.6.1 Fishery scenario

The use of the World Ocean products in the 20th century were characterized by an increasing trade in fish and other elements of higher trophic levels, with trade intensity doubling every 10–15 years. The maximum size of catch was in 1970 61 Mt. The per capita consumption of sea products varied between 7.2 to 11.8 kg. Numerous forecasts of the possible per capita use of the World Ocean products gave values of 12 kg/yr. Therefore assume a hypothetical dependence of the intensive fishing of nekton in each *i*th region:

$$\lambda_{GRi} = \begin{cases} \lambda_{G0i} - (\lambda_{G0i} - \lambda_{G\infty i})(t - t_0)/(t_{\lambda Gi} - t_0), & t_0 \leq t \leq t_{\lambda G_i} \\ \lambda_{G\infty i} & t > t_{\lambda Gi} \end{cases}$$

where λ_{G0i} is the fishing intensity at the moment t_0 (e.g., in 1970 $\lambda_{G0i} \approx 0.0286$ and in 1999 $\lambda_{G0i} \approx 0.0312$) and $\lambda_{G\infty i}$ is the maximum possible trade intensity reached at the moment $t_{\lambda Gi} \geq t_0$. The λ_{GRi} coefficient characterizes the fishing trade in the *i*th region and is considered in the demographic unit of the global model. A human being is considered one of the higher trophic levels with respect to nekton. It is assumed that in the multitude Γ there is the element s_H, for which the relationship is valid:

$$C_{rs_H} R_{s_H} = r \sum_{i=1}^{m} \lambda_{GR_i}$$

The contribution of the fishing industry I_i in the diet of *Homo sapiens* V_G is determined by the ratio of the volume of products fished out of the World Ocean to land-based agricultural products.

1.6.2 Scenario of the distribution of the soil–plant formation areas

An intensity of food production on land depends on the area covered with agricultural crops and on the crop productivity. Clearly, within the global model, while considering details of all processes and elements it is impossible to predict all the directions of agricultural development. Therefore all the processes of increasing productivity have been generalized, reflecting general trends. We introduce, for agricultural formation, the identifier k and consider that in any region, part of the territory can be occupied with cultivated lands:

$$\sigma_{ki} = \begin{cases} \sigma_{koi} + (\sigma_{k*i}) - \sigma_{k0i}(t - t_0)/(t_{kiS} - t_0) & t_0 \le t \le t_{kiS} \\ \sigma_{k*i} & t > t_{kiS} \end{cases}$$

$$H_{kiS} = \begin{cases} 1 + (H_{k*i} - 1)(t - t_0)/(t_{kBiS} - t_0) & t_0 \le t \le t_{kBiS} \\ H_{k*i} & t > t_{kBiS} \end{cases}$$

where $\sigma_{ki} \le \sigma_i$ is the area of land cultivated in the ith region and H_{kiS} is the indicator of changes in the agricultural productivity with respect to the time moment t_0:

$$(H_{kiS} = R_{ki}(\varphi, \lambda, t)/R_{ki}(\varphi, \lambda, t_0)$$

1.6.3 Investment scenario

According to our suppositions, in the ith region the area σ_{ki} under agricultural crops starting from the moment t_0, changes following the linear law from σ_{k0i} to σ_{k*i} in the time interval to t_{kiS}. The time for reaching the level σ_{k*i} depends on the amount of investments into agriculture with the inverse proportion coefficient t_{k*i} and on other factors affecting the time constant $t_{ki} : t_{kiS} = t_0 + t_{ki} + t_{k*i}/(q_{Vi}V_i)$. Possible changes in the agricultural productivity H_{kiS} are also approximated by a linear law where the value H_{k*i} shows how many times the productivity in the ith region can change from t_0 to t_{kBiS}. The reserves for increasing H_{ki} are rather large. Even the highly productive plants such as sugar cane, in the process of photosynthesis, consume only about 2% of the solar energy reaching the Earth surface annually. Other crops consume 1%, with other plants consuming even less. Plant physiology and agro-technology have broad possibilities for multiple increases of agricultural productivity and, hence, the food quantities available for *Homo sapiens*. Assume that to reach the level H_{k*i}, the population of the ith region has the time $t_{kBiS} - t_0$:

$$t_{kBiS} = t_0 + \bar{t}_{kBi} + t_{kBi}/(q_N V_i)$$

where the constituent $\bar{t}_{kBi}$ is independent of the agricultural investments, and t_{kBi} characterizes the efficiency of these investments.

According to the V_{Gi} food spectrum for *Homo sapiens* a certain role is played here by forest ecosystems, the general trend in changes of their areas being characterized by a negative derivative. On a global scale, the forest resources are constantly depleting, so that on the cusp of the 20th and 21st centuries the total area of forests constitutes $\sigma_L = 4184$ Mha, with a 31% density (about 28.3% land area).

Two principal processes are observed in the practice of forestry: deforestation and afforestation. We can describe the totality of these processes using the scenario of forest area changes:

$$\sigma_{Li} = \begin{cases} \sigma_{L0i} + (\sigma_{L*i} - \sigma_{L0i})(t - t_0)/(t_{Li} - t_0) & t_0 \le t \le t_{Li} \\ \sigma_{L*i} & t > t_{Li} \end{cases}$$

This dependence foresees that the area under forests in the *i*th region, until the moment t_{Li}, varies linearly from σ_{L0i} to σ_{L*i} and then remains constant. Values of the input parameters constitute the freedom of choice in simulation experiments.

To complete the formulation of the global scenario describing the level of food provision for *Homo sapiens*, consider the law of changing the stock-breeding productivity. Suppose that with the constant constituent k_{FSi}, it increases in proportion to agricultural investments q_{Vi} with the coefficient k_{F*i} and has a time varying constituent with an exponential law of change:

$$k_{Fi} = k_{F3i} + k_{F*i}q_{Vi} + (k_{F1i} - k_{F3i})[1 - \exp(-k_{F2i}(t - t_0)]$$

The direction of the anthropogenic activity is determined by the intensity of pollution generation, time of pollution utilization, cost of environmental cleaning, amount of investments into the renewal of resources and into the prevention of the environmental pollution, investments into industry and agriculture, rate of the natural resources expenditure, and search for new sources. The hypothetical trends of these processes should be set in order to realize the predictions with the use of the global model.

Let the intensity of pollution vary between k_{Z0i} at the time moment t_0 and k_{Z1i} at the moment t_{Z*i}. Similarly vary the time of waste utilization T_{Zi} and the cost of cleaning of unit pollution G_{ZVi}, respectively, between the values $T_{Z0i} = T_{Zi}(t_0)$, and $T_{Z1i} = T_{Zi}(t_{Zi})$, and $G_{2ZVi} = G_{ZVi}(t_0)$ and $G_{1ZVi} = G_{ZVi}(t_{ZVi})$. Since the estimates of these indicators can vary for many reasons and at different times, the simplest scenario for the experiment will be the following functional presentations:

$$k_{Zi} = \begin{cases} k_{Z0i} - (k_{Z1i} - k_{Z0i})(t - t_0)/(t_{Z*i} - t_0) & t_0 \le t \le t_{Z*i} \\ k_{Z1i} & t > t_{Z*i} \end{cases}$$

$$T_{Zi} = \begin{cases} T_{Z0i} - (T_{Z1i} - T_{Z0i})(t - t_0)/(t_{Zi} - t_0) & t_0 \le t \le t_{Zi} \\ T_{Z1i} & t > t_{Zi} \end{cases}$$

$$G_{ZVi} = \begin{cases} G_{2ZVi} - (G_{2ZVi} - G_{1ZVi})(t - t_0)/(t_{ZGi} - t_0) & t_0 \le t \le t_{ZGi} \\ G_{1ZVi} & t > t_{ZGi} \end{cases}$$

Since it is rather difficult to specify the spheres of human activity, the spectrum of investments are confined here to the functions U_{ZG}, q_v, and U_{MG}. The strategy of the investments distribution for each region is determined by two-step functions of time:

$$U_{MGi} = \begin{cases} U_{MG1i} & t_0 \le t \le t_{MGi} \\ U_{MG2i} & t > t_{MGi} \end{cases}$$

$$U_{ZGi} = \begin{cases} U_{ZG1i} & t_0 \le t \le t_{ZVi} \\ U_{ZG2i} & t > t_{ZVi} \end{cases}$$

where t_{MGi} and t_{ZVi} are the moments of the economic policy change in the ith region in the field of investments into the renewal of natural resources and prevention of pollution, respectively. Here the term "renewal of resources" denotes the processes favouring an increase in mineral resources. In particular, an increase in M can be brought about by conversion of mineral resources, thus transforming their significance.

The agricultural investments are one of the most important components of investments. To obtain a high productivity of cultural crops, it is necessary to completely satisfy their water and mineral matter needs. This means that high yields require great amounts of energy. For example, the production of protein food has an effectivity coefficient approximately equal to 0.1. In other words, 1 calorie of protein food can be produced with expenditure of about 10 calories of energy. This energy production requires some share of investments and consumption of fossil products of photosynthesis. Hence here the investments and energy expenditures are mutually dependent. The global model unit that describes the agricultural investments is constructed by setting the dependences of agricultural productivity on energy expenditures. The investment parameter of the control U_{BGi} is described by the two-step function:

$$U_{BGi} = \begin{cases} U_{BG1i} & t_0 \leq t \leq t_{BGi} \\ U_{BG2i} & t > t_{BGi} \end{cases}$$

where t_{BGi} is the moment of the change of the investment policy in agriculture in the ith region. The T_{Bi} parameter characterizing the time of assimilation of the agricultural investments in the ith region can also change at the moment t_{Bi}:

$$T_{Bi} = \begin{cases} T_{B1i} & t_0 \leq t \leq t_{Bi} \\ T_{B2i} & t > t_{Bi} \end{cases}$$

At present, humankind's need for energy is 90% satisfied by burning fossil products of photosynthesis (coal, oil, gas) and only 10% satisfied by hydroelectric power stations. In the future these relationships should change drastically due to mastering of new technologies of energy production and, in particular, increasing the efficiency of solar energy fixation. Here the processes of accumulation and transformation of solar energy, by improving and adapting technology to make use of World Ocean energy (waves, tides, ebbs, currents, etc.), can play an important role, so that the human race at some future moment will use new energy sources. Of course, this moment will be different for various regions, according to their levels of economic development. In the global model this is reflected by substitution of the initial value of the component M_i with a new M_{0GMi} value. The t_{GMi} parameter is the function of the multiplier m_{Gi} determined by the ratio of the natural resource supplies to their annual expenditure at a given time moment. It can be regulated in each region by different means with numerous aspects of human activity taken into account. Without dwelling upon details, take the following scenario of a possible change of the multiplier m_{Gi}:

$$m_{Gi}(t) = \begin{cases} m_{G0i}(t) - (m_{G0i} - m_{G\infty i})(t - t_0)/(t_{Mi} - t_0) & t_0 \le t \le t_{Mi} \\ m_{G\infty i} & t > t_{Mi} \end{cases}$$

Within the time interval $[t_0, t_{MGi}]$ in each region the cost G_{MGi} of natural resource renewal can vary linearly from the value G_{MG0i} to the value G_{MG1i} thus:

$$G_{MGi}(t) = \begin{cases} G_{MG0i} - (G_{MG0i} - G_{MG1i})(t - t_0)/(t_{MGi} - t_0) & t_0 \le t \le t_{MGi} \\ G_{MG1i} & t > t_{MGi} \end{cases}$$

The basic capital V_i $(i = 1, \ldots, m)$ invested into the development of industry, science, agriculture, construction, and other spheres of human activity eventually get worn out, resulting in decreasing efficiency. To parameterize the processes taking place, we introduce the term "the ageing time" T_{VG} (~40 yrs). An assessment of this parameter is a characteristic of each region. As before, assume that T_{VG} varies linearly from the value T_{VG0i} to the value T_{VG1i} within the time interval $t \in [t_0, t_{Vi}]$ and then remains constant:

$$T_{VGi}(t) = \begin{cases} T_{VG0i} - (T_{VG0i} - T_{VG1i})(t - t_0)/(t_{Vi} - t_0) & t_0 \le t \le t_{Vi} \\ T_{VG1i} & t > t_{Vi} \end{cases}$$

1.7 THE MAJOR SUDDEN STRATOSPHERIC WARMING AND THE OZONE HOLE SPLIT OVER ANTARCTICA IN 2002

1.7.1 The unprecedented event

In late September 2002, the ozone hole over the Antarctica was much smaller than in the previous six years and it had split into two separate holes (NASA/NOAA, 2002a; Varotsos, 2002c). It has been proposed that the plausible cause for this event was the prevalence of very strong planetary waves, which led to the appearance of a major, sudden stratospheric warming (SSW), an event that has never been observed before in Antarctica (Varotsos, 2002b, c). Subsequently, a NASA/NOAA press release of 6 December 2002 (NASA/NOAA, 2002b), also reported the prevalence of very strong planetary waves in Antarctica.

According to Varotsos (2002a, b) the basic polar vortex in the stratosphere of the southern hemisphere was step-by-step elongated up to the occurrence of a dramatic change, notably, the polar vortex split into two centres. In this respect, Varotsos (2002b) suggested that the amplitude of the zonal wave – number 2 (derived from the Fourier analysis of the isobaric height and temperature fields at 10 hPa) reached a peak when the polar vortex split into two circulations (reversal of the zonal mean wind), a characteristic feature for the occurrence of the unprecedented major SSW in the southern hemisphere. In the following, Varotsos and Kondratyev (2003) confirmed the mechanism proposed by Varotsos (2002b, c) and further suggested that this mechanism does not consider the need for the ozone layer to recover.

Varotsos (2003c) mentioned that some scientists argue that the destruction of ozone by chlorofluorocarbons (CFCs) is the direct and primary cause of delayed stratospheric warmings in Antarctica. In this respect, he suggested that if this was the major effect, the early major stratospheric warming of 2002 would not have occurred since CFC loading of the stratosphere has remained relatively stable in recent years (due to the Montreal Protocol and its amendments).

Furthermore, according to Varotsos (2003a), similar to the 10 hPa features, were found (from a Fourier analysis of the time-series of the 100 hPa) that the ozone hole split occurred not only in the stratosphere, but it also extended into the lower altitudes (upper troposphere).

In addition, Varotsos (2003d) noticed that an approximate 10-day component in zonal mean temperature and wind at high latitudes of the southern hemisphere, that seems to originate in the upper stratosphere, would be of crucial importance for the occurrence of this unprecedented event.

Moreover, Varotsos (2003e) proposed that strong changes in the sea surface temperature (SST) anomalies in the eastern equatorial Pacific (during 1950–2002) took place well before the occurrence of the "no ozone hole" episodes over Antarctica in 1988 and 2002. The effect of the strongest one would have been to have extremely increased planetary wave activity in the southern hemisphere leading to the major SSW in Antarctica.

Finally, Eskes *et al.* (2003) discussed the ozone forecasts and ozone analyses as produced before, during, and after the southern hemisphere ozone hole split event. The ozone hole of 2002 as measured by the Total Ozone Mapping Spectrometer (TOMS) showed anomalously high total ozone content (TOC). During the major SSW, the daily minimum of TOMS TOC was about 90 Dobson Units (DU) higher than in recent winters, and the area contained within the 220 DU contour (nominal ozone hole threshold value) was reduced to 10% of its typical size (Stolarski *et al.*, 2003).

Nevertheless, the present understanding of the dynamics of the stratosphere is still limited and most probably this unprecedented episode in Antarctica will further elucidate the causes and the mechanisms of SSWs. The next subsection illustrates the current attempts by various teams worldwide to clarify the present understanding of this peculiar event which took place in Antarctica in September, 2002.

1.7.2 The role of the tropospheric dynamics

Charlton *et al.* (2003) argued that the dynamics of the vortex split involved non-linear dynamics in the combined troposphere/stratosphere system, and that the concept of forcing from the lower atmosphere might be inappropriate. They also proposed that, as the vortex elongated, it became hydrodynamically unstable, and that this instability could have affected the whole of the upper troposphere and stratosphere. In fact, analyses of meteorological parameters throughout the winter showed that increased wave activity forced from the troposphere occurred as early as May (e.g., Varotsos 2002b; Varotsos and Kondratyev 2003; Varotsos 2003a, c, d; Newman and Nash, 2003; Scaife *et al.*, 2003).

Nowadays, there is growing evidence that stratospheric variability plays a crucial role in driving climate variability at the troposphere and at the Earth's surface on a range of timescales. Thus, the major SSW of 2002 that provoked large amplitude variation in the strength of the southern hemisphere polar vortex can induce anomaly in the tropospheric circulation, which could also be followed by coherent changes in Antarctic surface temperatures. In this respect, the examination of the temporal evolution of the tropospheric circulation following large amplitude variations in the strength of the southern hemisphere polar vortex in data from 1979–2001, and following the major SSW of 2002, performed by Thompson *et al.* (2003), showed that in both cases, anomalies in the strength of the southern hemisphere polar vortex precede similarly signed anomalies in the tropospheric circulation, which reflect a bias in the polarity of the southern hemisphere annular mode, and changes in surface temperatures.

Krueger *et al.* (2003) suggested that the analysis of winter 2002 reveals the dominant role of eastward traveling waves and their interaction with quasi-stationary planetary waves forced in the troposphere that finally lead to the sudden break-up of the vortex, which is in accordance with the earlier published results in Varotsos (2002b, 2003a).

Newman and Nash (2003) pointed out that the polar night jet was considerably weaker than normal, and was displaced further poleward than has been observed in previous winters, resulting from a series of wave events (propagated from the troposphere) that took place over the course of the winter, which preconditioned the polar night jet for the extremely large wave event of 22 September, 2002. They tabulated the dates of the most significant wave events in 2002, based on the 100-hPa eddy heat flux. Furthermore, they also enumerated the morphology of planetary wave energy deposition in the stratosphere, and speculated about a possible role of the tropics in the tropospheric wave forcing.

Kondragunta *et al.* (2003), reported that a 23-year time series of solar backscatter ultraviolet (SBUV) daily zonal mean total ozone values between 70°S and 80°S show record high values (~385 DU) between 25 September and 1 October, 2002. They have also proposed that due to the existence of an ozone depleted region between 0 and 24 km, the ozone destruction was present in the elongated but unsplit vortex in the lower stratosphere; ozone in the middle to upper stratosphere was not vertically aligned with ozone in the upper troposphere and lower stratosphere. This offset in the vertical distribution of ozone resulted in higher total column ozone and masked ozone depletion in the lower stratosphere.

1.7.3 The role of the planetary waves

The variability in 2002 is at least twice as high as in 1999, indicating a much stronger response of the stratosphere to wave activity in 2002, as discussed by Scaife *et al.* (2003), verifying the previously published results in Varotsos (2002b, 2003b).

Weber *et al.* (2003) examined the interannual variability in TOC and OClO (chloroperoxy radical) in terms of the atmospheric wave activity, by using Global Ozone Monitoring Experiment (GOME) observations for both hemispheres. The main finding was that the absolute winter eddy heat flux between 43°S and 70°S latitudes at 100 hPa correlates extremely well with spring-to-fall ratio of TOC polewards of 50°S and negatively correlates with the winter integrated maximum OClO column amounts. The unusual ozone ratio for austral winter/spring 2002 lays almost midway between typical values for Antarctica and those for recent cold Arctic winter/spring seasons. They have also found that GOME TOC (averaged from 50–90°S) during late September 2002 was ~40–50 DU larger than between 1995–2001.

Roscoe *et al.* (2003) have come to the conclusion that the vortex has probably not split before the final warming since Antarctic records began in the late 1950s, and almost certainly not in August or September. This conclusion appears to contradict some model results, of a decreasing likelihood for a vortex split given increasing greenhouse gases (GHGs). They have also found higher wavenumber-2 activity since 1979, on analysing the long-term radiosonde record from Halley Bay dating back to 1956.

Jadin *et al.* (2003) noticed that the relative ozone decrease (increase) in the eastern Antarctic is tightly associated with westerly (easterly) zonal wind anomalies near the southern tip of South America. In particular, unusual behaviour of the ozone hole in 2002 appears to be caused by great easterlies in this region. For the understanding of long-term changes of the atmospheric circulation and ozone layer, they proposed a simple mechanism of an interference of planetary waves excited by the standing orographic sources (Andes in the southern hemisphere and Rockies in northern hemisphere) and by the variable thermal sources depending on the SST anomalies with the well-defined spatial structure (dipole across Andes in the southern hemisphere and dipole across Rockies in the northern hemisphere).

Varotsos (2003e) suggested the necessity of the examination of the role of the SST anomalies in the eastern equatorial Pacific, the equatorial quasibiennial oscillation, and the El Niño phenomenon on the TOC variability over Antarctica. The main finding is that two extremely strong changes in the SST anomalies in the eastern equatorial Pacific (during 1950–2002) took place well before the occurrence of the "no-ozone-hole" episodes over Antarctica in 1988 and 2002. In particular, the strongest change in SST anomalies would have been to increase planetary wave activity in the southern hemisphere leading to the early major SSW in Antarctica that consequently resulted in the early breakdown of the 2002 Antarctic polar vortex. As to the equatorial quasibiennial oscillation, it appears that there was high (low) TOC over Antarctica. Finally, the maxima and minima in the seasonal SST anomalies in the eastern equatorial Pacific correspond approximately to times of occurrence of El Niño and La Niña phenomena, respectively, denoting however that the El Niño characteristics in 1988 and 2002 may not have been similar.

1.7.4 The role of the polar stratospheric vortex

Manney *et al.*, (2003) suggested that the anomalously strong tropospheric forcing could be viewed as the primary direct cause of the SSW. The warming was accompanied by rapid, large-scale transport of air from the subtropics to the polar region (e.g., Charlton *et al.*, 2003). Differences between 1999 and 2002 are not as dramatic at 500 K, consistent with previous findings that the vortex at this level remained intact and air within it was confined during the major SSW (e.g., Varotsos 2002b; Manney *et al.*, 2003). Manney *et al.* (2003) used long-lived (e.g., CH_4, H_2O, and N_2O) and idealized tracers (e.g., a tracer initialized with EL and one initialized with $\log(\theta)$) to show that enhanced poleward transport and mixing are the dominant transport processes during the SSW, with enhanced diabatic descent dominating in a core region of the vortex in the lower stratosphere that remained isolated throughout the warming.

Nedoluha *et al.* (2003) noticed that after the major SSW in September 2002, a sudden decline in polar stratospheric cloud (PSC) activity in 2002 occurred and rehydration of the lower stratosphere also began – but the rate of increase of water vapor was not exceptionally high. The major SSW was accompanied by rapid, large-scale transport of air from the subtropics to the polar region (e.g., Charlton *et al.*, 2003; Manney *et al.*, 2003). According to Randall *et al.* (2003) dynamical forcing early in the 2002 winter influenced the morphology of the stratosphere in a significant and unusual manner, possibly setting the stage for the unprecedented major SSW in late September.

Allen *et al.* (2003), performing diagnostics of transport and mixing, found that during the major SSW the lower stratospheric vortex remained intact, while the middle stratospheric vortex split into two pieces; one piece rapidly mixed with extra-vortex air, while the other returned to the pole as a much weaker and smaller vortex. This unprecedented early vortex split created many filaments and vortex remnants containing ozone depleted air, which were rapidly transported into the midlatitudes. In this respect, Charlton *et al.* (2003) noticed that the other cyclone and the dominant anticyclone moved eastward and poleward, the anticyclone continuing to strengthen while the cyclone decayed. The weaker cyclonic vortex disappeared completely by early October. The remaining cyclonic vortex remained relatively weak and finally broke down by the middle of November (Orsolini *et al.*, 2003).

Hoppel *et al.* (2003) suggested that in the altitude region where chemical ozone loss from chlorine-catalyzed chemistry occurs (below 600 K potential temperature) the 2002 ozone loss was similar to previous years up to the time of the major SSW. The ozone loss diminished after this time (about 1 week earlier than usual), resulting in up to 20% less chemical ozone loss within the vortex than in previous winters. This corresponds to partial column ozone values below 600 K inside the polar vortex that were ~25 DU larger in 2002 than in previous years. They also showed that the 21–60-km ozone partial column, derived from polar ozone and aerosol measurement (POAM) profile measurements, increased by ~150 DU during the major warming.

According to Varotsos (2003a, e), the SSWs in Antarctica observed before 1980 were considerably more prevalent. In 1968, a year of early SSW, the TOC over the South Pole in October was much higher than that in 1969, a year of late SSW. During 1964–1979, a delay in SSW relative to the time of SSW in 1978 occurred at the South Pole in only 6 out of 16 years. From 1979 through to 1985 (and also in 1986 and 1987), SSW delays relative to the 1978 SSW occurred in all years.

1.7.5 What do the models show?

It is of great importance to stress that the possibility of major warmings in the southern hemisphere is also supported by some GCM experiments – prior to September 2002 – and the occasional major SSW in the southern hemisphere were considered a model "defect" (Baldwin *et al.*, 2002).

Since long ago, the combination of wind and temperature fields could provide a reproduction of ozone observations by utilizing specific models. A chemical transport model (CTM) used by Sinnhuber *et al.* (2003) showed that the Antarctic TOC was significantly higher during the winter and spring of 2002 than during previous years, due to the increased transport during the winter 2002, while the estimated chemical loss of TOC of about 80 DU south of 60°S is comparable to earlier winters, although the chemical ozone depletion inside the polar vortex was terminated by the major stratospheric warming much earlier than in normal ozone hole winters.

The break-up of the polar stratospheric vortex in the northern hemisphere is an event that is known to be predictable up to a week or so ahead. For the split of the southern stratospheric vortex in late September 2002, however, the highly unusual flow conditions nevertheless exposed a computational instability of the forecast model (Simmons *et al.*, 2003). The observations and analyses from 1957 onwards in the 40-year European Re-Analysis (ERA-40) archive confirm the extreme nature of the 2002 event. Secondary vortex development by barotropic instability is also discussed in Simmons *et al.* (2003).

Grooss *et al.* (2003) performed a chemical simulation for this event using the Chemical Lagrangian Model of the Stratosphere (CLaMS), indicating a very fast chlorine deactivation that is triggered by the warming associated with the split of the vortex. They also found that the ozone depletion rates in the polar vortex parts decrease rapidly to zero, while outside the polar vortex high ozone depletion rates were observed, caused mainly by NO_x chemistry.

Konopka *et al.* (2003) by utilizing CLaMS-3d model concluded that the CH_4-distribution after the vortex split shows a completely different behaviour above and below 650 K. Above this potential temperature level, until the beginning of November, a significant part of the vortex is transported into the mid-latitudes up to 40°S. The lifetime of the vortex remnants formed after the vortex split decreases with altitude with values of about 3 and 6 weeks at 900 and 700 K, respectively.

Manney *et al.* (2003) presented a mechanistic model simulation initialized on 14 September, 2002 which reproduced the dynamical features of the 2002 Antarctic major SSW. Model results and analyses show a pattern of strong upward wave propagation throughout the warming, with zonal wind deceleration throughout the stratosphere at high latitudes before the vortex split, continuing in the middle and upper stratosphere and spreading to lower latitudes after the split.

1.7.6 Conclusions

The year 2002 "no-ozone-hole" episode does not denote a recovery of the ozone layer, but it will hopefully motivate the needed research in dynamics and chemistry that could lead to a more complete understanding of SSWs and global ozone changes that have occurred since the late 1970s.

Although some initial steps have already been taken and described above, there still remain some important questions, notably: Is the stratosphere well understood; what caused the southern hemisphere to exhibit very strong planetary waves and, subsequently, an unprecedented major SSW in 2002? Global ozone can be affected by changes in sea surface temperature via a number of processes: changes in convective activity over the ocean, modulation of the Hadley Cell circulation, and, most importantly, changes in the interaction of equatorial quasibiennial oscillation winds and extratropical planetary waves. Through teleconnections, this interaction affects the transport to polar regions, not only of ozone, but also of heat that – depending on its arrival time at high-latitudes – influences the destruction, or lack thereof, of ozone by CFCs.

2

The biogeochemical carbon cycle in nature

2.1 SOURCES OF AND SINKS FOR CARBON DIOXIDE IN THE BIOSPHERE

Carbon dioxide is circulating in the environment among its reservoirs listed in Table 2.1. The carbon fluxes in the form of CO_2 can be roughly described by the scheme shown in Figure 2.1. In general, the diverse compounds of carbon continuously form, change and decompose. Among all this diversity, natural and anthropogenic CO_2 fluxes are formed in the processes of respiration and decomposition of vegetation and humus, in burning of the carbon-containing substances, in rock weathering, etc. Part of the CO_2 dissolves in the World Ocean giving carbonic acid and products of its dissociation. The content of carbon in its reservoirs and estimates of its fluxes between them is the most important problem of the analysis of the global CO_2 cycle. Numerous schemes of this cycle drawn from analysis of global interactions of living organisms and their physical and chemical media as well as estimates of carbon supplies accumulated during the historical period serve as the basis for predictions of the CO_2 concentration dynamics in the Earth's atmosphere, which has recently been the subject of a hot dispute in connection with assessments of the role of CO_2 in climate warming (Itoh, 2002; Houghton *et al.*, 2001; Houghton and Yihmi, 2001; Macdonald *et al.*, 2002a, b; Nabuurs *et al.*, 2000; Patra and Maksyutov, 2002; Scanlon and Albertson, 2001; Schulze *et al.*, 2000; Spero and Lea, 2002).

The marine environment is subject to a variety of threats. Together with the possible future effects of climate change discussed here, these threats have such consequences as loss or degradation of biodiversity, changes in structure of trophic chains, and loss of habitats. According to Sherman (2003) 64 large marine ecosystems (LMEs) exist in the world in which detailed studies are taking place.

LMEs encompass coastal areas from river basins and estuaries to the seaward boundaries of continental shelves and the outer margins of the major coastal

Table 2.1. Global carbon reservoirs.

Reservoir	Amount of carbon (Gt C)
Atmosphere	720
World Ocean	38,400
Total inorganic carbon	37,400
Surface layer	670
Deep layers	36,730
Total organic carbon	1,000
Lithosphere	
Carbonate sedimentary rocks	>60,000,000
Kerogens	15,000,000
Land biosphere	2,000
Living biomass	600–1,000
Dead biomass	1,200
Biosphere of inland waters	1–2
Burnt fuel	4,130
Coal	3,510
Oil	230
Natural gas	140
Others (peat, etc.)	250

currents. These areas occupy more than $0.2 \cdot 10^6\ km^2$ of coastal waters. These waters are characterized by most of the global ocean pollution and overexploitation of their ecosystems. That is why the study of LMEs is of principal importance in addressing the marine ecosystem problems on a global scale. At present, the following set of indicators characterizing the LMEs state exist (Sherman, 2003):

- *Productivity indicator*. Photosynthetic activity, zooplankton biodiversity, and oceanographic variability.
- *Fish and fisheries indicator*. Biodiversity, finfish, shellfish, demersal species, and pelagic species.
- *Pollution and ecosystem health indicator*. Eutrophication, biotoxins, emerging disease, and health indices.
- *Socio-economic indicator*. Integrated assessments, human forcing, sustainability of long-term socio-economic benefits.
- *Governance indicator*. Stakeholder participation and adaptive management.

A modular strategy proposed by Sherman (2003) provides information for the various aspects of the state of LMEs. The productivity indicator supports data for the ecosystem parameters that are used within different items of the global model. Other indicators help to make the model items that parameterize the carbon fluxes within the marine ecosystem structure more precise.

An important stage in understanding the processes of CO_2 exchange between the biospheric reservoirs is a study of the laws of the development of various ecosystems

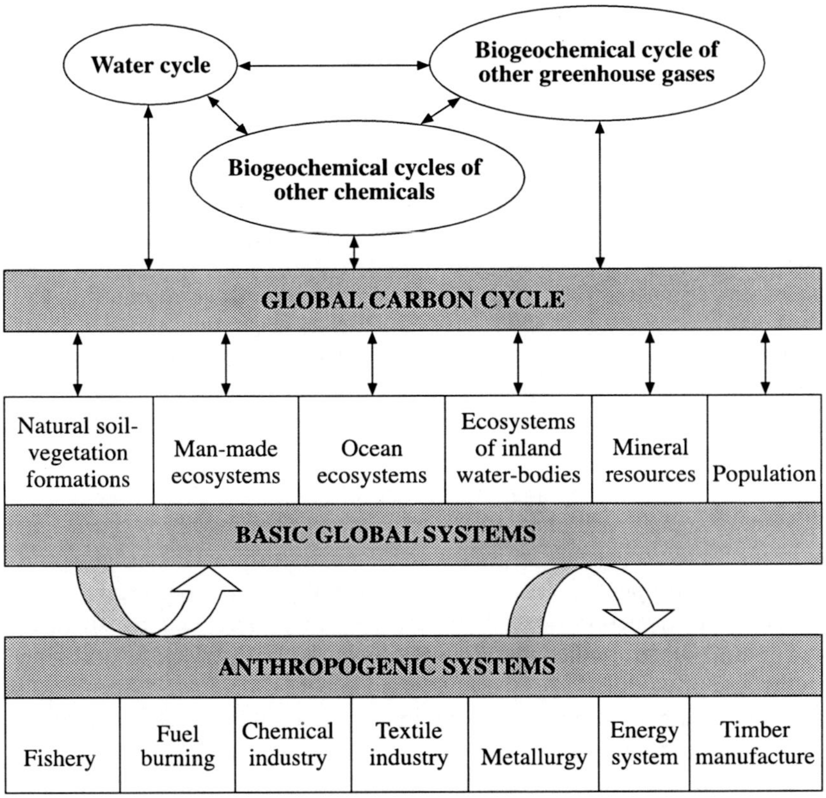

Figure 2.1. The conceptual scheme of the carbon cycle in the environment.

Table 2.2. Data on carbon supplies in the surface vegetation and in the 1-m soil layer (Bolin and Sukumar, 2000).

Biome	Area (10^6 km^2)	Carbon supplies (Gt C) Vegetation	Soil	Total
Tropical forests	17.6	212	216	428
Temperate zone forests	10.4	59	100	159
Boreal forests	13.7	88	471	559
Tropical savannahs	22.5	66	264	330
Temperate zone meadows	12.5	9	295	304
Deserts and semi-deserts	45.5	8	191	199
Tundra	9.5	6	121	127
Heavily moistened soils	3.5	15	225	240
Plough soils	16.0	3	128	131
Total	*151.2*	*466*	*2011*	*2477*

Table 2.3. CO_2 budget on the Earth and its exchange activity in the biosphere.

Reservoir	Carbon content as CO_2 (10^9 t C)	Component of the process in the biosphere	Exchange rate (10^9 t C/yr)
Atmosphere	697	Forests	10
Hydrosphere	35,420	Cultivated soils	4
Geosphere		Grass ecosystems	1
Inorganic carbonates	$18.3 \cdot 10^6$	Deserts	0.2
Inorganic non-carbonates	$6.8 \cdot 10^6$	Photosynthesis north of 45°	15
Coal, gas, oil	7,400	Human respiration	0.14
Substratum, detritus		Domestic animals respiration	0.5
In the ocean	3,220	Wild animal respiration	0.4
On land	710	Soil respiration	8.4
Biosphere			
In the ocean	10		
On land	124		

in the pre-industrial epochs, without any anthropogenic factors (Katz *et al.*, 1999; Katz, 2002). Natural carbon fluxes between the atmosphere, oceans, surface ecosystems, and inland water bodies are strongly variable both in space and time (from year to year and seasonally). Analyses of the ice core data from Greenland and the Antarctic have reliably shown variations in the atmospheric CO_2 levels in the past. Eight-thousand years ago the CO_2 concentration in the atmosphere constituted 200 ppmv. By the beginning of the pre-industrial epoch this estimate had varied between 275 ppmv and 285 ppmv (± 10 ppmv). By the year 1985 the concentration of CO_2 in the atmosphere reached $\sim$345 ppmv. But in 1998 it was already 366–367 ppmv (Bolin and Sukumar, 2000). The total amount of carbon in the atmospheric CO_2 is estimated at about $700 \cdot 10^9$ t C. The natural CO_2 budget is estimated at $\sim 150 \cdot 10^9$ t C emitted annually in the processes of respiration and decomposition and assimilated in photosynthesis both on land and in the ocean, as well as dissolution into the World Ocean.

Special emphasis has recently been placed on the circulation of organic and inorganic carbon in the water domain of the World Ocean whose mechanisms are closely connected with the CO_2 partial pressure dynamics in the atmosphere. With some assumptions, an increase of the atmospheric share of CO_2 is followed by an increase of the partial pressure of carbon dioxide in the surface layer of the ocean at about the same rate. In this case a mechanism is started for the interaction of the carbonic acid elements (HCO_3^-, $CaCO_3$, and CO_3^{2-}) which, depending on the relationship of its characteristics, limits or, on the contrary, stimulates the process of either assimilation or emission of CO_2 by the ocean. The formation of the carbon flux on the atmosphere–ocean border depends also on the phytoplankton production, relationship between the organic and inorganic shares of carbon, temperature, and hydrodynamic and other parameters of the water domain. At a certain level this flux also depends on the processes of vertical transport of carbon via descending

Table 2.4. Quantitative characteristics of the types of land vegetation formations (Bazilevich and Rodin, 1987).

Type of vegetation formation (see Table 2.5 for an explanation)	Formation area (10^6 km^2)	Annual production (kg/m^2/yr)	Phytomass (kg/m^2)	Dead organic matter (kg/m^2)
A	2.55	0.17	0.4	1.3
C	2.93	0.36	1.9	5.6
M	2.33	0.38	1.9	5.5
L	1.55	0.65	3.8	9.0
F	5.45	0.54	10.0	8.1
D	5.73	0.63	22.5	10.8
G	6.60	0.65	23.5	14.5
R	2.12	0.87	25.0	25.1
+	7.21	1.25	45.0	24.8
P	5.75	1.72	43.0	22.2
U	3.91	0.56	3.8	15.0
X	3.72	0.74	1.9	38.0
W	4.29	0.79	1.9	33.0
E	1.66	1.11	3.8	21.0
V	2.66	0.38	0.8	12.6
#	2.08	0.45	0.4	12.1
S	2.69	0.25	0.2	8.1
&	1.99	0.35	0.8	8.8
H	7.16	0.12	0.1	1.4
B	1.15	0.47	0.8	16.9
Q	3.54	0.76	1.9	24.0
Z	10.4	3.17	60.0	21.6
Y	7.81	2.46	60.0	20.5
N	9.18	1.42	10.0	15.1
J	17.1	1.35	0.1	2.0
T	11.5	0.18	0.4	4.9
K	0.38	0.18	0.4	4.9
I	0.9	1.96	45.0	21.6
*	14.6	0	0	0

dead organisms, becoming bottom sediments whose contribution to the so-called biological pump depends on the water depth. In this monograph an attempt has been made to describe these mechanisms on a formalized level. Of course, a search for critical factors for the impact on the difference of partial pressures of atmospheric and oceanic reservoirs of CO_2 requires a thorough and detailed observational study of the ecological and geophysical processes functioning in the oceanic system. In particular, the role of marine organisms in the process of transformation of calcium carbonate and its accumulation, that is, in the change of the domain acidity, has

Table 2.5. Identifier of the types of soil–plant formations following the classification after Bazilevich and Rodin (1967).

Type of the soil–plant formation	Symbol
Arctic deserts and tundras	*A*
Alpine deserts	*B*
Tundras	*C*
Mid-taiga forests	*D*
Pampas and grass savannahs	*E*
North taiga forests	*F*
South taiga forests	*G*
Subtropical deserts	*H*
Subtropical and tropical thickets of the grass-tugai type	*I*
Tropical savannahs	*J*
Solonchaks	*K*
Forest tundra	*L*
Mountain tundra	*M*
Tropical xerophytic open woodlands	*N*
Aspen–birch sub-taiga forests	*O*
Subtropical broad-leaved and coniferous forests	*P*
Alpine and subalpine meadows	*Q*
Broad-leaved coniferous forests	*R*
Sub-boreal and saltwort deserts	*S*
Tropical deserts	*T*
Xerophytic open woodlands and shrubs	*U*
Dry steppes	*V*
Moderately arid and arid (including mountain) steppes	*W*
Forest-steppes (meadow steppes)	*X*
Variably-humid deciduous tropical forests	*Y*
Humid evergreen tropical forests	*Z*
Broad-leaved forests	+
Subtropical semi-deserts	&
Sub-boreal and wormwood deserts	@
Mangrove forests	#
Lack of vegetation	*

been poorly studied. All these processes have different directions and power in the open ocean, estuaries, shallow waters, coastal zones, and river deltas.

To analyse the CO_2 dynamics in the biosphere, it is important to take into account the maximum possible number of reservoirs and fluxes as well as their spatial distribution. It is here that numerous global models of the carbon cycle differ. The present level of these studies does not allow one to answer the principal question about an extent of details of the database concerning the supplies and fluxes of carbon. Therefore many authors analysing the dynamic characteristics of the global CO_2 cycle, rather arbitrarily utilize fragments of

Table 2.6. Estimates of some characteristics of land and ocean ecosystems assuming that the content of carbon in a dry substance constitutes 45%. Notation: σ – area of the ecosystem (10^6 km^2), MPP – mean pure primary production (g C/m^2/yr), TPP – total pure primary production (10^9 t C/yr), MPB – mean plant biomass (kg C/m^2).

Ecosystem	σ	MPP	TPP	MPB
Tropical humid forests	17.0	900	15.3	20
Tropical seasonal forests	7.4	675	5.1	16
Temperate zone evergreen forests	5.0	585	2.9	16
Temperate zone deciduous forests	7.0	540	3.8	13.5
Boreal forests	12.0	360	4.3	9.0
Woodland and shrubs	8.0	270	2.2	2.7
Savannah	15.0	315	4.7	1.8
Temperate zone grass systems	9.0	225	2.0	0.7
Tundra and alpine meadows	8.0	65	0.5	0.3
Desolate dense shrubs	18.0	32	0.6	0.3
Stones, ice, and sand	24.0	1.5	0.04	0.01
Cultivated lands	14.0	290	4.1	0.5
Marshes and boggy territories	2.0	1125	2.2	6.8
Lakes and rivers	2.5	225	0.6	0.01
All land ecosystems	*148.9*	*400.6*	*48.34*	*6.26*
Open ocean	332	57	18.9	0.0014
Upwelling zones	0.4	225	0.1	0.01
Continental shelf	26.6	162	4.3	0.005
Algae and reefs	0.6	900	0.5	0.9
Estuaries	1.4	810	1.1	0.45
The whole ocean	*361*	*69*	*24.9*	*0.0049*
The whole biosphere	*510*	*144*	*73.2*	*1.63*

databases of the distribution of the carbon sinks and sources. These fragments are exemplified in Tables 2.1–2.6.

2.2 ANTHROPOGENIC SOURCES OF CARBON

A key component of the global CO_2 cycle are anthropogenic emissions into the environment. The principal problem studied by most of the investigators consists of the assessment of the ability of the biosphere to neutralize an excess amount of CO_2. It is here that all predictions of the consequences of the greenhouse effect are open to criticism. All the models of the global CO_2 cycle are based on scenarios that describe the dynamics of the extraction and burning of fossil fuels. Here, however, a natural need appears, for models to consider the energy–economy system which requires a detailed parameterization of the geopolitical structure of the world. So far, among the most widely used models of this type is a model developed by the International Institute of Applied Systems Analysis (IIASA), in which the globe is

Table 2.7. Characteristics of the rates of growth of the economic efficiency and population in the world regions following the IIASA scenario (Mintzer, 1987).

Region	P	Size of population (millions)		
		Years		
		2025	2050	2075
Australia	2.3	160	150	150
Africa	1.6	1,600	2,200	2,700
Canada and Europe	1.6	520	540	540
China	1.9	1,600	1,700	1,700
Latin America	1.9	720	850	900
Russia and FSU countries	1.3	470	500	510
Middle Asia	1.9	280	360	410
USA	1.2	290	290	290
South-East Asia	1.8	2,600	3,100	3,400

Notation: P – annual increment of labour productivity (%), FSU – Former Soviet Union.

divided into nine regions differing in levels of per capita energy consumption and other parameters. The regional structure of this model is shown in Table 2.7. With this scenario of the socio-economic structure one can attribute to it the strategies of the development of each region and assume possible consequences for the environment of future behaviour of individual regions. Most of the similar scenarios use such indicators as the ratio of acceleration of energy consumption. This parameter varies from 0.2% to 1.5% per year. Various combinations are considered when choosing a source of energy from oil, gas, nuclear, solar, hydroelectric, and solid waste. Naturally, one has to take into account the demographic, technical, political, and macroeconomic factors. The size of population in most of the scenarios is assumed to grow at a rate that would give population levels of 7.9 and 10.5 billion people in 2025 and 2075, respectively. If all these suppositions in the scenario are assumed to be true, one can calculate the anthropogenic emissions of CO_2 and other greenhouse gases (GHGs). Then it is necessary to determine the total temperature impact ΔT_{Σ} of these gases. Mintzer (1987) recommends the simplest additive algorithm to estimate this effect (see Chapter 6).

The anthropogenic constituent in the global CO_2 cycle causes changes in the reservoirs of the CO_2 sinks. Maximum changes are connected with urbanization, deformed structures of the soil–plant communities, and hydrospheric pollution. The rates of change of forest masses for pastures and cultivated lands are estimated at $0.05 \cdot 10^6\ \text{km}^2/\text{yr}$. Dense tropical forests are substituted with plantations at a rate of $10^5\ \text{km}^2/\text{yr}$. This process increases the rate of desertification ($\sim 5 \cdot 10^4\ \text{km}^2/\text{yr}$), which increases the amount of emitted carbon ($\sim$0.1 Gt C/yr).

The general pattern of the present level of anthropogenic CO_2 fluxes have been rather well studied. So, due to burning of solid and liquid fuels, about $20 \cdot 10^6$ t of CO_2 are emitted every year (at a ratio of 1:1). Burning of gas fuel contributes about

Table 2.8. Distribution of the scales of biomass burning.

Region	Sources of biomass burning (Mt of dry matter per year)						
	T	S	L	B	Y	A	C
Tropical America	590	770	0	0	170	200	7.5
Tropical Africa	390	2,430	0	0	240	160	9.3
Tropical Asia	280	70	0	0	850	990	3.3
Tropical Oceania	0	420	0	0	8	17	0
USA and Canada	0	0	0	0	80	250	0.5
West Europe	0	0	0	0	40	170	0.2
Temperate-zone forests	0	0	224	0	0	0	0
Boreal forests	0	0	0	56	0	0	0
Total	*1,260*	*3,690*	*224*	*56*	*1,438*	*2,017*	*21*

Notation: T – tropical forests, S – savannah, L – temperate zone forests, B – boreal forests, Y – household fuel, A – agricultural wastes, C – brown coal.

$4.5 \cdot 10^6$ t of CO_2 into the atmosphere. The contribution of the cement industry is estimated at $750 \cdot 10^3$ t of CO_2. Individual regions and countries contribute to these fluxes rather non-uniformly. Table 2.8 gives some estimates of such contributions due to biomass burning.

The biomass burning in the tropics is one of the main sources of input of minor gaseous components and aerosol particles to the troposphere. The share of the tropics is about 40% of the global land area and about 60% of the global primary productivity. The types of vegetation in the tropics are much more diverse compared to other regions. However at present the tropical forests and savannahs are being transformed into agricultural lands and pastures at a rate of about 1%/yr. This transformation is mainly caused by biomass burning which strongly affects the chemical composition of the atmosphere and, hence, the climate.

In the process of biomass burning, huge amounts of non-methane hydrocarbons (NMHC), NO_x, and many other gaseous components enter into the atmosphere. As shown by analysis of the data of satellite observations, the share of the tropics constitutes about 70% of the burnt biomass, about half of which is concentrated in Africa, with a maximum of biomass burning in the annual course (in the period of dry season) observed north of the equator. Savannahs and forests in the tropics also emit into the atmosphere a great amount of nutrient compounds. Widespread fires in savannahs and their strong impact on the environment were studied by Nielsen (1999) who performed an analysis of the special features of the spatial–temporal distribution of fires in the region during the field experiment EXPRESSO (Experiment for Regional Sources and Sinks of Oxidants in Central Africa) was carried out from the data of Advanced Very High Resolution Radiometer (AVHRR) carried by the *NOAA* satellites for the periods of dry seasons, from November 1994 to December 1997. The fires variability was described by three characteristics: (1) fire probability at a given point, at a given time moment; (2) probability of repeated fires at a given point during a certain time period; and (3) the spatial extent and the burning savannah temperature affecting the conditions at a given point.

Processing of satellite imagery has shown that a fire is not an accidental process. The fire probability increases, for instance, with increasing numbers of fires occurring in the neighbourhood of the point considered. A combined analysis of the characteristics of the spatial and temporal variability of fires has made it possible to substantiate 12 typical regimes of fires as well as the dependence of the special features of the fires on those of the vegetation cover. Alhough there is no doubt that, as a rule, the savannah fires are caused by humans and not by other factors, specific causes of fires as a function of human activity remain unclear. From the viewpoint of temporal variability, it is expedient to classify the fires taking into account the beginning of the fire season, the rate of their development, and duration of the fire season. In this context the following types of fires can be identified by the specific dynamics of their development: fast, late or long.

The contribution of fires in savannahs constitutes more than 40% with respect to the global level of biomass burning due to which the atmosphere receives minor gaseous components, such as NMHC, carbon monoxide, methane, etc., as well as aerosols. According to the available estimates for the period 1975–1980, 40–70% of savannahs were burnt every year and about 60% of such fires took place in Africa. In 1990 about $2 \cdot 10^9$ t of vegetable biomass were burnt, and as a result, 145 Tg CO got to the atmosphere, which constitutes about 30% compared with the anthropogenic CO emissions.

Forest fires are one of the most serious impacts on the global carbon cycle. Although forest fires can occur naturally, being caused by lightning strikes for example, human contributions to their occurrence are constantly growing. An occurrence of a forest fire due to a lightning strike is only possible if it strikes standing wood or, in the case of a thin forest, if it strikes moss or litter-covered soil. The electrical resistance of standing wood is known to be almost 100 times greater than that of the growing trees, and therefore when the lightning strikes a living tree, it does not even become charred. Therefore, the monitoring of fire danger in forests gives reliable estimates of the probability of lightning-caused forest fires. A more complicated problem is predicting the anthropogenic causes of forest fires. More than 90% of forest fires are known to occur within the 10-km zone of populated areas. Hence, the fire load on forests is strongly correlated with the spatial distribution of population density. Of course, the intensity and frequency of occurrence of fires depends on the climate dryness in a given territory, and on the forest density and health. A forest fire is dangerous not only because it is a source of pollutants for the atmosphere, but also because their consequences are dangerous. Fires change the forest microclimate, in particular, an illumination and heating of soil intensifies, and the hydrological regime of the territory changes. Moreover, within the territory of forest fires the bioproductive ability of biocoenoses deteriorates and, the role of the territory in the biogeochemical cycles changes. It is well known that in a region with a dry climate the fire-destroyed forests are not restored naturally, and are reforested. Therefore it is important to understand the laws of interaction of the forest fire and the biocoenoses within the territory. For instance, fires in the boreal forests contribute not more than 2% to carbon emissions into the atmosphere, but seriously affect the chemical processes in the high-latitude troposphere and the atmospheric radiative properties. This can have global climatic consequences.

In general, for different reasons, biomass burning is a complex anthropogenic source of atmospheric pollution and has global impact on the biosphere as a whole (Table 2.8). Estimates obtained by many authors show that the radiative forcing on the climate determined by aerosols from biomass burning constitute about $-1.0\,W/m^2$ (in the case of pure scattering aerosols the uncertainty of the estimates ranges between -0.3 and $-2.2\,W/m^2$).

2.3 CONCEPTUAL DIAGRAMS OF THE GLOBAL CARBON CYCLE

The reliability of the assessment of the role of CO_2 in the greenhouse effect formation depends on detailed consideration of the global biogeochemical carbon cycle dynamics of models and on the accuracy of the assessment of its characteristics. There are dozens of diagrams of the global carbon cycle in the form of CO_2 changes. We shall consider some of them to demonstrate their principal features and to understand the limits for necessary details of the simulation of the carbon cycle compounds, beyond which it is impossible to obtain any additional knowledge about this cycle and, hence, about the greenhouse effect due to CO_2. Note that all known diagrams of the global CO_2 cycle are divided into two classes: pointwise (globally averaged) and spatial (locally averaged). All the diagrams are similar in that the biosphere is divided into the atmosphere, oceans, and land ecosystems. Many diagrams divide carbon into organic and inorganic forms. As a rule, the time step of averaging all processes and reservoirs of carbon is assumed to be one year, and therefore the atmospheric reservoir is considered as homogeneously mixed (pointwise). The World Ocean and surface ecosystems are considerably detailed. This detailing is based on global databases for these reservoirs of carbon. As a rule, the final results of the diagram studies are either of methodical character or they predict the atmospheric CO_2 concentrations within the limits of a certain scenario of anthropogenic activity.

The scheme in Figure 2.2 (Bolin and Sukumar, 2000) gives an idea about the amounts of carbon supplies for its basic reservoirs. The estimates shown in this scheme differ significantly from the estimates by other authors. Nevertheless, their relationships and orders of magnitude coincide in most cases. As seen, the largest carbon supply is concentrated in the World Ocean, with the smallest supply in the atmosphere.

Natural processes determining the global carbon cycle dynamics have various timescales. Some of them, such as burial of the dead organic matter at the bottom of the oceans, have characteristic timescales of hundreds and thousands of years. Others (i.e., the biological carbon cycle on land), have periods of several dozens of years. Therefore consideration of the timescales of the carbon cycle in the biosphere is an important stage in studying of the dynamics of the CO_2 content in the atmosphere. Here it is also important to consider the fact that the characteristic time of

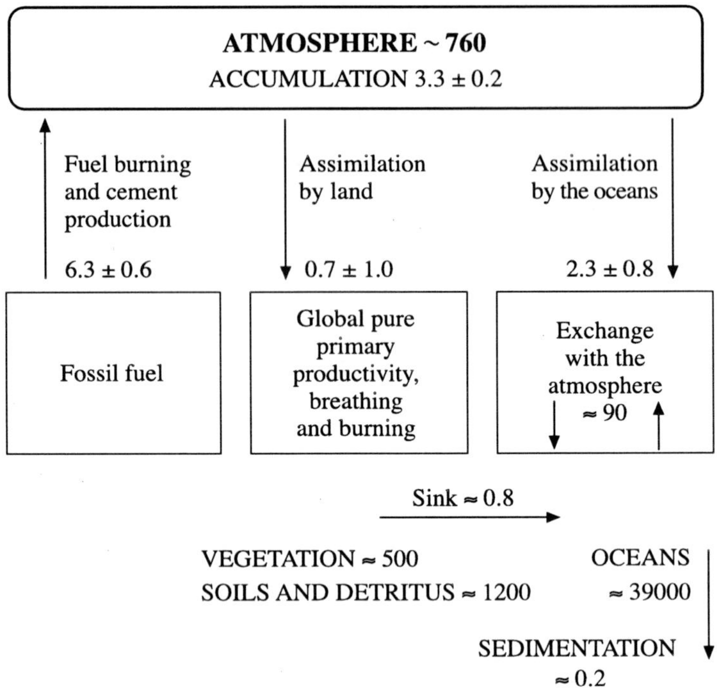

Figure 2.2. Global carbon cycle after Bolin and Sukumar (2000). Carbon supplies are given in Gt C, and its fluxes in Gt C/yr.

complete mixing of the atmosphere constitutes from several months to two years. Regular measurements at different stations of monitoring have shown that the CO_2 concentration varies considerably during one year. The difference between maximum and minimum estimates of the atmospheric CO_2 concentrations varies from 10 ppm at the South Pole to 15.10 ppm at northern hemisphere high-latitudes. This spatial non-uniformity is explained by the presence of large seasonally photosynthesizing vegetation communities in the northern hemisphere. Figure 2.3 shows some estimates of the latitudinal distribution of CO_2 in the atmosphere. An increased CO_2 content in the northern hemisphere atmosphere is closely connected with the impact of human activity through direct CO_2 emissions and also to the impact of the activity on vegetation cover. Almost 90% of the total carbon emissions due to organic fuel burning fall within the zone 30°N–60°N. It follows from the data in Figure 2.3 that the conceptual schemes of the global biogeochemical cycle of carbon dioxide should also consider the spatial non-uniformity of atmospheric processes (Kaminski *et al.*, 2001).

An important constituent of most of the conceptual schemes of the global carbon cycle is the structure of carbon fluxes in the World Ocean. As follows from Table 2.1, there is a certain information possibility to select in the oceans

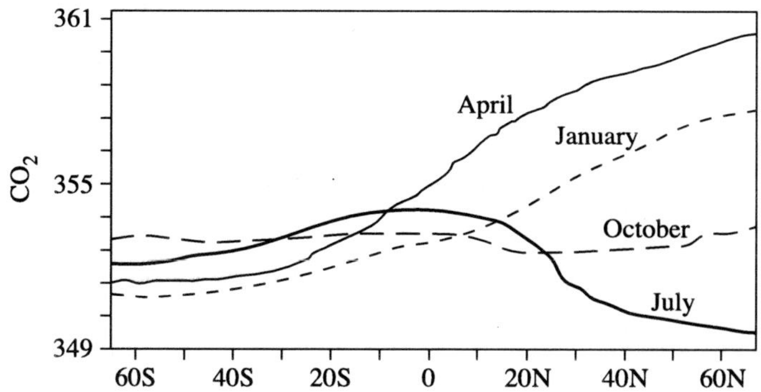

Figure 2.3. Latitudinal distribution of CO_2 concentration in the atmosphere in 1990 from the *TRENFS'1993* data (Boden *et al.*, 1994). The data are averaged over the months given in the curves.

several layers at various depths and to distinguish between the spatial non-homogeneity in the structure of the ocean surface. Most authors consider the vertical structure of the oceans to consist of two or three layers covering the photic layer and deep layers. Within the photic layer can be selected layers both above and beneath the thermocline. The spatial non-uniformity of the ocean is demonstrated via upwellings and latitudinal zones, with different rates and directions of oceans–atmosphere CO_2 exchange. More detailed schemes of the global carbon cycle take into account non-uniformities in the carbonate system of the oceans, which makes it possible to considerably increase the accuracy of the respective models. Two of the widely used schemes of this type are shown in Figures 2.4 and 2.5.

The role of the World Ocean in the global CO_2 cycle is mainly manifested through the process of its exchange at the atmosphere–ocean border. The intensity of gas exchange between the ocean and the atmosphere is determined by the dynamic characteristics of the turbulent layers of water and air near the interface. Here numerous physical schemes appear that reflect the situations of sea wave formation, as well as formation of foam and various films. As a result, carbon dioxide either dissolves in the ocean providing thereby the inflow of CO_2 needed for photosynthesis, or, is emitted from the ocean into the atmosphere. The cause of this binary situation on the air–water border is the difference between the partial pressures of CO_2 in the atmosphere and CO_2 dissolved in the water. This directed transport of CO_2 on the atmosphere–ocean border is more complicated. Its study requires expensive field experiments and a detailed classification of both synoptic and physico-geographical situations on the ocean surface.

In the programmes of studies of the global carbon cycle the emphasis is placed on the role of surface ecosystems in its utilization (Houghton *et al.*, 2001). In the process of photosynthesis, plants assimilate carbon dioxide and, whilst the decomposing dead plants emit carbon dioxide into the atmosphere. Thus a continuous CO_2 exchange between living and dead organic matter and the atmosphere takes place within the land biosphere. There are many conceptual diagrams that formalize this

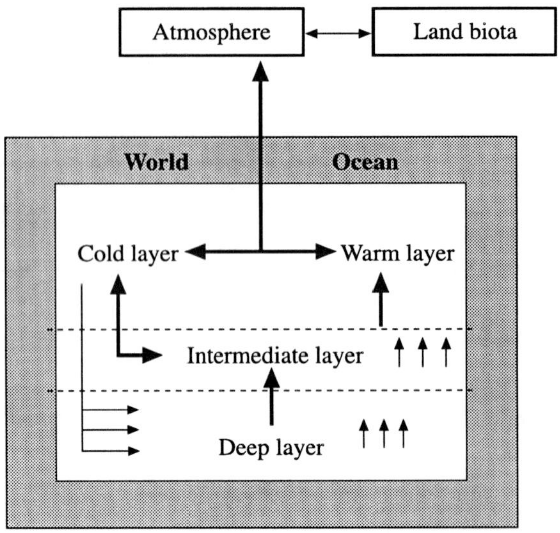

Figure 2.4. A scheme of the carbon reservoirs and fluxes in the models with a vertical structure of the World Ocean.

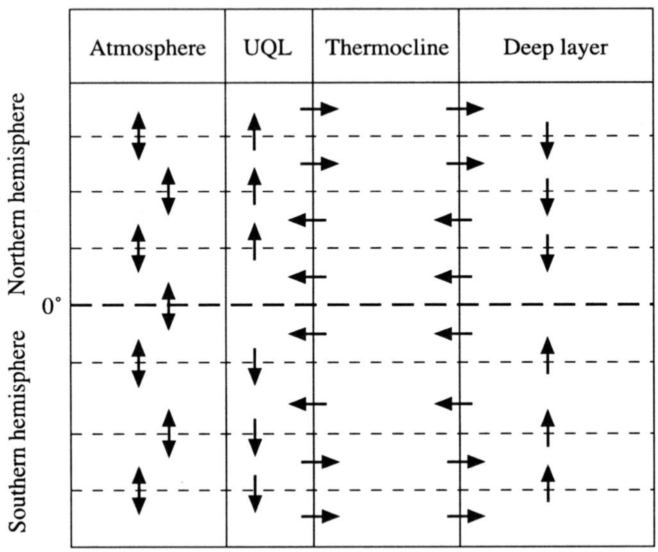

Figure 2.5. A scheme of the World Ocean global circulation as the basis for the description of carbon cycles. UQL is the upper quasi-homogeneous layer.

exchange and serve as the basis for global models of the carbon cycle. Two examples of such diagrams are given in Figures 2.6 and 2.7. Clearly, the accuracy of estimates of carbon fluxes in the terrestrial part of the biosphere is the function of detailed quantization of the types of soil–plant formations and accuracy of the

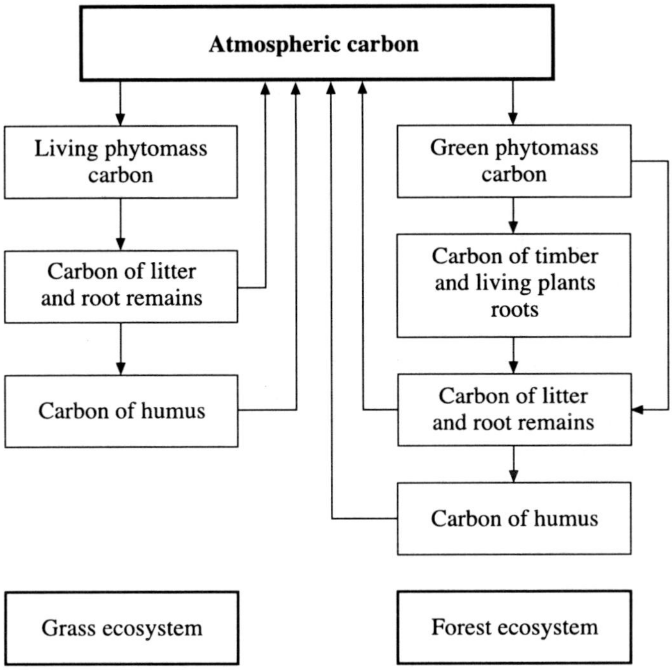

Figure 2.6. Diagram of the carbon fluxes in the "atmosphere–plant–soil" system using grass and forest ecosystems as an example.

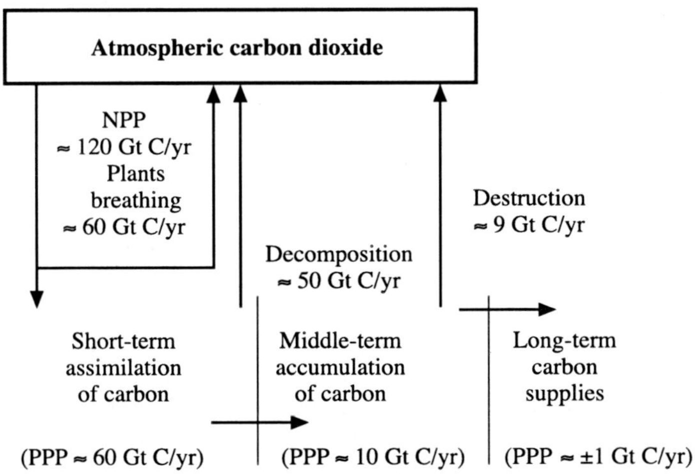

Figure 2.7. A scheme of the global carbon cycle in surface ecosystems after Watson *et al.* (2000). Notation: PPP – pure primary production, PPE – pure production of ecosystems; PBP – pure biome production, NPP – net primary production.

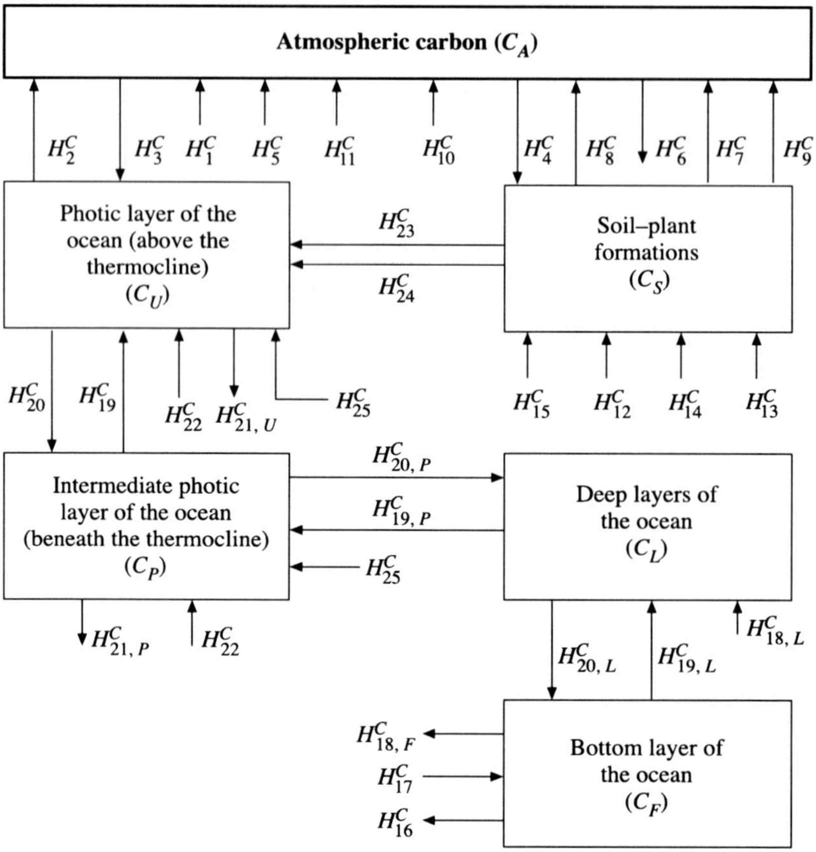

Figure 2.8. A block-diagram of the global biogeochemical cycle of carbon dioxide in the "atmosphere–land–ocean" system (*MGCDC-unit*). The CO_2 reservoirs and fluxes are described in Table 2.9 (Krapivin and Kondratyev, 2002; Krapivin and Potapov, 2002; Kondratyev *et al.*, 2002; Caspersen *et al.*, 2000; Charette *et al.*, 2001; Ciais, 1999a,b, 2002; Clerbaux *et al.*, 2001).

parameterization of the biocoenotic processes. In this regard world maps of vegetation and soils have been drawn and their areas estimated, data on the vital functions of soil micro-organisms have been accumulated, and technologies for facilitating operational monitoring of landscapes have been calculated. Unfortunately, there are no reliable estimates of the limits for considering details of soil–plant formations, which could provide the necessary accuracy for carbon flux estimation. To solve this problem, a global model of the carbon cycle is needed which would reflect a hierarchy of importance, based on expert estimates, constituents of the biosphere, and the processes of carbon transformation. An exemplary scheme of carbon fluxes is characterized in Figure 2.8. Numerical experiments with this model will make it possible to obtain comparative assessments of the consequences of complicating its

Table 2.9. Reservoirs and fluxes of carbon as CO_2 in the biosphere in a simulation model MGCDC of the global biogeochemical cycle of carbon dioxide as shown in Figure 2.8.

CO_2 reservoirs and fluxes	Identifier in the model	Average estimate of the reservoir (10^9 t) and flux (10^9 t/yr)
Carbon		
Atmosphere	C_A	650–750
Photic layer of the ocean	C_U	580–1020
Deep layers of the ocean	C_L	34,500–37,890
Soil humus	C_S	1,500–3,000
Emission due to burning		
Vegetation	H_8^C	6.9
Fossil fuel	H_1^C	3.6
Desorption	H_2^C	97.08
Sorption	H_3^C	100
Rock weathering	H_4^C	0.04
Volcanic emanations	H_5^C	2.7
Assimilation by land vegetation	H_6^C	224.4
Respiration		
Plants	H_7^C	50–59.3
Human	H_{10}^C	0.7
Animal	H_{11}^C	4.1
Emissions		
Decomposed soil humus	H_9^C	139.5
Plant roots	H_{15}^C	56.1
Vital functions		
Human population	H_{12}^C	0.3
Animal	H_{13}^C	3.1
Vegetation decay	H_{14}^C	31.5–50
Sedimentation to bottom deposits	H_{16}^C	0.1–0.2
Solution of marine sediments	H_{17}^C	0.1
Decomposition of detritus		
Photic layer	H_{22}^C	35
Deep layers of the ocean	H_{18}^C	5
Upwelling with deep water	H_{19}^C	45
Sinking with surface water due to gravitational sedimentation	H_{20}^C	40
Photosynthesis	H_{21}^C	69
Underground sink	H_{23}^C	0.5
Surface sink	H_{24}^C	0.5–0.6
Respiration of living organisms in the ocean	H_{25}^C	25

constituents, thereby allowing a sequential approach to calculating the level of optimal spatial quantization.

2.4 RESOURCES OF THE BIOSPHERE AND CARBON CYCLE

The dynamics of the global carbon dioxide flux is determined by natural and anthropogenic factors. Natural factors are formed by the process of evolution of the biosphere, and their dynamics depend on the interaction between natural ecosystems. The level of the anthropogenic forcing on the global CO_2 cycle is determined by the relationship of natural forces with numerous aspects of humankind's development – political, demographic, cultural, religious, economic, etc. All this anthropogenic diversity at present is limited by the resources of the biosphere, which will eventually determine development of humankind. The tabulated estimates of resources given in Chapter 1 lead to the conclusion that humankind's belief in its omnipotence, with its nature-destroying capacity, is limited by many circumstances. Humankind contaminates the environment either by using mineral resources or by changing the planet through changing one ecosystem for another. In both situations humankind, sooner or later, becomes at variance with its goals, and therefore limiting factors exist in the global dynamics of destruction – a depletion of resources bringing about a worsening of living conditions.

A human being as an element of nature, forming his or her environment, first of all, is interested in the sources for material production: mineral resources. With growing scientific–technical progress the rate of mineral resource consumption is constantly increasing, approaching critical levels. Although the late 20th and early 21st centuries are characterized by a broadened spectrum of mineral deposits used by industry, nevertheless, no alternative to oil, coal, and gas has been found. An increase of energy resources consumption is confirmed by the fact that almost half of the oil equivalent used by humankind ($\sim 90 \cdot 10^9$ t) has been consumed during the last 25–30 years.

By the end of the 20th century about $200 \cdot 10^9$ t of mineral deposits had been extracted. There are no reliable estimates of the global supplies of mineral resources. Therefore scenarios of carbon emissions into the atmosphere used by many authors should be considered conditional and disputable (Watson *et al.*, 2000; Houghton *et al.*, 2001). Nevertheless, there are undisputable relationships between volumes of emitted CO_2 and national production. This relationship varies between developed and developing countries by a factor of 30, which means a considerable reserve for calculating an optimal strategy for the formation of the profile of the global curve of CO_2 emissions into the atmosphere.

The World Ocean is one of the poorly utilized sources of mineral resources. Along with farming food products, humankind is gradually harnessing the mineral fuel supplies lying under the bottom of the oceans and around ocean shelves. Oil, gas, and coal are already being extracted in large volumes from marine deposits but there are both economic and technical limitations to this. On the whole, from preliminary estimates, from 180 gas basins discovered by geologists within the World

Ocean, the potential supplies of oil and gas are estimated, respectively, at $300 \cdot 10^9\,m^3$ and $150 \cdot 10^{12}\,m^3$. Thus here there also exists limiting parameters for scenarios of CO_2 emissions into the atmosphere.

The human activity in changing land cover is one of the factors of the anthropogenic forcing on the global carbon cycle dynamics that is difficult to assess. During the last century humankind strongly affected the global distribution of vegetation – one of the important sinks for atmospheric CO_2. This was caused by the human activity in engineering, construction, mining engineering, creation of new types of land formations, biological re-cultivation, etc. Especially dangerous for the environment are processes of deforestation and desertification. For instance, after the estimate by Watson *et al.* (2000) with the realization of the IPCC scenario of re-forestation, by the year 2050 an additional 60–87 Gt C (70% – tropical forests, 25% – temperate-zone forests, and 5% – boreal forests) may be removed from the atmosphere. An important parameter of this scenario is the rate of tree growth. For instance, if the forest grows at a rate of 3 t C/yr/ha, then 1 t C can be attributed to the effect of atmospheric CO_2 assimilation. In general, the control of land biocoenoses can markedly affect the biogeochemical carbon cycle. What is important is the manner in which this control is realized – planned or spontaneous. Humans create anthropogenic landscapes on the Earth's surface to improve their habitat. An appearance of man-made seas, recreation zones, cities, and other anthropogenic landscapes reduces the level of natural land cover evolution. Therefore, human activity which changes the landscapes can strongly affect the dynamics of atmospheric CO_2.

One of the sources of changing land cover is human agricultural activity. The impact of agricultural activity on the formation of the anthropogenic landscape propagates to the zones of the biosphere where the processes of transformation of inorganic masses intensify the food production. Here the ability of humankind to affect the global carbon cycle is also limited. Nobody knows the limits of the consequences of these transformations. To understand the level of such trends, note that, according to chronicles of the late 16th to early 17th centuries, the size of population was 1–2 people per square kilometer, 400 years later the population density increased by more than a factor of 50. During this time period the forested areas decreased by a factor of 3, being substituted mainly by agricultural crops and other cultivated landscapes. This trend still remains in the 21st century because, on the whole, the agricultural activity improves human habitats, although negative phenomena have also appeared. To improve the conditions for agriculture, changes are introduced in geological, geochemical, and hydrogeological processes, which eventually lead to irreversible changes in the global biogeochemical cycles of chemicals. These changes include the processes of wind and water erosion of soils, washing out of nutrients from the fertile soil layer and their consequent transport to the World Ocean, as well as formation of ravines, etc.

A very important stage in the framework of the global carbon cycle is the study of the role of the Arctic environment. A set of Arctic environmental problems is in the focus of many national and international plans to study the marine ecosystems of high latitudes. For example, the Strategic Plan of the Arctic Ocean environment study proposed by Canada and Iceland for discussion in the framework of the International Program for the Protection of the Arctic Marine Environment (PAME) includes the following subjects that warrant specific consideration and analysis:

1. Arctic marine transformation.
2. Understanding Arctic marine science.
3. Indigenous peoples interests and needs in the protection and utilization of Arctic marine resources.
4. Offshore Arctic oil and gas development.
5. Ecosystem- and risk-based approaches to conserving and managing Arctic flora and fauna.

3

Surface ecosystems and carbon cycle

3.1 LAND ECOSYSTEMS AND THEIR GLOBAL DYNAMICS

The growing global size of population and the associated growing forcing of human activity on the environment and ecosystems have become not only the main threat to further sustainable development of civilization in the context of global ecological safety but also reflects a dangerous disorder in the normal functioning of various systems of life support (Caspersen *et al.*, 2000; De Lucia *et al.*, 1999; Grigoryev and Kondratyev, 2001; Grogan *et al.*, 2001; Kojima, 1998; Leach, 2002; Lueker *et al.*, 2001; Mander and Jongman, 2000; McNulty, 2002; Morgan, 2002; Nepstad *et al.*, 2002; Quay, 2002; Räisänen, 2001). An alarming symptom of barbaric impacts on nature was the bombarding of Yugoslavia by NATO (Kondratyev, 2001a,b). In connection with the key role of the ecosystems in the processes of natural regulation of environmental properties, of principal importance is an analysis of the available data on the global dynamics of ecosystems and an assessment of possible trends. Important information on these problems can be found, in particular, in recent publications by many authors (Hackler *et al.*, 1999; Ito and Oikawa, 2002; Seppelt and Voinov, 2002; Gbondo-Tugbawa and Driscoll, 2002; Lee *et al.*, 1999; Lischke *et al.*, 2002), among which the most informative is the report prepared by the Institute of World Resources (Washington, DC) (*A Guide to World Resources 2000–2001*) supported by the UN Programme of Development, UNEP, and the World Bank. This report emphasizes a close relationship between the global ecosystems and the global population – a symbiosis unique and extremely sensitive to external forcings. These presentations have been thoroughly substantiated by Gorshkov *et al.* (2000) in the form of a concept of biotic regulation of the environment. Monographs by Kondratyev (1998b, 1999a–c) and new publications (Ernst, 2000; Kondratyev *et al.*, 2000; Kondratyev and Krapivin, 2001a, b) contain an analysis of the key aspects of the global changes from the viewpoint of functioning of the global systems of life support and requirements to an adequate ecological monitoring.

Global natural and regulated ecosystems play an important role as factors of the environmental dynamics ranging from micro-scales (e.g., soil bacteria) to the whole planet. They are vitally important sources of drinking water, food, timber, paper, and other means of life support. As has been mentioned earlier (Kondratyev, 2000a; Kondratyev and Demirchian, 2000; Kondratyev, 1998a), an urgency of the problem is that the world, on the whole, has already drawn near to the limits of ecosystem impact. Exceeding these limits is fraught with the irreversible destruction of the global systems of life support. It is suggested from some indicators that these limits have already been exceeded, though the present enthusiasm for apocalyptic predictions is, so far, unfounded (this especially refers to so-called "global warming").

The extreme complexity of the problem discussed is that it is necessary to explain (and, as far as possible, to predict) the dynamics of the interactive "society–nature" system (the society should be placed first here since its functioning determines its impact on nature) with its numerous feedbacks, non-linear nature, and "surprises". Unfortunately, the present stage of studies of the "society–nature" system can be considered not more than initial and preliminary. This even refers to a simple description of the present condition of nature (global ecosystems), which results from the observational data deficit with an apparent abundance of some observational means (especially expensive space-borne means). Therefore the report of the Institute of World Resources (Washington, DC) is in many respects incomplete, being concentrated only on consideration of five types of ecosystems (the share of land surface is given in brackets, except for the Antarctic and Greenland which are occupied by the respective ecosystem): agricultural (28%), coastal regions (22% within a 100-km band), forest (22%), freshwater (<1%), and grass (41%). An abandonment of the World Ocean is, of course, a serious, though justified (in view of information deficit) flaw. Special attention should also be paid to the soil ecosystems (Wofsy *et al.*, 2001; White *et al.*, 1999; Tangley, 2001; La Deau and Clark, 2001).

These ecosystems are very important for the solution of various problems of human life support and regulation of environmental conditions (Table 3.1).

3.1.1 Characteristics of the anthropogenic impacts on ecosystems

The authors of the report of the Institute of World Resources (Washington, DC) (*A Guide to World Resources 2000–2001*) have briefly summarized the anthropogenic impacts on the ecosystems during the civilization development, beginning from an intensive application of the irrigation systems during the Shumer civilization, which led to soil salination, to the present global processes of atmospheric pollution and ozone layer depletion (Kondratyev and Varotsos, 2000). The data of Table 3.2 characterize the most important aspects of anthropogenic impacts on the ecosystems today.

Examples of destructive (and even catastrophic) impacts on ecosystems and their economic consequences are numerous (see for details Grigoryev and Kondratyev, 2001). The collapse of the cod catch in 1990 in the sea regions of Canada made about

30,000 fishermen unemployed, and in the region of Newfoundland brought about serious economic difficulties in 700 settlements. Material losses in China reaching $11.2 \cdot 10^9$ US\$/yr have resulted from a deficit of drinking water due to polluted river and sub-soil waters. In India, commercial forest felling and the transformation of the deforested lands into agricultural lands have not only changed the traditional way of people's lives but also caused a deficit of wood fuel and timber to the detriment of 275 million people in rural areas.

As for estimates of the consequences of the global anthropogenic impacts, the situation with water resources is a good example: about 28% of the global population have no access to pure drinking water; every year about 5 million people die because of low quality of drinking water and anti-sanitary conditions; and about 90% of all wastes in developing countries go to rivers, lakes, and coastal regions of the seas. Intensified emissions of CO_2 into the atmosphere have caused considerable changes in the global carbon cycle (Demirchian and Kondratyev, 1999; Kondratyev and Demirchian, 2000).

The most important fact is that the levels of impact on the ecosystems have become global scale. About 75% of marine fish populations have either decreased due to violation of the permissible amounts of catch or come close to the population threshold of their survival. Intensive forest felling has almost halved the forested areas, and the construction of various economic infrastructures has caused a fragmentation of forest cover (Kaiser, 2000; Katul and Albertson, 1999; Keller *et al.*, 2001). About 58% of coral reefs are seriously affected by fisheries, tourism, and pollution. Almost 65% of arable lands have partially lost their fertility. The scales of economic usage of ground waters exceed the rate of their natural recovery by at least $160 \cdot 10^9$ m^3/yr. In most cases the anthropogenic load on the ecosystems has intensified.

It is well known that the main causes of ecosystem degradation are the increasing size of the population and increased needs for natural resources as well as enhancing loads on the environments (Table 3.2). Concrete detailed data characterizing the present global situation can be found in numerous publications (Danilov-Danilian and Losev, 2000; Kondratyev, 1999a–c, 2000a; Ernst, 2000; Watson *et al.*, 2000; Kondratyev, 1998a,b). Note that an extremely important feature of the growing scales of consumption is the strong geographical non-uniformity in consumption reflecting the socio-economic contrasts in the world (Table 3.3).

3.1.2 The present state of individual ecosystems and the trends of their variability

Now we shall discuss the principal results obtained within the framework of the Pilot Analysis of the Global Ecosystems (PAGES). The main difficulties of this analysis are connected with the lack of available information. The PAGES project concentrates on the consideration of three types of indicators of impacts on ecosystems dynamics: (1) anthropogenic loads (increasing size of population, growing level of resources consumption, pollution, and over-exploitation of natural and controlled ecosystems); (2) the spatial extent of ecosystems (size, shapes, localization, and geographical distribution); and (3) production of

Table 3.1. Ecosystems and their functions of life support.

Ecosystem	Human life support	Regulation of the state of the environment and human life support
Agricultural	Food (agricultural crops) Fibre (agricultural crops) Genetic diversity of agricultural crops	Maintaining the functions of watersheds (infiltration, regulation of run-offs, and partial protection of soils) Provision of habitat for birds, pollination processes, and functioning of soil organisms. Formation of soil organics. Accumulation of carbon from the atmosphere. Provision of population employment.
Coastal	Fish and mollusks. Fish as food for animals. Algae (as food and raw material for industry). Salt. Genetic resources.	Some impact on storms (mangrove thickets, "barrier" islands). Habitat for wild fauna and flora. Biodiversity maintenance. Solution and processing of wastes. Provision of harbours and routes for water transport. Habitation for humans. Provision of population employment. Provision of places for recreation and aesthetic satisfaction.
Forest	Wood for construction. Wood fuel. Water for drinking and reclamation. Non-wood products (wine, bamboo, leaves, etc.) Food (honey, mushrooms, fruit, etc.). Genetic resources.	Removal of atmospheric pollutants and emission of oxygen. Nutrient cycles. Maintaining the functions of watersheds (infiltration, water cleaning, regulation of run-off, and stabilization of soil properties). Biodiversity maintenance. Accumulation of carbon from the atmosphere. Some weakening of extreme impacts of weather. Soil formation. Provision of habitat for humans and wild animals. Provision of places for recreation and aesthetic satisfaction.

Freshwater	Water for drinking and reclamation. Fish. Hydroelectric energy. Genetic resources.	Regulation of water regime. Solution and removal of wastes. Cycles of nutrient salts. Biodiversity maintenance. Provision of aquatic habitat. Provision of water transport routes. Provision of population employment. Provision of places for recreation and aesthetic satisfaction.
Grass	Domestic animals. Water for drinking and reclamation. Genetic resources.	Maintaining the functions of watersheds (infiltration, water cleaning, regulation of run-offs, stabilization of soil properties). Cycles of nutrients. Removal of atmospheric pollutants and formation of oxygen. Biodiversity maintenance. Soil formation. Accumulation of carbon from the atmosphere. Provision of habitat for humans and animals. Provision of places for recreation and aesthetic satisfaction.

Table 3.2. Anthropogenic impacts on ecosystems.

Ecosystems	Anthropogenic impacts	Causes of impacts
Agricultural	Use of agricultural lands for urban constructions or industrial purposes. Pollution of inland waters due to nutrients, run-offs, and silting. Impoverishment of water resources as a result of irrigation. Reduction of soil fertility due to erosion, change of agricultural practice, or decreased nutrient input. Weather change.	Increasing size of population. Increasing needs for food and industrial products. Urbanization. Governmental subsidies for agriculture. Poverty and unemployment. Climate changes.
Coastal	Exhaustion of fish resources (exceeding the permissible catch levels). Transformation of wetlands and coastal habitats. Water pollution by agricultural and industrial sources. Fragmentation and destruction of natural barriers by tides and reefs. Intrusion of alien species. Possible sea level rise.	Increasing size of population. Increasing needs for food and the growth of tourism. Development of urbanization and of recreational places. Governmental subsidies to fisheries. Inadequate information about the state of ecosystems, especially from the viewpoint of fishery conditions. Poverty and unemployment. Uncoordinated ecological policy in coastal regions. Climate changes.

Forest	Transformation or fragmentation of forest ecosystems as a result of either agricultural use or urbanization. Biodiversity losses due to fires and deforestation, emissions of stored carbon, and pollution of the atmosphere and water basins. Acid precipitation due to industrial pollution. Intrusion of alien species. Over-exploitation of water resources in their economic use.	Increasing size of population. Increasing requirement of wood. Governmental subsidies connected with deforestation and use of timber. Inadequate estimation of losses caused by industrial pollution of the atmosphere. Poverty and unemployment.
Freshwater	Over-exploitation of water sources in their acceptable levels. Breaking the acceptable levels of fish catch in inland water basins. Building of dams in the interests of irrigation, creation of hydroelectric power stations, and flood control. Pollution of water basins due to various factors of economic water use. Intrusion of alien species.	Increasing size of population. General deficit of water and natural non-uniform distribution of water resources. Governmental subsidies to maintain water use. Inadequate estimate of losses due to water basins pollution. Poverty and unemployment. Growing needs for hydro power stations.
Grass	Transformation or fragmentation due to either agricultural use or urbanization. Biodiversity losses due to fires, emissions of stored carbon, and atmospheric pollution. Reduction of soil fertility and water pollution by domestic animals. Unacceptable scales of the use of animals for entertainment.	Increasing size of population. Growing needs for agricultural products – especially for meat. Inadequate information on the state of ecosystems. Poverty and unemployment. Accessibility of grass ecosystems.

Table 3.3. Contrasts in per capita consumptions. Notation: *S* – total per capita consumption (US$, 1997); *F* – fish (kg, 1997); *M* – meat (kg, 1998); *G* – grain (kg, 1997); *P* – paper (kg, 1998); *R* – fossil fuels (kg, equivalent oil, 1997); *A* – cars (per 1000 people, 1996).

Country	*S*	*F*	*M*	*G*	*P*	*R*	*A*
USA	21,680	21.0	122.0	975.0	293.0	6902	489.0
Singapore	16,340	34.0	77.0	159.0	168.0	7825	120.0
Japan	15,554	66.0	42.0	334.0	239.0	3277	373.0
Germany	15,229	13.0	87.0	496.0	205.0	3625	500.0
Poland	5087	12.0	73.0	696.0	54.0	2585	209.0
Trinidad/Tobago	4864	12.0	28.0	237.0	41.0	6394	94.0
Turkey	4377	7.2	19.0	502.0	32.0	952	55.0
Indonesia	1808	18.0	9.0	311.0	17.0	450	12.2
China	1410	26.0	47.0	360.0	30.0	700	3.2
India	1166	4.7	4.3	234.0	3.7	268	4.4
Bangladesh	780	11.0	3.4	250.0	1.3	67	0.5
Nigeria	692	5.8	12.0	228.0	1.9	186	6.7
Zambia	625	8.2	12.0	144.0	1.6	77	17.0

economically important products such as agricultural crops, wood, fish, etc. A considerable drawback of each and every one of these indicators is that they do not contain information about the thresholds of an ecosystems' ability to perform its functions of life support.

Figure 3.1 presents expert assessments of the state of ecosystems and their functioning as life support systems including their respective trends. The results of the assessments shown in Figure 3.1 testify, on the whole, to considerable anthropogenic forcings on ecosystems (especially during the 20th century) that have led to considerable changes in their ability to perform the life support functions. Although in some cases (e.g., food production) the ecosystem productivity has substantially increased, in providing the required level of production, in other cases (drinking

	Agricultural	Coastal	Forests	Water	Grass
Food/fiber	↘ 2	↘ 3	↗ 2	↑↓ 2	↘ 3
Water quality	↘ 4	↓↑ 3	↘ 3	↘ 4	6
Water amount	↘ 3	6	↘ 3	↘ 3	6
Biodiversity	↘ 4	↘ 3	↘ 4	↘ 5	↘ 3
Carbon supplies	↑↓ 3	6	↘ 3	6	↘ 2
Recreations	6	? 2	6	6	↘ 2
Protection of coasts	6	↘ 4	6	6	6
Wood fuel	6	6	? 3	6	6

Figure 3.1. Expert estimates of the levels of anthropogenic forcing on ecosystems and their consequences. Notation: increasing ability of ecosystem to perform life support functions (↗), a reduced ability (↘), mixed trends (↑↓), unknown trends (?). The numbers characterize the levels of ecosystem life support functioning compared with the situation 20–30 years ago: 1 – excellent; 2 – good; 3 – satisfactory; 4 – weak; 5 – unsatisfactory; 6 – no estimate.

water quality, biomass preservation, etc.) there has been a degradation of the ecosystem functioning. Let us now consider these results in detail (see also Kondratyev, 1998b, 1999a–c; Ernst, 2000).

3.1.2.1 Food production

The state of the present agricultural ecosystem responsible for 95% of protein due to cultivated plants and animals and 99% of calories consumed by humans is contradictory. On the one hand, the agricultural yields have increased, but on the other hand, in most countries the quality of the agricultural ecosystems has worsened. So, for instance, a decrease of the natural soil fertility is recorded on 65% of agricultural land. In spite of this, a wide application of fertilizers, irrigation, and various new technologies (seeds, pesticides, etc.) have more than compensated for the impact of deteriorating conditions of agricultural ecosystems, and this trend will remain, apparently, in the near future (for how long it is unclear).

Much more problematic is the fishing situation. The present state of coastal marine ecosystems can be described as no better than satisfactory and the situation is worsening. In 28% of the areas of water bodies in the most intensive fishing regions, fish supplies are either decreased due to over-fishing or are only now beginning to recover from previous over-fishing periods. In 47% of water basins the fishing levels are close to the maximum permissible biological levels. Unfortunately, the information on the potential supplies of biomass in the ocean is contradictory. The general laws of distribution of the organic life in the World Ocean and in various seas have been more or less studied. Phytoplankton, the source of productivity for other trophic levels of the ocean, is known to develop mainly in the upper 100-m layer with 65% of the zooplankton biomass and >90% of nectone being located between the surface and 500-m deep. The benthos animals habitat is mainly found at a depth up to 200 m (about 80%). Considering productivity, the World Ocean is non-uniform. Less than 25% of its area is occupied by ecosystems with plankton biomass above 200 mg/m^3. More than 50% of the area is occupied by unproductive ecosystems (<50 mg/m^3).

The situation in freshwater basins is rather contradictory. On the one hand, there is excessive fishing, but on the other hand, natural fish supplies are constantly being replenished through fish-breeding. The scale of fish-breeding in ponds is constantly increasing. On the whole, an increasing dependence of fish catch on the aquacultural production and a decrease of natural fish supplies threatens the population of the developing countries, where aquacultural production is absent.

3.1.2.2 Amount of water

The water resources of the biosphere have been systematized in detail from the viewpoint of both supplies and quality. The main water fluxes between the oceans and land as well as water volumes of rivers, lakes, sub-soil, and soils have been estimated. On the whole, the global freshwater supplies are not great. Annually all the rivers bring to the World Ocean about $50 \cdot 10^3$ m^3 of water, which is only 1.5 times greater than the volume of the water in Lake Baikal or the American Great Lakes.

Continental water supplies are distributed non-uniformly and not in proportion to the population density (Gleick, 1993).

The construction of dams and other forms of river course regulation have seriously changed conditions both for people and for the water ecosystems. The level of economic use of river water has reached, on average, about half the available resources (with this situation strongly differing in various countries). Building of dams and other engineering constructions has led to a considerable fragmentation of about 60% of the largest river systems in the world. For instance, as a result of this fragmentation, the time needed for river waters to reach their respective seas has tripled on average. The regime of inland waters has changed due to deforestation and the drying of wetlands. This has especially manifested itself in the tropics where the forests are a key factor of water regime dynamics. The global area of freshwater wetlands has almost halved, which has had an effect on water supplies and flood regimes.

The main causes of the ongoing deterioration of the quality of natural waters are chemical pollution of water basins and input of nutrients, a reduced ability of the ecosystem to perform water filtration, and intense soil erosion due to land use mismanagement (Watson *et al.*, 2000). Pollution by nutrients resulting from the run-off of fertilizers into river water is a serious problem for all agricultural regions in the world. This kind of pollution causes eutrophication of waters in lakes and near-shore regions of the sea (this is especially observed in the Mediterranean and Black Seas, in the north-western part of the Gulf of Mexico, and in many large lakes). During the last two decades a dangerous florescence of water basins took place due to the input of nutrients. In many cases the ability of freshwater and marine coastal ecosystems to maintain the required quality of natural waters has been broken. Although, on the whole, in industrial countries the quality of natural inland waters has somewhat improved, the situation in developing countries is quite different, especially near large cities and in industrial regions. The worsening quality of water most noticeably effects the poorer populations who do not have access to pure drinking water.

3.1.2.3 Carbon cycle

To avoid repetitions, we remind the reader that only plants and soil micro-organisms favour both the removal of carbon dioxide from the atmosphere and its storage in their tissues, consequently slowing down the anthropogenic increase of CO_2 concentrations in the atmosphere. It is important to emphasize in this context that an intensified exploitation of ecosystems as sources of food and other means of life support negatively affect the ability of the ecosystems to accumulate carbon. This takes place, in particular, when forests are substituted with agricultural ecosystems (or even with secondary forests) with a much weaker potential for carbon accumulation. This change in land use creates a considerable source of CO_2 emissions entering the atmosphere – up to 20% compared to the total anthropogenic emissions of CO_2.

Despite these circumstances, the ecosystems still remain a considerable reservoir of carbon that play an important role in the formation of its global cycle. Tropical and boreal forests where carbon is accumulated, respectively, in the vegetation cover and in soils (especially in peatbogs) are the main reservoirs of carbon on land. At present 38–39% of carbon are concentrated in the forests and 38% in the grass ecosystems. The share of carbon in the agricultural ecosystems is 26–28% (Stern *et al.*, 1998; Smith and Shugart, 1993; Smith, 2002; Scholes and Noble, 2001; Schlesinger, 1999; Rosenberg and Izaurralde, 2001; Bazzaz, 1998).

Between 1850 and 1998 the concentration of CO_2 in the atmosphere increased from 285 ppm to 366 ppm, that is by about 30%. This has mainly resulted from fossil fuel burning, but changes in land use, that have led to carbon emissions into the atmosphere by the vegetation cover and soils, have also contributed significantly. About 33% of the carbon accumulated in the atmosphere during the last 150 years, has been caused by deforestation and changes in land use. The emission process of carbon dioxide and other greenhouse gases (GHGs) into the atmosphere is interactive with respect to climate change. So, for instance, the climate warming in the Arctic could lead to accelerated decomposition of gigantic permafrost masses in the ecosystems of tundra and taiga (permafrost melting should be followed by powerful emissions into the atmosphere of an important greenhouse gas – methane). An important aspect of the problem of increasing CO_2 concentrations is the effect of fertilization – stimulating the growth of some species of vegetation in conditions of increased CO_2 concentrations. In particular, in middle and high latitudes one can expect a 5% increase in crop yield but at low latitudes (mainly in Africa) a 10% decrease in crop yield is possible.

As for the global nitrogen cycle, the impact of human economic activity has turned out to be even more substantial and biologically significant. For most of the species of natural vegetation a deficit of nitrogen is the basic limiting factor for their development (therefore nitrogen fertilizers are used). A growing increase in the use of fertilizers, fossil fuel burning, and deforestation have resulted in ecosystems using nitrogen in amounts significantly exceeding natural levels. This has caused serious problems of eutrophication (especially for freshwater and coastal ecosystems) because an excess of nitrogen stimulates the development of "parasitic" algae which produce a deficit of oxygen effecting aquatic organisms, including fish populations.

3.1.2.4 Water cycle

As has been mentioned above, the level of anthropogenic impacts on water systems has become rather substantial. This is illustrated, for instance, considering that almost half the river run-off is used for economic (mainly agricultural) needs, a figure which may reach 70% by the year 2025. The transport of fresh water from rivers and other reservoirs to agricultural fields ensures an increase of the crop yields, but in this case various natural ecosystems and water users located downstream suffer serious damage. In many cases water does return to the rivers but is strongly polluted and, as a rule, unfit for further consumption. The best example of such processes is a change of the water regime in the Aral Sea basin.

3.1.2.5 Biodiversity

The biodiversity losses during the last century have acquired threatening scales and concern to practically all ecosystems (mainly as a result of the loss of habitats). For example, forested areas have probably decreased in size (on a global scale) by 50%. Some forest ecosystems (like the dry tropical forests of Central America) have disappeared completely. In many countries up to 50% of mangrove plantations have been lost, areas of wetlands have almost halved, and in some regions the size of territories covered with grass has decreased by 90%. Only the tundra, arctic, and marine deep-water systems show little change, although in these cases substantial anthropogenic impacts also occur sometimes.

Even when the initial areas of the ecosystems have been preserved, many species are threatened with pollution, over-exploitation, intrusion of "alien" species, and degradation of habitats. From the viewpoint of biodiversity, the freshwater ecosystems have suffered most. For instance, about 20% of the freshwater fish species have been exterminated, and many other species are under threat of extinction. The forest, grass, and coastal ecosystems face serious problems of survival. The biodiversity is threatened with frequented diseases of marine organisms, intensive florescence of water bodies, and a considerable reduction of amphibian population. Apart from the respective losses for medicine (due to a decreasing volume of raw materials for drugs), banks of genes, and eco-tourism, a reduction of biodiversity also means a threat to ecosystem productivity, integrity, and resistance to various external forcings.

Finally, it is clear that there are now numerous signs of anthropogenically caused degradation of ecosystems resulting in their reduced ability to function as systems of life support and controllers of environmental properties. The data discussed above shows that all the ecosystems considered suffer serious and increasing anthropogenic forcings. Intensive processes of transformation of natural ecosystems take place due to deforestation, pollution (including, nutrients), dam construction, and biological intrusions of alien organisms. Anthropogenic forcings result in considerable transformations of the global biogeochemical cycles, whose normal functioning is a guarantee of ecosystem safety.

So far, the negative trends of the state of global ecosystems do not vividly threaten the high levels of production of various products and services. The food and fibre production has never reached such high levels as today, and dam construction has ensured an unprecedented regulation of the river run-off providing water for the interests of agriculture. However, the progress in creating the material well-being is fraught with a long-term prospect of reduction of productivity of global ecosystems and their reduced ability to regulate environmental quality. The use of new technical means and technologies such as synthetic fertilizers, improved means of fishing, and irrigation systems has masked a decrease of bioproductivity of natural ecosystems, which ensures the required growth of food and fibre production. As for the long-term prospects (and it is in this context that sustainable development is defined), one should consider the reduced functioning of natural systems of life support, which manifests themselves, for instance, in the form of a dangerous decrease of biodiversity, reduced quality of drinking water, intensive emissions of greenhouse gases into the atmosphere, and many other negative phenomena.

In the light of the World Summit on Sustainable Development (WSSD) in Johannesburg (2002) ("Rio+10"), of key importance is a substantiation of priorities whilst considering the problem of the functioning of the "society–nature" interactive system. The acuteness of the situation becomes especially apparent when we consider the limited success of the UN Second Conference on Environment and Development (Rio de Janeiro, 1992), the Special UN Session "Rio+5", and the complete fiasco of the 6-year series of annual conferences of the representatives of the governments signing the UN Framework Convention on Climate Change (FCCC) that served as the basis for the Kyoto Protocol. Speculative exaggerations (sometimes reaching the level of apocalyptic global predictions) characteristic of the concept of "global warming" have diverted attention (and to a considerable extent the financial resources) away from the most significant problem of global ecosystems safety within conditions of an increasing size in the global population and an intensifying of anthropogenic forcing on natural systems of life support. In this regard the authors of *A Guide to World Resources 2000–2001* draw attention to the availability of the "ecosystem approach" founded on the fact that the use of natural resources should be based on a consideration of the ability of the global natural systems of life support to function normally.

The ecosystem approach has the following main features:

- Complex (systematic) character of analysis of interactions within the "society–nature" system;
- not fragmentary but integral analysis of ecosystem functioning;
- consideration of all the spatial–temporal diversity of the processes with the indispensable priority of long-term prospects;
- a detailed consideration of interactive processes in the "society–nature" system; and
- justification and accounting of the limits of the normal functioning of ecosystems, and in this regard, permissible anthropogenic loads.

The data in Table 3.4 illustrate the differences between traditional and ecosystem approaches to the use of natural resources as applied to forestry. Of course, the advantages of the ecosystem approach are not in doubt. The problem is, however, to what extent the conditions of the socio-economic development of a country permit a successive realization of the ecosystem approach. The negative manifestations of the dynamics of various ecosystems reflect vividly great difficulties (sometimes insurmountable) in realizing the ecosystem approach. In this regard the participants of the PAGE project have formulated the following general recommendations:

- to stimulate the development of science and observational means;
- to realize and quantitatively estimate the value of functions performed by ecosystems;

Table 3.4. Differences between traditional and ecosystem approaches to studies of natural resources with forestry as an example.

	Traditional approach	Ecosystem approach
Goal	Maximum possible production of consumer goods. Maximizing the value of produce. Maintaining the state of the forests or the use of forest products at levels equal to or below those ensuring the growth or restoration of the forests.	Maintenance of forest ecosystems as an interconnection component of the natural medium with provided stable production of consumer products. Preserving the possibility of choice in the future. Providing a persistent productivity of ecosystems taking into account forest aesthetics and the social acceptability of forestry practices.
Scale	Works on the level of forest stands within the boundaries of the states or property.	Works on the levels of ecosystems and landscapes.
Role of science	Consideration of problems of forestry management as the applied science.	Forestry management with scientific data and social factors taken into account.
Role of management	Special attention to end products (production of goods and services). Main problems: provision of needs for timber, recreation, environment protection, and fodder. Ensuring the required industrial production. Priority of wood production. Desire to avoid a wood deficit. Consideration of forests as systems of wood production. Assessment of the economic efficiency of forestry.	Emphasis on the safety of forest ecosystems with an account of required soil properties, biological diversity, and ecology on the whole in the interests of raising the level of production of goods and services. Ensuring the solution of problems in protecting nature. Complex approach to solution of the production problems considering similar importance of the problems of nature protection. Desire to avoid losses of biodiversity and soil degradation. Approach to forests as safe natural systems. Assessment of economic profits and social acceptability.

- to stimulate a broad discussion on the goals, scientific foundations, and practical importance of the ecosystem approach;
- to provide the participation of all needed specialists in the solution of the problems of the ecosystems management.

The scientific understanding of the laws of ecosystem functioning and, moreover, the possibility of predicting ecosystem dynamics are still in the initial phases of development. To a great extent this is explained by inadequate complex information about the ecosystems and their functioning. So, for instance, the problem of the global carbon cycle has attracted serious attention for many years (Kondratyev, 1998b, 1999c; Demirchian and Kondratyev, 1999; Kondratyev and Demirchian, 2000; Watson *et al.*, 2000), nevertheless, even the system requirements have not been formulated yet for complex observational data, which would make it possible to also obtain adequate information about the laws of the carbon cycle (of course, this requires a combined use of both ground-based and satellite observations). In this regard, the UN Secretary-General Kofi A. Annan justly noted:

> It is impossible to justify an efficient ecological policy if it is not based on reliable scientific information. Though in many spheres a considerable progress has been reached in obtaining the observational data, there remain considerable gaps in our knowledge. In particular, so far, no complete global assessment of the state of the main ecosystems existing over the globe has been obtained. A response to this necessity was the planned Millenium Ecosystems Assessment (MEA) as an important international initiative aimed at mapping the "health" of our planet. This initiative has been backed by many Governments, as well as by the United Nations Environmental Programme (UNEP), United Nations Development Programme (UNDP), Food and Agriculture Organization (FAO), and United Nations Educational, Scientific and Cultural Organization (UNESCO). I call upon the Governments – UN members to give the needed financial support in assessing the state of ecosystems for the last Millenium and take an active part in the programme.

Realization of the MEA programme started in 2001 and marked the first step toward an assessment of the state of global ecosystems, as well as development and realization of measures needed to preserve the ecosystems on the planet at a time when the human need for natural resources was growing. Furthermore, there appeared the acute problem of preserving not simply the quality of the environment but also normal functioning of life-support systems on the Earth, without which further development of civilization is impossible.

3.2 THE ROLE OF FOREST ECOSYSTEMS

Over many years, the problem of the impact of atmospheric carbon dioxide on the global climate has been discussed both by scientists and politicians. Some people believe that humankind will inevitably change the climatic situation on the Earth due to an enhanced greenhouse effect, which will change living conditions (probably, for the worst). Therefore, it is necessary to reduce industrial emissions of CO_2. Others, agreeing with the consequences of the greenhouse effect, deny the strategy put forward by the Kyoto Protocol and believe that the recommended reductions (quotas) will lead to an aggravation of the economy in many regions of the globe, without solving the problem of the greenhouse effect, but further worsening the global ecological situation. The opponents to the Kyoto strategy think that the greenhouse effect can only be prevented by the correct management of the structure of surface land cover and by introducing a strict control of the World Ocean pollution (Kondratyev, 1998b; Kondratyev and Demirchian, 2001a,b). In this regard, the Intergovernmental Panel on Climate Change (IPCC) at the 8th Session in June 1998 in Bonn and at the 14th Session in October 1998 in Vienna prepared a special report on the role of the strategy of using the surface land cover (forests, in particular) in the global balance of CO_2. This report discusses the problems of interaction between the anthropogenic activity in the field of surface land cover reconstruction and the distribution of CO_2 and other GHGs in the biosphere. An assessment was given of various scenarios following from the Kyoto Protocol and concerning the problem of the impact of human society on the surface land cover structure in general and on forested territories, in particular. A brief analysis of this report is given below.

3.2.1 The greenhouse effect and forest ecosystems

Item 3.1 and Appendix 1 of the Kyoto Protocol foresee a limitation and then a reduction of GHGs emissions during the period up to 2008–2012. Before this time some problems should be solved to assess the role of the use of the Earth's surface. In particular, among these problems is the problem of the formalized description of the processes of changes of the Earth's land cover structure, such as afforestation, forest reconstruction, deforestation and the associated carbon supplies. Understanding the meteorological processes as functions of GHGs refers to one of the key problems of humankind in the first decade of the third Millenium. Only an adequate knowledge of the meteorological phenomena of various spatial–temporal scales in conditions of supplies of CO_2 and other GHGs will enable us to make an accurate and constructive decision in the field of global environment protection.

The dynamics of the surface ecosystems depends on interactions between biogeochemical cycles, which during the last decade of the 20th century suffered from an anthropogenic modification. This especially refers to the cycles of carbon, nitrogen, and water. The surface ecosystems, in which carbon remains in living biomass, decomposing organic matter, and soil, play an important role in the global CO_2 cycle. Carbon exchanges between these reservoirs and the atmosphere take place through photosynthesis, respiration, decomposition, and burning. Human interference in these processes takes place through changing the structure of the Earth's land cover, pollution of water basin surface and soil areas, as well as through direct emissions of CO_2 into the atmosphere (Rojstaczer *et al.*, 2001; Qi *et al.*, 2001; Potter and Klooster, 1999).

The role of various ecosystems in the formation of carbon supplies to the bio-

Table 3.5. Carbon supplies in vegetation and in a 1-m soil layer (Watson *et al.* 2000).

		Carbon supplies (Gt C)		
Biome	Area (10^9 ha)	Vegetation	Soil	Total
Tropical forests	1.76	212	216	428
Temperate zone forests	1.04	59	100	159
Boreal forests	1.37	88	471	559
Tropical savannahs	2.25	66	264	330
Temperate zone meadows	1.25	9	295	304
Deserts and semi-deserts	4.25	8	191	199
Tundra	0.95	6	121	127
Marshes	0.35	15	225	240
Sown areas	1.60	3	128	131
Total	*15.12*	*466*	*2011*	*2477*

spheric reservoirs determines the rate and direction of changes in regional meteorological situations and in global climate. An accuracy of assessment of the level of these changes depends on the reliability of the data on the surface ecosystem inventory. Table 3.5 data show that a considerable scattering of the estimates of carbon supplies in various types of vegetation suggests the conclusion that it is important to classify the surface ecosystems more specifically.

The anthropogenic constituent of the global carbon budget, beginning from the mid-19th century, increases the scale of its effect practically on every natural component. From 1850 to 1998 about 270($\pm$60) Gt C were emitted as CO_2 into the atmosphere due to fuel burning and cement production. About 136($\pm$55) Gt C entered the atmosphere as a result of anthropogenic reconstruction of surface land cover. This has led to an increase of atmospheric CO_2 by 176($\pm$10) Gt C, that is, the partial pressure of carbon dioxide in the atmosphere has increased from 285 to 366 ppm (by 28%). In other words, during this 148-yr period 48% of carbon emitted to the atmosphere remained there and was not assimilated by any land surface or ocean ecosystem (230($\pm$60) Gt C were assimilated).

Some idea about the global carbon budget can be obtained from the data in Table 3.6. This table shows that the rates and trends of carbon accumulation in the surface ecosystems are rather uncertain. However, it is clear that the surface ecosystems are important assimilators of excess CO_2. Understanding the details of such assimilation is only possible through modelling the process of the plant growth, that is, considering the effect of the nutrient elements of soil and other biophysical factors on plant photosynthesis.

According to Table 3.5, the forest ecosystems and associated processes of natural afforestation, forest reconstruction, and deforestation should be studied in detail. The same has been emphasized in Items 3.3 and 3.4 of the Kyoto Protocol, where the necessity to determine national and international strategies of forest management is emphasized. In a forest range, the volume of the reservoir for CO_2 coming from the atmosphere is a function of the density of its canopy, and in a given

Table 3.6. Characteristics of the mean annual CO_2 budget (Watson *et al.*, 2000; Lee *et al.*, 1998).

Characteristic	Estimate (Gt C/yr)	
	1980–1989	1990–1998
(1) CO_2 emission due to fossil fuel burning and cement production	5.5 ± 0.5	6.3 ± 0.6
(2) CO_2 accumulation in the atmosphere	3.3 ± 0.2	3.3 ± 0.2
(3) Assimilation by the oceans	2.0 ± 0.8	2.3 ± 0.8
(4) $(1) - [(2) + (3)]$	0.2 ± 1.0	0.7 ± 1.0
(5) CO_2 emission due to changes in the use of land resources	1.7 ± 0.8	1.6 ± 0.8
(6) $(4) + (5)$	1.9 ± 1.3	2.3 ± 1.3

time period, a change of this volume is determined by the level and character of the dynamic processes of the transition of a given type of forest to another state. The causes of this transition can be natural, anthropogenic, or a mixture of both. Biocoenology tries to create a universal theory of such transitions, but so far, there is only a qualitative description of the observed transitions. As mentioned in the Kyoto Protocol, of importance is the correct definition of the notions "afforestation, forest reconstruction, and deforestation". Afforestation means to forest a land area used before (for 20–50 yr and longer) for other purposes. Usually this term determines the process of natural succession at the expense of propagation of forest over other territories without human interference. The process of forest reconstruction is defined as planting trees. Deforestation is a substitution of the forest territory for another ecosystem. Thus two opposite processes are possible in the forest ecosystem dynamics that can be controlled by both nature and people. Each of these processes has its versions characterized by special dynamics of the vegetation over a given territory. Of special status is the process of foresting a territory where historically trees had never grown. In this case this territory immediately becomes important in CO_2 dynamics.

Table 3.7 illustrates the impact of the afforestation/deforestation processes on carbon supplies following the FAO scenario (Watson *et al.*, 2000) where the forest is a land area not less than 0.5 ha, with trees more than 5 m high with a canopy covering more than 10% of the area. Deforestation is determined as a change of the surface cover with its canopy covering less than 10% of the area, as well as a change of the class of the forest with negative consequences (e.g., a decrease of productivity). Afforestation is the planting of trees over the area where trees had never grown. Note that "natural broadening" (i.e., propagation of forest over the agricultural territories without human interference) due to the FAO scenario is also referred to as the process of afforestation. Finally, forest reconstruction is a direct planting of trees on the territories earlier covered with forest.

The technology of considering the scenarios of the type given in Table 3.7 makes it impossible to choose the scenario to be recommended for use. An approach

Table 3.7. Assessment of the calculated change of the mean annual carbon supply for the afforestation/deforestation scenario (Watson *et al.*, 2000).

			TR		*FR*	
Region	*RF*	*AF*	*A*	*B*	*A*	*B*
Boreal	35	0.4–1.2	0.5	0.1	−18	−185
Temperate	60	1.5–4.5	2.1	1.9	−90	−501
Tropical	120	4–8	13.7	2.6	−1644	−1352

Notation: *A* – deforestation; *B* – afforestation; *RF* – change of average carbon supply after deforestation, t C/ha; *AF* – average rate of CO_2 assimilation at afforestation, t C/ha/yr; *TR* – change of area (10^6 ha/yr) resulting from the deforestation-to-afforestation transition; *FR* – forecast of changes in carbon supplies (10^6 t C/yr) in 2008–2012 after the FAO scenario.

suggested in the Kyoto Protocol is oversimplified for reliable assessments of the CO_2 dynamics as a function of numerous natural and anthropogenic parameters.

3.2.2 Global cycles of GHGs

The fact that differences in the estimates of the global carbon cycle by various authors reaches 100%, demonstrates an impossibility, with the present knowledge, to reliably simulate the global CO_2 dynamics and, hence, identify objective causes to make decisions (recommendations) similar to those of the Kyoto Protocol.

Carbon dioxide, which dissolves in the oceans and is assimilated by vegetation, leaves the atmosphere through a series of complicated mechanisms of the biospheric cycle of substances and energy. In the oceans such a mechanism is the carbonate system which, with a certain combination of environmental parameters, draws off the atmospheric CO_2 and sends it to the bottom sediments. The fate of CO_2 fixed on land depends on which ecosystem and which reservoir of carbon is acting as the assimilator (e.g., living biomass or soil). Carbon fixed in a reservoir in a year or less (for leaves and roots) either returns to the atmosphere or remains in the reservoirs with characteristic cycle times of decades to centuries (for trunks, stems, branches, organic matter). As a result, a pure carbon flux forms between the surface ecosystems and the atmosphere, introducing a slight instability between assimilation in photosynthesis and the various processes of carbon return to the atmosphere. Plants, soil micro-organisms, biochemical processes, and animals contribute to this instability (Post *et al.*, 2001; Pacala *et al.*, 2001; Niles and Schwarze, 2001; Ni, 2001; Nabuurs and Sikkema, 2001). Climate change and human activity contribute to this process through transforming the structure of the surface ecosystems and polluting the environment. From the available estimates, pure assimilation of CO_2 by the surface biosphere constituted 0.2–0.1 Gt C/yr during 1980–1989 and 0.07–0.6 Gt C/yr during 1990–1998 (see Table 3.6). At the same time, because of anthropogenic interference into the structure of surface land cover, the pure emission of CO_2 reached 1.7–0.8 Gt C/yr and 1.6–0.8 Gt C/yr, respectively. On the whole, variations of the global carbon cycle are estimated at ±1 Gt C/yr, and a clear correlation is

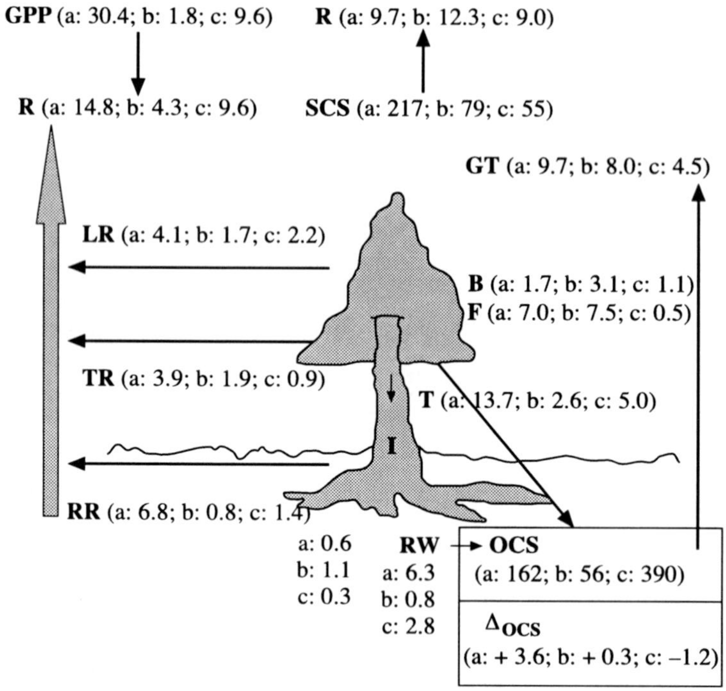

Figure 3.2. Annual supplies (t C/ha) and fluxes (t C/ha/yr) of carbon for tropical (a), temperate (b), and boreal (c) forests (Watson *et al.*, 2000). Notation: RR – roots respiration; TR – trunk respiration; LR – leaves respiration; T – transport of carbon from the surface part of the tree to its roots; B – pure increment of surface biomass; GPP – gross primary production; F – falling of leaves and dead biomass; I – pure increment of root biomass; RW – root withering; R – respiration; OCS – organic carbon of soil; Δ_{OCS} – pure increment of OSC; SCS – surface carbon supply; GT – gross transpiration.

observed between variations and phenomena such as El Niño or volcanic eruptions (Hernandez *et al.*, 2001; Jones and Cox, 2001). The total carbon flux assimilated by plants from the atmosphere reaches approximately 120 Gt C/yr. Part of the carbon rapidly returns to the atmosphere due to transpiration (~60 Gt C/yr) and decomposition of dead organic matter (~10 Gt C/yr). Forests contribute most to this balance (see Table 3.5). As seen from Figure 3.2, various forests have quite different internal and external supplies and fluxes of carbon. Hence, the estimates of carbon fluxes averaged over large territories do not even give approximate patterns of changes of its supplies to biospheric reservoirs. Therefore numerous carbon circulation models based on these data cannot be an objective instrument for decision making.

A similar conclusion is valid for other GHGs, such as CH_4 and N_2O (see Tables 3.8 and 3.9) (Isidorov and Kondratyev, 2001). The concentration of methane increased from 700 ppbv in the pre-industrial epoch to 1700 ppbv

Table 3.8. Global estimates of CH_4 sources (Prather *et al.*, 1994).

CH_4 source	10^6 t CH_4/yr	10^9 t C/yr
Domestic animals (ferments, wastes)	110 (85–130)	0.6 (0.5–0.7)
Rice plantations	60 (20–100)	0.3 (0.1–0.6)
Biomass burning	40 (20–80)	0.2 (0.1–0.5)
Natural heavily moistened soils	115 (55–150)	0.7 (0.3–0.9)

Table 3.9. Global estimates of N_2O sources (Prather *et al.*, 1994).

N_2O source	10^6 t N_2O/yr	10^9 t C/yr
Cultivated soils	3.5 (1.8–5.3)	0.9 (0.5–1.4
Biomass burning	0.5 (0.2–1)	0.1 (0.05–0.3)
Domestic animals (horned cattle and pastures)	0.4 (0.2–0.5)	0.1 (0.05–0.13)
Natural soils in tropical latitudes		
Wet forests	3 (2.2–3.7)	0.8 (0.6–1)
Dry savannahs	1 (0.5–2)	0.3 (0.1–0.5)
Natural soils		
Forests	1 (0.1–2)	0.3 (0.03–0.5)
Steppes	1 (0.5–2)	0.3 (0.1–0.5)

nowadays. About $550 \cdot 10^6$ t CH_4/yr are emitted into the atmosphere from various sources. From 1980–1990, the CH_4 concentration in the atmosphere increased at a rate of 13 ppbv/yr, but in 1990 and in 1996 the rates were 8 ppbv/yr and 4 ppbv/yr, respectively. The lifetime of methane is about 9 years. The contribution of anthropogenic sources of methane into its total flux in the atmosphere is estimated at 70%. Clearly, this share consists of not only direct emissions of methane from industry and agriculture, but also the contributions from the anthropogenic transformations of surface land cover.

The sources of another important GHG, N_2O, are also closely connected with the structure of the surface land cover. The global emission of N_2O is estimated at $14 \cdot 10^6$ t N_2O/yr, with the anthropogenic contribution to this reaching 50%. This estimate is strongly affected by agricultural lands, tropical forests, and especially by soil moistening. The microbiological processes in soils are the primary sources of N_2O.

The land use affects directly the processes of GHG exchange between the surface ecosystems and the atmosphere. Changes in land use, such as forest clearing to broaden the extent of agricultural fields or for building/expanding settlements cause synchronous (or slightly delayed) changes in the surface cover structure and in the distribution of carbon supplies to the surface ecosystems. These actions are connected with the necessity to increase food production, to create new areas for human habitation, and recreation. Therefore the problem of balanced regulation of the processes of land use becomes urgent. To solve this problem using the approaches suggested by the Kyoto Protocol is impossible. It is necessary to put into practice some methods of eco-informatics which make it possible to study the "nature–society" system as an integral system based on a complex global model that describes a totality of biogeochemical, biogeocoenotic, climatic, demographic, and socio-economic processes with their binding with spatial and temporal scales. The idea of such a model has been described in a monograph by Krapivin and Kondratyev (2002). One of the versions of the global model described by Krapivin (1993) uses a spatial geographical grid $4^\circ \times 5^\circ$ with detailed forest ranges, following the scheme in Figure 3.2. According to this model, the estimates of CO_2 assimilation drastically change the results of the scenario modelling given in Watson *et al.* (2000). Respectively, both prediction and conclusions are changing. Of course, it is necessary to take into account a variety of possible scenarios (Nabuurs *et al.*, 2000; Matsumura, 2000; O'Neill and Oppenheimer, 2002; Pielke, 2001a; Kok and Verweij, 1999).

3.2.3 Afforestation and deforestation

Let us return briefly to the problem of anthropogenic changes of the structure of forest ranges. In the Kyoto Protocol its solution is connected with the problem of definition of such notations as forest, afforestation, tree planting, deforestation, biome, surface cover, land use, degradation, etc. Anyhow, a question arises here: to what extent is the use of either definition justified at making a decision and how does the result of its realization depend on their quality? Not only does the Kyoto Protocol not answer this question, it does not even raise it. In this respect, Table 3.10 is demonstrative and serves as the basis for drawing various conclusions. In fact, a huge range of uncertainty exists here due to which of the processes of afforestation and deforestation cannot be estimated uniquely. Clearly, a single correct way to solve the principal problem and the respective partial problems is to make a detailed inventory of the forest ranges and to use the obtained data in the global model. The specific features of this inventory should be determined in an adaptive regime of using the model, gradually changing the spatial–temporal scales, starting from the known model with a mean-annual time step and the geographical grid $4^\circ \times 5^\circ$.

3.2.4 Conclusion

The problem discussed above is urgent from the viewpoint of the global changes of the environment both ecologically and economically. Combining these aspects, it is necessary to find an efficient mechanism for global forestry management. This problem cannot be solved independent of other nature protection problems. Apparently, even on a local level a decision about changing the Earth's cover should be made on the basis of assessments of the global consequences in the future. The technology of making such well considered decisions was proposed in Ugolnitsky *et al.* (2000) and was called Geoinformation monitoring system (GIMS) technology.

Table 3.10. Directions of carbon supplies changes with different forms of soil cover reconstructions (Watson *et al.*, 2000).

Type of land cover reconstruction	Direction of changes in carbon supplies					
	Biomass		Forest cover/wood			
	A	*U*	*S*	*L*	*Q*	*M*
Cultivated land → forest	↑	↑	–	↑	↑	↑
Uncultivated land → forest	↑	↑	–	↑	?	↑
Forest → cultivated land	↓↓	↓	↓	↓	↓	–
Forest → pasture	↓↓	↓	↓	↓	?	–

Notation: *A* – above ground; *U* – under ground; *S* – short-lived; *L* – long-lived; *Q* – organic matter; *M* – wood production.

Coming back to the Kyoto Protocol, note that despite numerous quite acceptable conclusions and scenarios, recommendations to introduce quotas on GHG emissions cannot be considered scientifically substantiated (Gorshkov and Makaryeva, 1999). After all, even the most optimistic assessments of the role of the forest ecosystems in CO_2 assimilation from the atmosphere fluctuate between 10 and 27%.

To solve the problem of the use of forests, coordinated with the dynamics of global changes, is only possible with the use of GIMS technology including the model for forest dynamics, which describes changes in the structure of forest areas and temporal variations as a result of natural growth of plantations and a complex network of external forcings.

From the viewpoint of humankind it would be worthwhile and advantageous to concentrate the efforts of experts in different sciences into dealing with the problem of global modelling, in a more organized manner (Kondratyev *et al.*, 2000).

3.3 THE IMPACT OF CARBON DIOXIDE ON PRIMARY PRODUCTION

The relationship between the global CO_2 cycle and surface vegetation is manifested through the dependence of primary production and the rate of decomposition of dead biomass on atmospheric temperature and CO_2 concentration. The temperature dependence is manifested most significantly in northern latitudes, where mean-annual temperature variations can reach high levels, and the vegetation period of plants varies from two to seven months. An impact of atmospheric CO_2 on the growth of plants depends on many factors. There are two basic types of plants if classified by the form of their reaction to changes in the partial pressure of atmospheric CO_2. The first, most widespread type of plants (type C_3) is characterized by photorespiration realized due to ferments, which can simultaneously assimilate and emit CO_2 and O_2. This process has the so-called compensation point Γ, when the balance of all functions of the plant with respect to CO_2 concentration (C_a) is optimal. This point is characterized by the value $\Gamma \cong 50$ mmol/mol at 25°C, and grows with increasing temperature, is proportional to the value $(C_a - \Gamma)$ up to the level 1000 mmol/mol, and its efficiency of the initial use of light increases with increasing CO_2 in proportion to $(C_a - \Gamma)/(C_a + \Gamma)$ (Goudrian *et al.*, 1990).

Plants of another type (C_4), such as high tropical grass (maize, sugar cane, millet, and sorghum), assimilate CO_2 from the atmosphere independently of the O_2 concentration, so that Γ remains practically constant at a low level $\cong$5 mmol/mol. These plants react weakly to changes in the concentrations of carbon dioxide.

Numerous laboratory studies of the reactions of both types of plants to changing values of C_a testify to a wide scattering of the quantitative estimates of variations in photosynthesis for the type C_3 plant. On average, the plants react to changes of CO_2 concentrations with a 1-month lag; a CO_2 doubling can cause an increase of the photosynthesis rate from 10 to 100%. Further increase of C_a up to 400% leads to an effect of photosynthesis saturation for some plants, that is, an addition to the rate of photosynthesis can reach 20%. In some cases (e.g., for *Setaria lutescens*) photosynthesis is oppressed (Bazzaz, 1998). Practically, all the plants of the type C_4 at present values of C_a are in a state of photosynthesis saturation.

The effects of CO_2 on plants are many, being manifested through changes of the nutrient regime of photosynthesis. Table 3.11 exemplifies this effect. Of course, the elemental composition of the plant's structure varies. It includes C, O, H, N, S, P, K, Ca, Mg, Fe, and the exchange processes in the "plant–atmosphere" system includes chemical compounds, such as CO_2, CH_4, H_2O, H_2S, NH_3, and NO_2. A living plant consists of 50–95% water, and the remaining part, the so-called dry substance, includes 70–98% of an organic substance, which will burn. In other words, each

Table 3.11. Changing content of the nutrient element in trees as a result of a 2-year impact of changed CO_2 concentrations for *Acer pseudoplatanus* (*A*) and *Fugus sylvatica* (*F*) (Bazzaz, 1986).

Chemical element	CO_2 concentration (mmol/mol)					
	370		520		670	
	A	*F*	*A*	*F*	*A*	*F*
N	1	1	1.09	1.11	1.17	1.21
P	1	1	1.10	1.33	1.20	1.50
K	1	1	1.15	1.21	1.27	1.33
Ca	1	1	1.17	1.13	1.25	1.23
Mg	1	1	1.14	1.22	1.29	1.22
Mn	1	1	1.00	1.08	1.07	1.25
Fe	1	1	1.43	1.17	1.54	1.50

plant on the Earth plays its own role in the global biogeochemical cycle of CO_2 and other chemical elements. Therefore all the existing models of the CO_2 cycle based on rough classifications of the soil–plant formations are incorrect, and their reliability can hardly be assessed from the available databases on the vegetation cover and its parameters.

Quite special are the tundra and forest–tundra biocoenoses, whose share is about 4% of the land surface. Their role in assimilation or emission of CO_2 is of a seasonal character. Tundra with its marshes, water basins, and lakes is a source of CO_2 for the arctic atmosphere (Lafleur *et al.*, 2001). The soils of the arctic tundra play a special role in this process. In the period of snow melting they emit carbon monoxide (CO), and above the soil surface the CO partial pressure can reach 100 ppm with a mean-annual value of 0.05 ppm (Kelley, 1987). The CO concentration in air bubbles produced in water bodies and lakes with decaying biomass is estimated at 5–20 ppm. In the spring, near the bottom of the tundra plants the CO partial pressure in the air reaches 40 ppm, markedly decreasing by the end of summer. As a result, in spring, the CO_2 partial pressure in the near-surface atmosphere of the arctic tundra can reach 2100 ppm. What happens in winter with the CO_2 exchange on the border "atmosphere–tundra" is practically unknown. But there are data on the photosynthetic and respiratory activity of the tundra vegetation, from which it follows that this activity is preserved even with illumination of 5–7 W/m^2 day and a temperature below zero (Tiezen, 1973). This means that the tundra vegetation in late summer and early winter can serve as a sink for atmospheric CO_2. The sink of CO_2 from the atmosphere due to assimilation by the tundra vegetation is estimated at 146 g/m^2/day (Kelley, 1987; Eliasson *et al.*, 1999). Therefore to assess the role of soil–plant formations in the global cycle of CO_2, it is necessary that the model has a sufficiently detailed spatial and objective classification of the different types of soil–plant formation. Unfortunately, there is no information of the form that could be introduced into the model, and therefore the published estimates of the role of the soil–plant formations considered in the global models of the CO_2 cycle cannot be seen as reliable. Although, for instance, the estimates of the surface vegetation role (after the classification by Bazilevich and Rodin (1967)) in CO_2 assimilation over the territory of the FSU countries given in the monograph by Krapivin and Kondratyev (2002) do seem realistic.

One of the promising results of modelling the effect of "fertilization" due to changing concentrations of atmospheric CO_2 has been discussed by Alexandrov and Oikawa (2002) who, based on the Tsukaba Biosphere Model (TsuBiMo), obtained a globally-distributed estimate of the contribution of various types of vegetation to the change of CO_2 flux on the "atmosphere–land" border. To compare this result with the Global Simulation Model (GSM) data, Figure 3.3 presents two curves of the dependence of pure primary production (PPP) of the surface vegetation on changes of the atmospheric CO_2 partial pressure. The difference between these curves shows that a consideration in the GSM of additional feedbacks compared with the simple TsuBiMo specifies this dependence of PPP on CO_2 partial pressure and recommends a more thorough structure of GSM units. Nevertheless, it should be borne in mind that despite an apparent simplicity of the parametric descriptions assumed in the

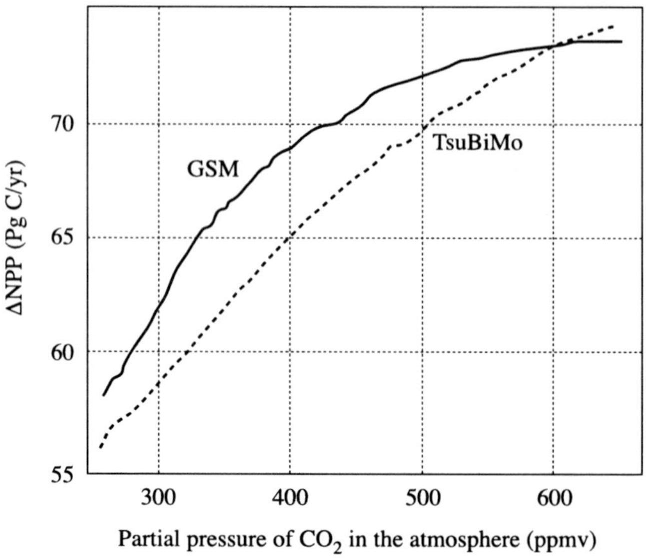

Figure 3.3. Comparison of two models of the global carbon cycle to assess the response of pure primary production of vegetation to changing concentrations of atmospheric CO_2. Notation: GSM – global simulation model (Kondratyev *et al.*, 2002); TsuBiMo – Tsukaba Biosphere Model (Alexandrov and Oikawa, 2002).

TsuBiMo, the result obtained with its use is reliable due to a detailed description of the spatial structure of vegetation cover (50 × 50 km). Therefore the strategy of modelling of the global carbon dioxide cycle should follow a paradigm of the complete description of all direct interconnections and feedbacks with the use of a set of functional dependences, the choice of which is made in the process of an adaptive-evolutionary numerical experiment.

3.4 CARBON EXCHANGE PROCESSES ON THE "LAND–ATMOSPHERE" BORDER

The interaction between two carbon reservoirs – the atmosphere and the land – is expressed by carbon fluxes formed through ecological, geophysical, and geochemical processes, including photosynthesis, respiration of plants and animals, decomposition of dead organic matter, vegetation and fuel burning, volcanic emanations, rock weathering, etc. Which of these processes prevails depends on many factors. Therefore in the scheme of Figure 2.8 and in Table 2.1 various processes are taken into account.

The most important aspect in studying the global carbon cycle is the contribution of the interaction between the surface vegetation and the atmosphere. This dependence is based on the fact that all plants create their biomass by assimilation of atmospheric constituents C, O_2, H, N, S, P, K, Ca, Mg, Fe, among which the main elements are carbon, oxygen, nitrogen, and sulphur. Clearly, in a detailed analysis of the process of photosynthesis one should take into account the kinetics of CO_2, CH_4, H_2O, H_2S, NH_3, and NO_2. A minimum requirement with ensured CO_2 assimilation is the availability of CO_2, H_2O, light, chlorophyll, and proper environmental conditions (temperature and humidity). Therefore the complex assimilation formula can be written as follows:

CO_2	$+ H_2O$	+ 675 kcal	$\rightarrow C_6H_{12}O_6$	$+ O_2$	$+ H_2O$
moles 6	moles 12	solar energy	mol 1	moles 6	moles 6
246 g	216 g		180 g	192 g	108 g

This formula can be used to calculate the balance between the plants and the atmosphere only for the CO_2 exchange, but cannot be used for water, since it is a limiting factor for photosynthesis. The plants use much more water because of transpiration (unconsidered in the formula). A more detailed pattern of variations of the carbon exchange can be restored from the data in Table 2.9. In the global models the process of carbon assimilation should be detailed carefully to avoid the violation of the balanced description of other processes. Usually this is realized through introducing required corrections (Krapivin and Kondratyev, 2002). For instance, possible losses in the balance relationship for photosynthesis are taken into account. These losses are assumed to be 20–30%, that is, on average, 6 mol CO_2 give 0.75 mol of glucose.

It is also necessary to consider the spatial heterogeneity of the Earth land cover which differs in biomass density and intensity of organic matter formation. On average, 90% of the total biomass ($\approx$830 Gt C) are forest ranges ($\approx 50 \cdot 10^6\ m^2$), 50% of this value constituting tropical forests ($\approx 24.5 \cdot 10^6\ m^2$), and only 10% ($\approx$84 Gt C) refer to bushes, savannahs, meadows, deserts, semi-deserts, marshes, and cultivated lands. The process of organic matter formation is highly inhomogeneous: forests at 33 Gt C/yr, the remaining vegetation at 20 Gt C/yr. These heterogeneities lead to mosaic patterns of bioproductivity and therefore should be taken into account in the synthesis of the model.

The relationship between the CO_2 global cycle and surface vegetation dynamics is manifested through the dependence of primary production and the rate of the dead biomass decomposition on atmospheric temperature and CO_2 concentration with the binary digitization of the types of plants into two classes being taken into account.

Let $R_\kappa(\varphi, \lambda, t)$ be the photosynthesis production for vegetation of the type κ at a latitude φ and longitude λ at a time moment t. Then the CO_2 flux from the atmosphere into the living biomass can be described with a simple model:

$$H_6^C(\varphi, \lambda, t) = C_{23} R_\kappa(\varphi, \lambda, t)$$

where the coefficient C_{23} reflects the efficiency of the mechanism of the photosynthesis response and, on average, is estimated at $C_{23} \cong 0.549$. Bjorkstrom (1979) proposed an approximate assimilation of CO_2 by vegetation with the formula:

$$H_6^C = k_b\left(1 + \beta \ln\left[\frac{C_A}{C_A^*}\right]\right) C_k$$

where the parameter $\beta \in [0, 1]$ is the measure of the ability of the vegetation system to react to an increase of the atmospheric CO_2 partial pressure, C_k is the carbon content in the biomass of the kth type of vegetation, k_b is the coefficient of proportion depending on the temperature and type of vegetation, and C_A is the concentration of carbon dioxide in the atmosphere in the pre-industrial period.

Various authors estimate the flux $H_6^C(\varphi, \lambda, t)$ ranging from 16.7 to 35 Gt C/yr. This negligible scatter of the estimates enables one to reliably estimate the coefficients in approximations for H_6^C. A more detailed description of H_6^C requires a construction of an additional unit of the model, taking into account the relationship between CO_2 concentration and functioning of the surface biomes on a given territory. These specifications have been made in the publications by Krapivin and Vilkova (1990), Nefedov and Tarko (1993), and Krapivin *et al.* (1996). To specify the function R_κ, empirical dependences are used additionally, and are exemplified in Tables 3.12 and 3.13. The database of the modelling system contains similar information as well as data on the parameters of the soil–plant formations. Of course, the global data on the CO_2 balance in the biosphere are rather contradictory and formed with a delay and averaged over long time periods (see Table 3.6). Such data are exemplified in Table 2.5 and Figure 3.4. In the publication by Vilkova *et al.* (1998) the regression formulae are given which enable one to calculate productivity $F(T_a, W)$, humus supplies $H^0(T_a, W)$, and phytomass supplies $B^\circ(T_a, W)$ as a function of atmospheric temperature T_a (°C) and precipitation W (mm/yr):

$$\begin{aligned}
F(T_a, W) &= 4.25 \cdot 10^{-4} T_a^3 - 8.76 W^3 - 1.99 T_a^2 W + 4.29 T_a W^2 + 2.29 T_a^2 \\
&\quad + 19.05 W^2 - 8.79 T_a W + 4.56 T_a - 14.16 W + 4.18 \\
H(T_a, W) &= -5.16 T_a^3 - 161.4 W^3 - 9.41 T_a^3 W + 6.79 T_a W^2 - 9.47 T_a^2 \\
&\quad + 199.51 W^2 - 4.37 T_a W + 7.47 T_a - 44.17 W + 4.93 \\
B(T_a, W) &= -9.02 T_a^3 + 225.79 W^3 + 1.11 T_a^2 W - 29.39 T_a W^2 - 5.87 T_a^2 \\
&\quad - 511.72 W^2 + 41.29 T_a W - 11.37 T_a + 356.97 W - 62.94
\end{aligned}$$

The interaction of vegetation and the atmosphere is characterized by the CO_2 flux H_7^C, resulting from the process of respiration. Therefore, if T_k denotes the losses of the gross production by vegetation of the type k in the process of respiration, then $H_7^C = C_{26} T_k$. In the first approximation we can take $T_k = \gamma B_k$, where B_k is the vegetation biomass, the product γC_{26} reflects the share of organic carbon emitted per unit time by the vegetation surface cover into the atmosphere.

The functioning of the atmosphere–land border in the process of CO_2 exchange also includes the fluxes H_9^C, H_{14}^C, and H_{15}^C, which play an important role in the carbon balance of the biosphere. Bjorkstrom (1979) used the following relationships to

Table 3.12. Dependence of the annual production ($kg/m^2/yr$) on the mean global temperature and total precipitation, $F(T_a, W)$.

Precipitation	Atmospheric temperature T_a (°C)											
W (mm/yr)	−14	−10	−6	−2	2	6	10	14	18	22	26	30
3125							3.4	3.5	3.7	3.8	3.9	4.0
2875							3.2	3.3	3.5	3.6	3.7	3.8
2625							3.0	3.2	3.3	3.4	3.5	3.6
2375							2.8	2.9	3.0	3.1	3.2	3.3
2125							2.5	2.6	2.7	2.9	2.9	3.0
1875						1.6	2.3	2.3	2.4	2.5	2.6	2.7
1625				0.4	0.6	1.3	2.0	2.1	2.1	2.2	2.3	2.4
1375		0.2	0.3	0.4	0.7	1.1	1.7	1.9	1.9	2.1	2.1	2.0
1125	0.2	0.3	0.3	0.4	0.6	1.0	1.6	1.8	1.9	1.8	1.8	1.7
875	0.2	0.3	0.4	0.5	0.8	0.9	1.5	1.4	1.3	1.3	1.2	1.2
625	0.3	0.3	0.5	0.6	0.9	0.9	0.9	0.8	0.8	0.7	0.7	0.7
375	0.4	0.4	0.5	0.7	0.6	0.6	0.6	0.5	0.5	0.5	0.4	0.4
125	0.1	0.3	0.3	0.2	0.2	0.2	0.2	0.2	0.2	0.1	0.1	0.1

Table 3.13. Dependence of the humus content (kg/m^2) in a 1-m layer of soil on the mean annual temperature and total precipitation $H(T_a, W)$.

Precipitation	Temperature T_a (°C)										
W (mm/yr)	−14	−10	−6	−2	2	6	10	14	18	22	26
3000							21.9	21.8	21.2	21.1	19.8
2750							22.4	22.4	22.4	21.2	20.3
2500							22.5	22.5	22.4	21.3	20.6
2250							22.6	22.6	22.5	21.5	20.7
2000							22.7	22.7	22.7	21.5	20.9
1750							22.8	22.7	22.7	21.6	21.1
1500						15.6	22.8	22.8	22.7	21.6	21.3
1250				6.0	6.1	16.2	23.2	23.0	22.9	21.7	21.4
1000		5.0	5.0	6.5	6.1	16.9	25.8	24.1	23.1	22.5	22.2
750	1.0	5.5	6.0	11.0	21.5	35.1	25.1	24.0	23.1	23.1	22.3
500	1.0	6.1	7.4	9.1	19.1	58.3	45.1	23.3	19.2	14.3	18.2
250	1.1	6.1	7.5	11.0	13.5	14.3	10.1	5.2	3.4	1.1	1.0

describe these fluxes in his model of the biospheric balance of CO_2:

$$H_9^C = B_\kappa/\theta_9 \qquad H_{14}^C + H_{15}^C = B_\kappa/\theta_{14,15}$$

where θ_9 is the characteristic time of the carbon cycle in the soil and $\theta_{14,15}$ is the characteristic time of the transition of the living biomass carbon to the soil

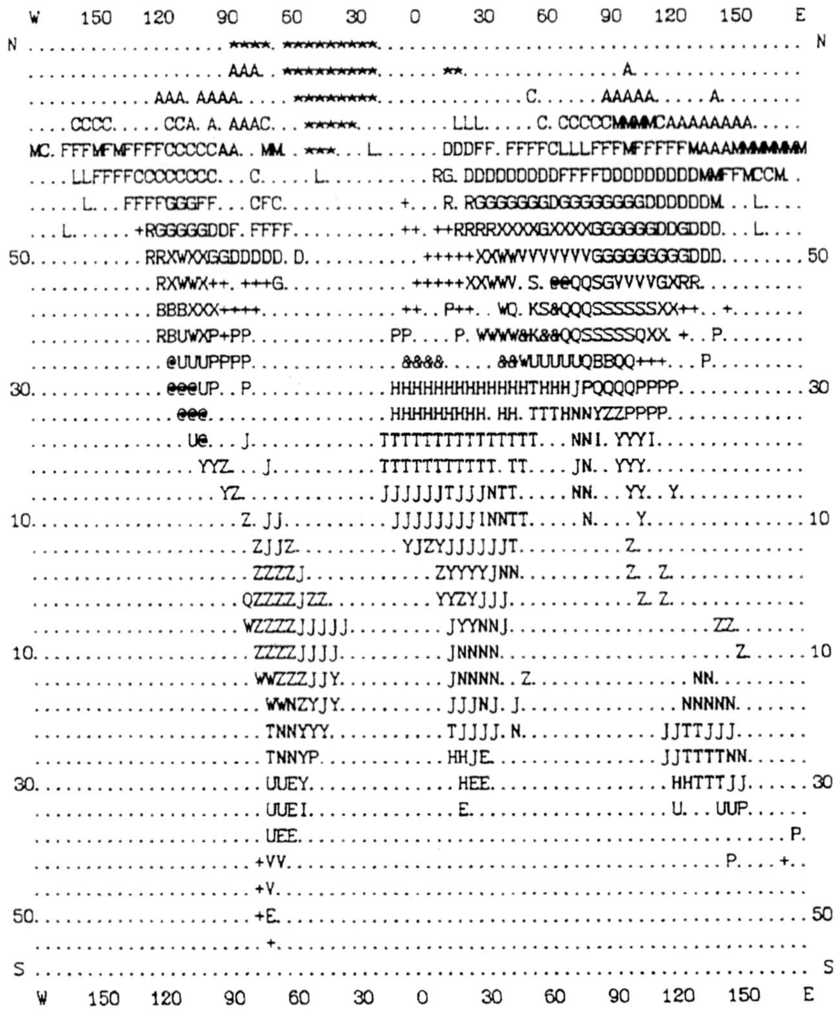

Figure 3.4. The spatial distribution of soil–plant formations over the geographical grid $4^\circ \times 5^\circ$. Notations are given in Table 2.5.

(about 1000 years). If R_Q denotes the rate of humus decomposition and M_k is the rate of vegetation decay, then:

$$H_9^C = C_{30} R_Q, \qquad H_{14}^C = C_{18} M_\kappa, \qquad H_{15}^C = C_{15} B_\kappa$$

With the carbon supply in the litter equal to 421.1 t/km^2, from which 12.95 t/km^2 are annually washed out into the soil, with 82.5% of this amount remaining in the upper soil layer up to 8 cm thick, we obtain $C_{18} = 0.31$.

The flux H_9^C has a more complicated dependence on the environmental parameters such as soil temperature and humidity. In the global model H_9^C is assumed to be a linear increasing function. The non-linear effects are considered due to heterogeneity of the types of soils manifested through the processes of oxidation of the organic matter. The most inert system is peat covering about $4.3 \cdot 10^6\,\mathrm{km}^2$ and containing 860 Gt of organic carbon. In a stable condition an organic substance in the soil in the process of oxidation emits CO_2 and accumulates the same amount of carbon during the process of plant decay:

$$(H_9^C = H_{14}^C + H_{15}^C + H_{12}^C + H_{13}^C)$$

For peat and tropical forests this balance is not observed. The soil in tropical forests emits CO_2 at a rate almost twice as high as the rate of CO_2 input into the soil by dead plants.

For the model of the atmosphere–soil CO_2 exchange to be complete, it is necessary to take into account geophysical and demographic aspects of the formation of additional carbon fluxes. They include volcanic emanations (H_5^C), rock weathering (H_4^C), vital functions of animals (H_{11}^C, H_{13}^C) and humans (H_{10}^C, H_{12}^C), as well as vegetation burning (H_8^C). Although nowadays some of these do not play a substantial role in the total balance of CO_2, an account of these fluxes is still worthwhile for a greater adequacy of the model response in conditions of stress simulation experiments. In the models the flux H_5^C is usually assumed a function of time, and spatial coordinates φ and λ. Fluxes $H_i^C\,(i = 10, \ldots, 13)$ are considered to be proportional to the sizes of population G and animals F:

$$H_{10}^C = C_3 G, \qquad H_{11}^C = C_2 F, \qquad H_{12}^C = C_{22} G, \qquad H_{13}^C = C_{21} F$$

With respect to the flux H_4^C, in the process of weathering of silicate rocks the rate of CO_2 extraction from the atmosphere is negligible compared with a similar process for carbonate rocks. Therefore consider the contribution of such rocks to the flux H_4^C. In equilibrium conditions the relationship $[Ca^{2+}][HCO_3]^2/p_a = \text{const}$ is valid. Usually $2[Ca^{2+}] = [HCO_3]$, and therefore:

$$\Delta[HCO_3]/[HCO_3] = (\Delta p_a/p_a + 1)^{1/3} - 1$$

The flux H_8^C reflects a fragment of the anthropogenic interference to the global cycle of carbon dioxide. The formation of industrial CO_2 can be described rather precisely as follows:

$$H_1^C + H_8^C = \gamma_0[\exp(rt)]^6 \quad \text{where } r \approx 0.029\,\mathrm{yr}^{-1}.$$

The surface cycle of carbon is, of course, more complicated in real conditions. Therefore many fluxes of carbon in Figure 2.8 should be considered as a combination of fluxes characterized by the scheme of Figure 3.5. A more detailed surface carbon cycle can be considered by dividing the body of plants and soil into their constituents. This process requires a deeper knowledge of the biogeocoenotic parameters, which on global scales is only possible in the form of average estimates of phytomass, biomass of the trunks and roots, supplies of organic matter in the litter, and humus.

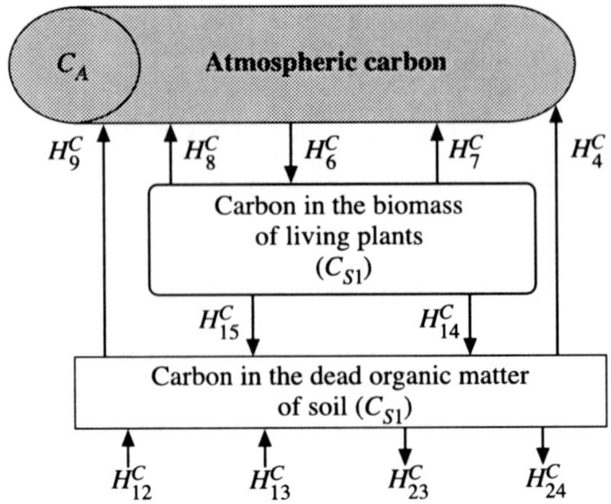

Figure 3.5. The scheme of carbon fluxes in the model of the atmosphere–vegetation–soil system. Notations are given in Table 2.9.

3.5 SOME ASPECTS OF BIOCOENOLOGY IN CONNECTION WITH THE ATMOSPHERIC GREENHOUSE EFFECT

To solve the problem of modelling the global greenhouse effect whilst taking an account of the role of the surface ecosystems, it is necessary to use methods, technologies, methodologies, and algorithms of eco-informatics (Krapivin and Kondratyev, 2002). Eco-informatics is the science of natural and anthropogenic systems and their interactions. It is characterized by the processes of differentiation and integration required, depending on the field of investigation. In the sphere of biocoenology these aspects can be clearly manifested, when only those types are selected from a variety of the surface ecosystems, which are typical of the assumed spatial grid of digitization of the Earth's surface. Unfortunately, at present there are no fixed terms, which would uniquely identify the sub-systems of the environment. Many authors put different meaning into such terms as the biosphere, ecosystem, biogeocoenosis, biome, geosystem, and the environment. From the viewpoint of informatics, the development of methods of data processing and models of the functioning of natural–anthropogenic systems such as divergences are not important, since they do not affect the formalized constructions of informatics. These divergences can manifest themselves only at the stage of interpretation of the results of data processing or modelling (Yokozawa, 1998; Kondratyev *et al.*, 2002a).

Interactions between informatics and biocoenology have been demonstrated mainly in the spheres of numerical modelling and monitoring. Since the nature of the biocoenology systems is determined by several abiotic and biotic factors, the use of the informatics methods in biocoenology requires a more detailed classification of these factors (Tikunov, 1997). One of the important factors is the climatic effect on a

Table 3.14. Ecological classification of the types of soil cover attached to the characteristic climatic zones.

	Specific features of the zone	
Climatic zone	Type of soil	Type of vegetation
1. Equatorial	Brown clay	Evergreen forests
2. Tropical	Red clay	Deciduous forests or savannah
3. Sub-tropical	Grey soil; stony, desert, and salt soils	Desert vegetation
4. Mediterranean	Brown soils	Wood vegetation
5. Warm temperate climate	Red or yellow forest soils	Temperate evergreen forests
6. Cold temperate climate	Brown and grey forest soils	Broad-leaved deciduous forests
7. Continental	Black brown, and grey soils	Steppes
8. Boreal	Ashen-grey soils	Taiga
9. Polar	Tundra	Undersized vegetation

special character of the distribution of plants. Each type of plant reacts with a time lag to temperature changes. The problem of modelling this dependence has been studied by many authors. It is a problem of succession of the vegetation cover depending both on climate changes and on anthropogenic forcing resulting from fires and deforestation. Different ideas and methodologies of correlation studies in biocoenology are caused by instability of climate processes. Therefore an application of stochastic mechanisms to analyse a hierarchy of the cycles of climate change is one of the goals persued in eco-informatics.

To use its methods, eco-informatics needs a certain level of formalization. For instance, Walter and Breckle (1985) introduced a zonal classification of the types of climatic situations characterized in Table 3.14. This classification simplifies the procedure of the spatial description of biocoenotic processes in the case of their modelling and in the formation of the biocoenology data-base. Of course, this example is one of numerous versions of attempts to formalize the environmental structure.

The classification aspect of biocoenology is important for the use of quantitative methods in the formalized description of the dynamics of various parameters of the surface ecosystems. The models of the cycles of some chemical constituents of the biosphere described by Krapivin and Kondratyev (2002) would be impossible to produce without a division of the soil–plant formations into homogeneous categories. The best known classification of the soil–plant formations is the classification by Bazilevich and Rodin (1967) given in Figure 3.4 and Table 2.5. Other classifications can be found through the Internet in the database of the World Resources Institute.

The level of detailed description of surface ecosystems is determined by many circumstances. The experts in concrete spheres of biocoenology study various sides of the functional behaviour of the ecosystems components. The problem for eco-informatics is to combine current knowledge with new knowledge that is the product of the synthesis of information fluxes from various scientific directions. So, for instance, phytocoenology studies plant physiology, the processes of plant competition for light, nutrients, and space. Here one cannot do without the methods of eco-informatics which make it possible, at a formal level, to formulate the behaviour of one plant or a community of plants.

The theory of complex system survivability enables us to formulate a model of the organization of phytocoenoses, based on the hypothesis of the usefulness of ecologo-coenotic strategies behind vegetation community development. For the first time, the so-called syn-ecological addition introduced by Rabotnov (1987) makes it possible to consider the process of the plant survival not as a mutual adaptation of concrete species but in a more general way, as a coexistence of species with different adaptive strategies. Each plant or their group in its competition for life can vary the elements of its strategy: change the distribution of roots over the soil horizons, change the ratio of the above ground and below ground parts of the plant, and change the requirements of the ecological niche. Therefore by using the theory of complex system survivability we should be able to gain a deeper understanding of the mechanisms of succession in vegetation communities.

The biocenological aspect of eco-informatics is especially manifested in the spheres of anthropogenic effects on the environment – landscape ecology and agricultural ecosystems. The methods of eco-informatics make it possible to formalize the scheme of the interaction of factors that form the landscape and lead to an appearance of potential vegetation in place of that destroyed by humans. There are formalizations in the landscape ecology, the use of which can help in the synthesis of the respective models of the landscape succession. For instance, there exists a 4-level hierarchy of landscape units: eco-top, micro-, meso-, and macro-flora. However, there is a disagreement in the subject sphere among various authors on typifying the territorial heterogeneities in vegetation communities of various ranks caused by internal landscape bonds. Therefore the use of system methods of eco-informatics with its possibilities of numerical experiments will help standardize the notions of landscape ecology with human interference taken into account.

In agro-ecology, methods of eco-informatics are introduced to develop the technologies to assess energy fluxes in the controlled ecosystems, which would ensure the necessary levels of their energy support. It is important that the specific energy expenditures in industrial agriculture are higher by an order of magnitude than in natural ecosystems. Humans have considerably reduced the species variety in agri-ecosystems to raise the productivity of useful species by managing the energy fluxes. Therefore the respective corrections should be introduced into the biogeochemical cycles, for instance, by introducing corrections in the Table 2.5 data. A variety of agro-ecosystems, not considered in the global biospheric models, results in the reliability of simulation modelling being substantially reduced.

3.6 THE SYSTEM APPROACH IN BIOCOENOLOGY

Practical problems of the rational use and reproduction of the biospheric resources require a quantitative study of the internal nature of the organisms constituting its communities (biocoenoses). This is necessary in studies of the World Ocean productivity and the structure and functioning of the communities residing in it, as well as in assessing the state of recreation zones. The use of only biochemical methods to solve these problems is insufficient in principle. It is necessary to apply biocoenotic methods. The latter, in contrast to biogeochemical methods, have no solid quantitative basis provided by physics and chemistry.

The quantitative basis of biocoenology is cybernetics, including the theory of complex systems. This theory manifests itself in biocoenology through numerous models of concrete biocoenoses aimed at solving the problems of predicting their dynamics and at finding the technologies to control these dynamics. In this respect the Lotki–Volterra models are rather popular as well as their numerous modifications that introduce into the model a factor limiting some functions of biocoenosis through environmental parameters. Unfortunately, all these models are only of theoretical character and serve to study a simplified phase structure of an idealized biocoenosis. Attempts to complicate the model to enable consideration to be given to the real processes leads to complex systems of differential equations in partial derivatives, of which an assessment of the coefficients requires serious field investigations (Bratus *et al.*, 2001; Novozhilov, 2002; Bratus *et al.*, 2000). Such a situation in biocoenology turned out to be the rule and not an exception. An attempt to express the ecological concepts as formulae or other formalized constructions is closely connected with the necessity for field measurements. Hence, the numerical modelling in biocoenology serves also to justify the necessity of measurements.

Biological systems have a hierarchical structure. Unicellular organisms are its lowest level, and the biosphere its top level. Biosystem A is determined by its structure $|A|$ and behaviour $\bar{A}$. The ecological law would not be broken if we suppose that a biosystem has some goal $\underline{A}$. There is no doubt that the strategic goal of any biosystem is either to survive or to favour the survival of a higher biosystem. However, the tactical goals of a biosystem favouring a realization of the strategic goal are not apparent.

Biosystems are open systems. It means they interact with the medium B, which also includes other biosystems. The biosystem–medium interaction consists of an exchange between them of substance, energy, and information. A set of the biosystems requirements to the medium is called its niche. Biosystems with similar niches can be considered as the same form of life. If the requirements of the biosystem to the medium are described by values of some parameters prescribed within certain ranges, the niche can be determined as some region in the respective multidimensional parametric space. Each element of the biosystem can also be described by some set of parameters, and in this case the bonds between the elements of the biosystem are presented as the dependence between the respective parameters. In other words, the biosystem A can be described by N parameters $\bar{x}(t) = \{x_i(t), i = 1, \ldots, N\}$, depending on time t. The structure $|A(t)|$ and behaviour $\bar{A}(t)$ of the biosystem are functions of these parameters. Therefore $A(t) = F(\bar{x}(t))$. Hence, the trajectories of the system A in time correspond to some curve in the N-dimensional Euclidean space.

For instance, the model of the system A is called an abstract formation $A_M(t) = F_M(\bar{x}^M(t))$, depending on M components of vectors $\bar{x}(t)\{\bar{x}^M(t) \in \bar{x}(t), M \leq N\}$ and considering not all bonds existing between them. Apparently, at $M \to N$ a discrepancy between the trajectories of the system itself and its model should decrease. An extent of discrepancy is chosen in the form of any natural criterion, for instance, in the form of a maximum difference between respective coordinates of the trajectories or some of their functions.

Note that the modelling of biocoenoses and its spatial distribution is the present-day equivalent to classical experimental studies. The principal obstacle here is the multi-dimensionality of the phase space, which is impossible to overcome. Therefore one should not expect more from numerical modelling than it can give. The well-developed numerical models perform a mainly explanatory function making it possible to simulate some processes in nature on an accelerated timescale and, without interference into real processes, to assess the effect of variations of individual characteristics on the dynamics of the environmental systems. Important models of biocoenoses are close, as a rule, in complexity to the modelled objects, and therefore often loose their explanatory character. Solving the appearing problems is connected with the use of the model only as an element of a more complicated construction discussed in Kondratyev *et al.* (2000) GIMS technology (Geographical Information System (GIS)+Model).

Let the biosystem's niche $A(t)$ be described by L parameters $\{c_j(t), L \ll N\}$ lying within the limits $a_j \leq c_j(t) \leq b_j (j = 1, \ldots, L)$. Then the biosystem environment bond can be described by a system of equations:

$$G_j(\{x_i(t)\}) = c_j(t), \quad (j = 1, \ldots, L)$$

All the systems whose parameters meet these relationships form the class Ξ.

Let us introduce some goal functional on the trajectory of the system $A(t)$:

$$V = f(\{x_i(t)\})$$

Let the parameter V reach a maximum at $\{x_{i,0}(t)\}$ on the trajectory $A_{\text{opt}}(t)$. Dependences of N parameters of an optimal system on L of its characteristics is a solution for this extreme problem:

$$x_{i,0} = H_i[c_1(t), \ldots, c_L(t)]$$

which enables us to reduce the model's dimensionality. By forming a multitude of tactical goals of the system $\Xi = \{A_1, \ldots, A_r, \ldots, A_m\}$ and establishing correlations $A_{r,\text{opt}}(t) = g(A_r), \underline{A}_r \in \Xi$ either experimentally or at an expert level, we obtain a limited set of possible optimal models whose trajectories can be compared with the trajectories of $A(t)$. The inverse problem can be solved using the given test of proximity:

$$\underline{A}_{r,0} = g^{-1}(A_{r,0,\text{opt}}(t))$$

As for the determination of the multitude of tactical goals, on a level of population they are depredation, parasitism, mimicry, etc. Determination of these goals on a level of biocoenoses is a more difficult problem. Here the problem is connected with understanding the categories of purposefulness, expediency, and reasonableness, as manifestations of biosystem efficiency. The problem is somehow simplified when biocoenosis is a part of the economic system.

3.7 MODELS IN BIOCOENOLOGY

Numerical biocoenology has become an independent scientific direction due to the works by many authors who combined in their studies the potentials of ecology and numerical modelling (Friend, 1998; Legendre and Legendre, 1998). This combination has resulted in a reorientation of many principles of the traditional ecology, which created a kind of information interface between ecology and other sciences and raised the role of complex interdisciplinary studies. The sphere of interests of this new direction includes the problems of complexity and generality in the development of ecological laws.

The recently developed models cover the problems such as a parameterization of the vegetation cover succession, description of the production process as a function of the environmental factors (Wirtz, 2000), modelling the competition for common resources (Alekseev *et al.*, 1992), modelling the natural–anthropogenic systems (Polishchuk, 1992; Ugolnitsky, 1999), modelling the spatial–temporal dynamics of the ecological systems, and many other problems (Roo-Zeilinska and Solon, 1997; Peritunen, 1998).

Many authors solve the problem of modelling a succession of vegetation communities based on their points of view on succession processes. There are two fundamental interpretations of these processes based on the concepts of endo-ecogenesis and exo-ecogenesis. The most stable ideas of the character of the dynamic phenomena in phytocoenoses is an idea of a single process of vegetation development under the influence of both external and internal factors. In this case the process of succession covers such factors as soil development, layer-by-layer division of the vegetation community, change of the height of plants and their accumulated biomass, microclimate formation, change of the species diversity, and relative stability of the community. The classification of the forms of vegetation community dynamics and its affecting factors is important for further synthesis of the numerical model. Vasilevich (1983) introduced the following categories of successions: endogenic (autogenic), exogenic (allogenic), continuous, and post-disruptive (post-catastrophic). In the case of endogenic succession the source of changes is located in the vegetation itself and through changes in the medium which affects it. There are no stages in continuous successions. In the course of post-disruptive succession, changes in the community take place from unstable to stable conditions. Exo-ecogenesis is characterized by prevailing external factors.

A new development in modelling of succession is the use of population dynamics. So, Hulst (1979) proposed three types of the models of succession. The first type is a model of succession with response but without competition, that is, there is an effect of reaction of the vegetation community to changes in environmental conditions. Succession is described by the usual differential equation:

$$\frac{dN_i}{dt} = \frac{r_i N_i (k_i - N_i)}{k_i}$$

where N_i is the number of individuals of the ith species, r_i is the rate of exponential growth, and k_i is a maximum of population density.

The second type of succession model is based on suppression of one species by another in the process of intrusion. In this case the coefficient of competition is used:

$$\alpha_{ij} = \frac{\dfrac{\partial [dN_i/dt]}{\partial N_j}}{\dfrac{\partial [dN_i/dt]}{\partial N_i}}$$

and the model itself is written as:

$$\frac{dN_i}{dt} = \frac{r_i N_i \left(k_i - \sum \alpha_{ij} N_i\right)}{k_i}$$

The third type of model of succession is of an alternative character. If in the first two models the population density is a function of the size of a previous species of vegetation, here a maximum of population density k_i depends on time.

Production of organic matter by a vegetation community depends on physiological processes in the organisms of the plants, resources of solar radiation, moisture, elements of mineral nutrition, temperature, and many other internal and external factors. Modelling of the production process takes into account photosynthesis, respiration, matter motion, transpiration, division of cells, etc. There are numerous correlations between these factors and production of plants. Lieth's formulae are used most often (Lieth, 1985). To estimate the annual production of plants R_p (g/cm^2/yr) as a function of mean-annual temperature T (°C) and mean annual precipitation r (mm), the following estimate can be used:

$$R_p = 3000 \min \{(1 + \exp(1.315 - 0.119T))^{-1}, (1 - \exp(-0.000664r))\}$$

In a more general case the vegetation biomass density P on a given territory is modelled by the balanced equation:

$$\frac{dP}{dt} = \min \{\rho_P, R_{PP}\} - M_P - T_P$$

where ρ_P is a maximum production/biomass P/B-coefficient of the type of vegetation considered, R_{PP} is the vegetation productivity for the P/B-coefficient values less than ρ_P for the present values of the P/B-coefficient, and M_P and T_P are parameters of decay and respiration expenditure per unit time, respectively.

The use of parametric descriptions of productivity in biogeocoenotic models is important as an addition to the monitoring of vegetation communities.

Unfortunately, the existing database and knowledge do not give any synthesized information about a huge amount of empirical correlations representing dependences of productivity and other functional components of vegetation communities.

Quite special in the numerical biogeocoenology is the problem of model description of the competition between individual plants. On the whole, the mutual impact of plants is described by the so-called competition indices (Bogatyrev, 1988). A freely growing plant is a plant whose functioning is independent of resource and is determined by only species characteristics and external parameters. An interaction between plants starts with their influence on the same resource and with the mutual intrusion on territories occupied by free growth. Biomes have more complicated and diverse interrelationships and, as a result, they differentiate niches and habitats. According to Mirkin (1986), who introduced five types of strategies for plants to function in the environment (K – violents, S – ecotopic patients, S_k – phytocoenotic patients, R – true explerents, R_k – false or phytocoenotic explerents), a systems approach is possible to model the description of competition and survival. All these strategies, to a differing degree, are inherent in individual species, and therefore the synthesis of the model of survival is determined by a concrete set of requirements of a plant to the niche parameters. Usually a plant's niche is an aggregation of the space it occupies, resources it uses, and additional conditions, for instance, the presence of pollinators, phytofuges, etc.

Each plant or some totality of plants N interacts with the environment H (including other plants) exchanging some amount of resources V_N for an obtained volume of resources W_N. The structure $|N|$ and the behaviour $\underline{N}$ of plants or their communities are in an equilibrium state with the structure $|H|$ and behaviour $\underline{H}$ of the environment, and this state is described with a minimax criterion:

$$W_{N,\text{opt}} = \max_{\{|N|,\underline{N}\}} \min_{\{|H|,\underline{H}\}} W_N(V_N, |N|, \underline{N}, |H|, \underline{H})$$

$$W_{H,\text{opt}} = \max_{\{|H|,\underline{H}\}} \min_{\{|N|,\underline{N}\}} W_H(V_N, |N|, \underline{N}, |H|, \underline{H})$$

The goals N_S and H_S of vegetation continuum N and environment H, respectively, can be opposite, partially coinciding, or indifferent. Due to the developed optimal strategy, plants of different kinds adjust to the levels of (V, W)-exchange that are possible on a given territory. Because of the wide differentiation in niches, there exist multilevel vegetation communities. In other words the factor of niche differentiation is multifaceted. It covers such special features of various kinds of plants as the location of roots at different depths, different periods of blossoming, different response of plants to climate changes, different needs for mineral nutrition of species, etc. In general, a plant's niche can vary in time both due to the dynamics of the plant itself and in connection with the drifting of the adjacent niches. The diversity of possible situations brings forth a series of models representing the functioning of vegetation communities.

One of the characteristic examples of the models of vegetation communities is the discrete model by Galitsky (1985). Let us consider it in more detail. The equation that describes a change in the plant's biomass B is written as:

$$\frac{dB}{dt} = k(t)(\alpha B^k + (B/B_f)^\gamma B_f) - \alpha B^k$$

where $k(t) = \min\{A/A_f(t), 1\}$ is the indicator of the plant's provision within a territory, A is the size of the available territory for the plant's growth, A_f is the size of the territory needed for free growth, B_f is the biomass of a freely growing plant, α is the characteristic of the item of expenditure in the balance (on the biomass maintenance), and γ is the index of the deficit of territory.

The gist of the Galitsky model's functioning is that the biomass of each plant located in a cell of its own in a common mosaic changes independently of other plants in the community until either this plant or any of its neighbours dies from lack of territory. At the moment of dying the adjacent plants rush to distribute the vacant territory, and after changing the parameter A, the equation of the model is valid again. As a result, as individual plants die, the original mosaic changes on a local scale, gradually covering larger areas. The criterion of dying off is simple: a plant is considered dead if its biomass becomes below some threshold level B_{dead} $(B_{\text{dead}}(t) = \varepsilon B_f(t), 0 < \varepsilon < 1)$.

3.8 MODELLING THE FOREST ECOSYSTEMS

The forest ecological system is an important constituent of the global continuum of soil–plant formations on the Earth (see Table 2.5). Therefore the development of models that would describe the forest dynamics is a principal stage of the synthesis of the global model. Of course, the importance of the models of the forest ecosystems is greater, since forest resources and their use affect the economic potentials of the regions (Ugolnitsky *et al.*, 2000).

3.8.1 Modelling the growth of a tree as an element of the forest biocoenoses

Many authors describe the dynamics of total biomass of an individual plant in the form of a logistic equation:

$$\frac{dB}{dt} = k(B_{\text{lim}} - B)B$$

where B is the biomass, B_{lim} is the threshold amount of biomass, and k is the coefficient reflecting the effect of environmental factors and the plant's type on its growth.

To describe the dynamics of the tree growth, the simplest functions are often used, such as Terazaki's equation $B = a\exp(-b/t)$, Koller's function $B = ax^b c^{-ct}$, and the approximation by Korsun' and Bakman $B = a\exp(b\ln t + c\ln^2 t)$, etc. Bichele *et al.* (1980) were first to propose a more complicated model of tree growth described by the system of ordinary differential equations:

$$dB_j = K_\Phi - k_R R_j - V_j + K_M$$

where K_Φ is an increment of biomass due to photosynthesis and respiration of the fresh assimilated components of a plant ($j = 1, 2, 3, 4$); M is the dry biomass of the whole plant; Φ_i is an amount of carbon dioxide assimilated during 24 hours by the ith organ of the plant ($k_F \Phi_i$ is an equivalent dry biomass, k_F is the coefficient of equivalence); R_i is an amount of carbon dioxide emitted during 24 hours by the ith organ of the plant ($k_R R_i$ is an equivalent dry biomass, k_R is the coefficient of equivalency); V_i is the dry biomass of part of the ith organ; α_{ij} is the share of "fresh" assimilates created during 24 hours in the ith organ of the plant, moving during the same period of time into the jth organ of the plant; β_{ij} is the exchange of "old" assimilates during 24 hours between the ith and jth organs, per gram of dry biomass of the whole plant; K_M is an increment connected with the redistribution of old assimilates, and $k_R R_j$ is the biomass expenditure on respiration.

Of course, the coefficients of the model described above are functions of the environmental parameters. An account of the effect of the environment on the dynamics of the tree growth is realized, as a rule, within the principle of the Libikh limiting factors. Among the factors of the environment are photosynthetic active radiation (PAR), carbon dioxide, temperature, humidity, as well as mineral salts in the soil. By the principle of limiting factors, a plant grows until the "resources" of some factor reach their limit. An introduction of this principle is connected with many uncertainties, and therefore in concrete situations it must be tested with the use of experimental estimates.

More developed models of the tree growth, instead of a set of empirical parameters whose uncertainties are often a great barrier to practical application of the models, include the functional dependences of the elements of tree growth on environmental factors. Among the most widespread and often applied is the Libikh criterion – the growth of a plant is confined to an element, whose concentration is at a minimum. By this criterion a plant or its organs grow until the resources of some element, for instance, PAR, moisture, temperature, carbon dioxide, or mineral salts are exhausted. There are other criteria testifying to a weighted influence of all environmental factors, these weights are characteristic indicators of the plant type.

The process of tree growth can be presented as a result of interaction of three processes of growth: change of the leaf (needle) mass, growth of the trunk, and development of the root system. This process is schematically shown in Figures 3.2 and 3.6.

Denote the biomass of the canopy, trunk, and roots of the tree as $x1$, $x2$, and $x3$, respectively. Then the balance equations of a tree model are written as:

$$x_i(t + \Delta t) = x_i(t) + \varepsilon_i(t)\{R_x - T_x\}\Delta t \quad (i = 1, 2, 3)$$

where Δt is the characteristic time step, R_x is the gross productivity of a tree due to photosynthesis, T_x is the biomass expenditure on respiration, and ε_i is the share of new biomass moving into the biomass of the ith organ of the tree.

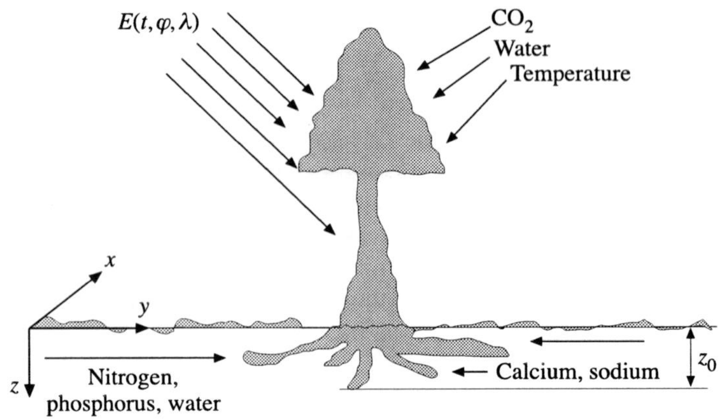

Figure 3.6. The scheme of the model of the growth of a tree as an element of the forest biogeocenosis. Notations: z_0 – thickness of the root system layer; E – solar radiation.

The ε_i parameters are the functions of time as well as other parameters of the environment. Following the principle of maximum survival, a supposition can be made that the principle of maximum primary productivity is valid:

$$\tilde{Y}^*(t+\Delta t) = \max_{\{\varepsilon_i(t)\}} Y\{x_i(t) + \varepsilon_i(t)Y[x_i(t), E(t,\varphi,\lambda), C_A, T_A]\Delta t, E, C_A, T_A\}$$

3.8.2 Modelling the water regime of plants

One of the important limiting factors of tree growth is water, the movement of which in the soil and tree body determines the dynamics of the tree biomass change. Following Kirilenko (1990), denote the soil as a homogeneous porous layer $[0, z_g]$, where z_g is the depth of ground water. Let us introduce the following notation: W – relative soil moisture (kg H_2O/m^3/[kg soil/m^3], ρ – water density, and ρ_s – volume density of soil. Let us introduce a system of coordinates (x, y, z), where the z-axis is directed down from the soil surface $(z = 0)$. Then the movement of water through the soil can be described by Darsi's 3-D equation:

$$V = -\frac{K[\nabla(\Phi_s - \rho g z)]}{\rho g} \tag{3.1}$$

where $V(t, x, y, z)$ is the rate of water flow, $K(t, x, y, z)$ is the hydraulic conductivity of water, g is the acceleration due to gravity, and $\Phi(t, x, y, z)$ is the water potential of soil.

The equation of the soil moisture dynamics is:

$$\frac{\partial W}{\partial t} = -(\rho/\rho_s)(f + \text{div } V) \tag{3.2}$$

where $f(t, x, y, z)$ is the sink function.

Dependences of the water potential of soil and its hydraulic conductivity on soil moisture is written as:

$$K = K_s^* W^m \qquad \Phi_s = \Phi_s^* W^{-n}$$

where

$$K_s^* = K_{FM} W_{FM}^{-m} \qquad \Phi_s^* = \Phi_{MG} W_{MG}^n \qquad m = 2n + 1$$

where W_{FM} is the total water capacity of soil, K_{FM} is the soil conductivity corresponding to the total water capacity, and m and n are constants.

The system of equations (3.1) and (3.2), with assumed notations, are rewritten as:

$$\frac{\partial \omega}{\partial t} = -\frac{\rho}{\rho_s}\frac{K_{FM}}{W_{FM}} \operatorname{div}\{\mu\omega^n \nabla\omega + i\omega^n\} + \frac{f}{W_{FM}}$$

$$V = K_{FM}(\mu w^n \nabla\omega + i\omega^n)$$

Where $w = n\Phi_{FM}/(\rho g)$ and i is the unit vector directed along the z-axis.

In the case of homogeneous processes in the plane (x, y) the problem is simplified:

$$\frac{\partial \omega}{\partial t} = \frac{-\rho}{\rho_s}\frac{K_{FM}}{W_{FM}} \frac{\partial}{\partial z \left\{ \dfrac{\mu\omega^n \partial\omega}{\partial z} + \omega^n \right\}} + \frac{f(t,z)}{W_{FM}}$$

$$V(t,z) = K_{FM}\left(\frac{\mu\omega^n \partial\omega}{\partial z} + w^n \right)$$

Boundary and initial conditions are given in the form:

$$\omega(t, z_g) = 1, \qquad V(t, 0) = P(t) - e(t), \qquad \omega(t_0, z_g) = \omega^0(z)$$

where $P(t)$ is precipitation intensity, $e(t)$ is the intensity of evaporation from the soil surface, and $\omega^0(z)$ is the initial soil moisture.

The f function describes the intensity of water absorption by a tree's root system:

$$f(t,z) = \xi_0[\Phi_s(t,z) - \psi_0(t)]\frac{ds}{dt}$$

where $\xi_0 = \xi_k\sigma_0$ is the conductivity of water entry to the root system, $\xi_k = r_k^{-1}$ is the specific conductivity of water entry to the root, r_k is the specific resistance of water entry to the root, σ_0 is the general area of the sucking roots of the tree per unit soil surface area ($\sigma_0 = \beta_k m_k$), m_k is the mass of roots beneath unit soil area, β_k is an empirical constant, ψ_0 is the water potential within the root system, and $s(z)$ is the function of the vertical distribution of the root system $[s(0) = 0, s(z_0) = 1]$.

To complete the synthesis of the tree's water regime, parameterize the processes of water flow in the plant and transpiration. Assume that from the root system the water flows by the trunk to the canopy (branches, leaves) and then evaporates into the atmosphere (Lai *et al.*, 2000; Lal *et al.*, 1998). Divide the above ground part of the tree of height H into n equal layers: $H = n\Delta z$. The water moves due to the difference between water potentials ψ_i and ψ_{i-1} across the boundaries of the ith layer overcoming the resistance of the xylem vessels:

$$r^*_{ks}(z) = (S^*_{ks}(z)\xi_{ks})^{-1}$$

where ξ_{ks} is the specific conductivity of the xylem and S^*_{ks} is the cross section of the tree at a height z. As a result, with the area of the cross section supposed to be vertically constant, we obtain the following formula to describe the rate of water flow in the trunk across the ith layer:

$$v^*_{i-1,i} = -\xi_{ks}S^*_{ks}\Delta z^{-1}(\psi_i - \psi_{i-1} + \rho g\,\Delta z)$$

According to Kirilenko (1990), the intensity of transpiration from the ith layer can be described with the formula:

$$\gamma^*_i = \frac{d^i}{r^i_L} S^{*i}_L$$

where d^i is the deficit of saturation of an absolute air humidity in the atmosphere, S^{*i}_L is the leaf area of the ith layer, $r^i_L = r^i_{st} + r_a, r^i_{dt}$ is the stomatal resistance, and r_a is the air boundary layer resistance.

To complete the description of the model after Kirilenko (1990), consider a standing wood of the density ρ_F. The model equations are written as:

$$\frac{d^2\varphi}{d\eta} = Uv(\eta)\varphi(\eta), \qquad \eta \in [0, \alpha] \tag{3.3}$$

the boundary conditions are:

$$\begin{cases} \varphi'_0 = -[G(\tilde{\varphi}_s - \varphi_0) + F] \\ \varphi'_\alpha = \qquad -F \end{cases}, \frac{d\varphi}{d\eta} = -F \quad \text{for } \eta \in [\alpha, 1] \tag{3.4}$$

At $\varphi_0 \leq 0$, the system of equations (3.3) and (3.4) becomes:

$$\begin{cases} \dfrac{d\varphi}{d\eta} = -F \\ \varphi_0 = \tilde{\varphi}_s \end{cases}$$

with the following notation: $\eta = z/H, \varphi(\eta) = 1 + \psi(\eta H)/\psi^*, \psi^*$ is the leaf's water potential, $1/D_1$ is a minimum mouth resistance, $U = Hr_{ks}D_1 d\beta m_L/\psi^*$, $F = \rho g H/\psi^*$, $G = Hr_{ks}\xi_0$, $\alpha = a/H, \beta$ is the leaf index, $h = -z$, and

$$a = \begin{cases} H & \psi(H) \geq \psi^* \\ h & \psi(h) = -\psi^*,\ \psi(H) < -\psi^* < \psi(0) \\ 0 & \psi(0) \leq -\psi^* \end{cases}$$

3.9 MODELLING THE ENERGY FLUXES IN THE ATMOSPHERE–PLANT–SOIL SYSTEM

The energy exchange in the atmosphere–plant–soil (APS) system plays an important role in the formation of climatic processes on every spatial and temporal scale, and therefore should be considered in the models of the greenhouse effect to increase the reliability of the estimates. The problems that appear include the development of methods to measure the energy fluxes in the air–surface layer as well as their modelling. A special aspect of these problems concerns the development of technologies of estimation of energy fluxes in the APS system on global scales to take these estimates into account in the global climate systems. There are well-developed methods of remote sounding of a wide set of climatic parameters including energy fluxes (Yakovlev, 1998, 2001; Krapivin and Kondratyev, 2002). However, in the remote sounding of the APS system a complicated problem arises of an assessment of the screening impact of vegetation over the wavelength regions used in the monitoring system.

Beginning in the 1980s, studies have been carried out on developing a model of photosynthesis of a vegetation bud. It was shown in Farquhar *et al.* (1980) how a biological model of the CO_2 assimilation by leaves is connected with fermentative kinetics and the electronic properties of the chloroplast, which made it possible to rather plausibly describe the process of photosynthesis for the plants of class *C*. Then Ball (1988) and Collatz *et al.* (1990, 1991, 1992), using a model technology and observational data on the process of vegetation bud conductivity in photosynthesis, derived a semi-empirical model of the vegetation bud functioning with an account of changes of the environmental temperature, humidity, CO_2 concentration in the atmosphere, and intensity of assimilation of the vegetation species itself.

The problem of the consequences of the distribution of the photosynthetic potential of leaves and canopy of plants with respect to illumination has been considered and a criterion has been developed to determine a common potential corresponding to maximum photosynthetic abilities of the plant in CO_2 assimilation.

Sellers (1985, 1987) developed methods demonstrating the vertical integration of leaf canopies along with models of the light scattering and absorption, photosynthesis and conductivity in a vegetation bud. These studies revealed an interaction between spectral characteristics of a vegetation species (spectral vegetation index – SVI) and the important functional relationships that regulate the processes of photosynthesis and evaporation from leaf canopies. Apart from this, an accomplished analysis revealed a connection between the share of the PAR absorbed by the vegetation canopy with the respective characteristic coefficient of vegetation species (simple ratio vegetation index – SRVI). Here we consider SRVI and relative difference (normalized difference vegetation index – NDVI) as SVI. Spectral characteristics of vegetation species SRVI and NDVI are based on the remote sounding of the reflected emission in the visible and near-IR regions and determined as follows:

$$\mathrm{SN} = \frac{a_N}{a_V} \qquad \mathrm{NDVI} = \frac{a_N - a_V}{A_N + a_V}$$

where a_N and a_V are hemispheric reflections and radiance in the visible and near-IR wavelength regions, respectively.

The parameters a_N and a_V are measured directly above the leaf canopy. Therefore the effects of absorption and scattering of emissions by the atmosphere are not considered here. For space-borne remote sensing a method has been developed to correct the experimental data called FASIR (Fourier-adjusted, solar zenith angle-corrected, interpolated and reconstructed data). A theoretical dependence (close to linear) has been established between PAR and SRVI confirmed later by experimental measurements in the region of Kansas (Sellers *et al.*, 1992, 1996a–c). A minimum and maximum of FPAR (the fraction of PAR absorbed by the green portion of the vegetation canopy) are determined experimentally, and the final relationship becomes:

$$\mathrm{PAR} = \frac{(\mathrm{SRVI} - \mathrm{SRVI}_{i,\mathrm{min}}) \cdot (\mathrm{PAR}_{\mathrm{max}} - \mathrm{PAR}_{\mathrm{min}})}{(\mathrm{SRVI}_{i,\mathrm{max}} - \mathrm{SRVI}_{i,\mathrm{min}})} \tag{3.5}$$

where $\mathrm{PAR}_{\mathrm{max}} = 0.950$ and $\mathrm{PAR}_{\mathrm{min}} = 0.001$, but in general, $\mathrm{PAR}_{\mathrm{max}}$ and $\mathrm{PAR}_{\mathrm{min}}$ depend on the type of vegetation; $\mathrm{SRVI}_{\mathrm{min}}$ and $\mathrm{SRVI}_{i,\mathrm{min}}$ can be calculated from the NDVI, based on experimental data.

Sellers *et al.* (1996*b*) use the Π parameter reflecting an amount of vegetation cellulose taking part in the process of photosynthesis. The Π parameter changes from 0 (lack of vegetation) to $\approx$1.5 for bushy species rich in photic vegetation cellulose and can be approximately estimated as:

$$\Pi \approx \frac{\mathrm{PAR}}{\bar{k}}$$

where $\bar{k}$ is the coefficient of attenuation for PAR averaged over time.

Apart from spectral characteristics, Sellers *et al.* (1996c) consider the photic characteristic (green leaf area index L_g) and the photic share (green part N) for a concrete vegetation species. Many authors derived dependences of L_g on FPAR for species growing separately and in groups. The result can be presented in the form:

$$L_g = (1 - F_{cl}) L_{\mathrm{g,i,max}} \frac{\log(1 - \mathrm{PAR})}{\log(1 - \mathrm{PAR}_{\mathrm{max}})} + F_{cl} \frac{L_{g,i,\mathrm{max}} \mathrm{PAR}}{\mathrm{PAR}_{\mathrm{max}}}$$

where F_{cl} is the share of bushy vegetation determined for each of the vegetation classes L_s.

The photic share N of vegetation species can be obtained from the following considerations. The plant canopy consists both of leaf matter (L_g) participating in the process of photosynthesis and of the trunk cellulose L_s (0.076 for forests, 0.05 for meadows and pastures) as well as a dead fraction L_d. The dead cellulose is supposed to remain in the leaf canopy for about a month before leaf falling or other means of extraction. Thus the photic share of vegetation cellulose is determined as:

$$N = \frac{L_g}{L_T}$$

where $L_T = L_s + L_d + L_g$, L_d is that fraction of vegetation cellulose that is dead $= L_{g_{n-1}} - L_{g_{n1}}$, $L_d \geq 0$, L_g is the photic fraction determined from PAR, L_T is the total territorial characteristic of a vegetation species, and L_s is the fraction of the trunk cellulose.

3.9.1 The model of photosynthesis of leaf canopy

In the model of the biosphere (simple biosphere model-2 – SiB2) derived by Sellers *et al.* (1996a) there are equations to determine temperatures, humidity, and evaporation for the vegetation canopy and 3-layer soil cover. Apart from this, a model of radiative transfer was synthesized (soil–leaf canopy–atmosphere) and also a model of photosynthesis. The model of the leaf canopy photosynthesis considers a 1-layer vegetation species, surface and root layers, as well as the zone of re-charging. Equations of the SiB2 are based on the concept of flux exchange in the system "soil–leaf canopy–atmosphere". For this purpose estimates of resistance r and conductivity g in inverse proportion ($r = 1/g$) are introduced. The flux in this case is determined by analogy with the electrostatic form:

$$\text{Flux} = \frac{\text{Potential difference}}{\text{Resistance}}$$

Sellers *et al.* (1996b, c) consider the fluxes of sensible and latent heat transport through evaporation of water vapour from a vegetation canopy λE_c and the surface soil layer, as well as the CO_2 flux (A_c and R_{SOIL} for soil). The potential differences connected with the fluxes are presented in the form of differences of temperatures, pressures, and partial pressures of CO_2 for heat transport, latent heat transport, and CO_2, respectively. In the model of photosynthesis of interest are CO_2 fluxes (A_c and R_{SOIL}) and the flux of latent heat exchange λE_c of leaf canopy and its respiration constituent λE_{ct}.

The model of the leaf canopy photosynthesis derived by Collatz *et al.* (1990) for vegetation species of classes C_3 and C_4 together with the bud model (Ball, 1988) have been used to derive the SiB2. In this model, photosynthesis and conductivity are closely connected (Sellers *et al.*, 1996a–c).

According to Collatz *et al.* (1991), there are three factors limiting plant photosynthesis intensity. These are an efficiency of the photosynthetic ferment system w_c (the so-called limitation after Rubisco), an amount of PAR absorbed by cellulose chlorophyll w_e, and an ability of a vegetation species to assimilate and emit to the environment the products of photosynthesis w_s. As a result, the value of photosynthesis intensity A is a minimum of the three limitations enumerated above:

$$A \leq \min(w_c, w_e, w_s)$$

where A, w_c, w_e, and w_s are measured in mol/m^2/s.

The physiological limitation of assimilation w_c is a function of the ferment supplies of species, in other words, it is an ability of cellulose to realize the biochemical reactions. The ferment reserve of the plant is described in the model with the $V_{\max}$ parameters, a maximum catalytic supply of ferments (Rubisco) taking part in photosynthesis. The w_c value depends also on the leaf canopy temperature and soil moisture. Collatz *et al.* (1990) suggest the following expression for w_c for vegetation classes C_3 and C_4:

$$w_c = V_m \left[\frac{c_i - \Gamma^*}{c_i + K_c(1 + O_2/K_o)} \right], \text{ for } C_3$$

$$w_c = V_m, \text{ for } C_4$$

where V_m is a maximum catalytic supply of ferments (Rubisco) (mol/m^2/s), c_i is the partial pressure of CO_2 in the leaf (Pa), O_2 is the partial pressure of O_2 in the leaf (Pa), Γ^* is the point of CO_2 compensation (Pa) $= O_2/S$, S is a special interaction between CO_2 and O_2, K_c is the Michaelis–Menten constant for CO_2 (Pa), and K_o is the constant of O_2 suppression (inhibition).

The V_m parameter is expressed through V_{max}, the function of dependence on the soil temperature and moisture. The S, K_c, K_o parameters are functions of temperature.

A limitation of assimilation intensity w_e is described as follows:

$$w_e = (F_\pi \cdot n)\varepsilon_3(1 - \omega_\pi) \left[\frac{c_i - \Gamma^*}{c_i + 2\Gamma^*} \right] \text{ for } C_3$$

$$w_e = (F_\pi \cdot n)\varepsilon_4(1 - \omega_\pi) \text{ for } C_4$$

where F_π is a PAR vector (W/m^2), n is the normal to the leaf surface, $\varepsilon_{3,4}$ is quantum efficiency of CO_2 assimilation (mol), and ω_π is the dispersion coefficient for PAR.

The third limitation w_s is written as:

$$w_s = \frac{V_m}{2} \text{ for} C_3 \qquad w_s = \frac{V_m}{2} \text{ for } C_4$$

where p is atmospheric pressure (Pa).

The photosynthesis intensity A as a minimum of functions w_c, w_e, and w_s is estimated in the simplest way, but observations show that the transition from one limitation to another is not that spasmodic. Therefore Collatz *et al.* (1992) proposed a method of combining the contributions of all the three limitations in the form of two equations, to be solved with the least values:

$$\beta_{ce} w_p^2 - w_p(w_c + w_e) + w_e w_c = 0$$

$$\beta_{ps} A^2 - A(w_p + w_s) + w_p w_s = 0$$

where β_{ce} and β_{ps} are coefficients of correlation and w_p is a "smoothed" minimum between w_c and w_e (mol/m^2/s).

Theoretically β_{ce} and β_{ps} can have values from unity (lack of relationships between limitations) to zero. Real estimates vary from 0.8 to 0.99.

Pure assimilation of CO_2 by a plant A_n is equal to $A_n = A - R_d$, where R_d are CO_2 losses resulting from respiration connected with the content of carboxylase in cellulose: $R_d = f_d V_m$, where $f_d = 0.015$ for C_3 and $f_d = 0.025$ for C_4.

The connection of the Collatz model of photosynthesis with the Ball semi-empirical model of vegetation bud conductivity is written as:

$$g_s = m\frac{A_n}{c_s}h_s p + b$$

where g_s represents the vegetation bud conductivity (mol/m^2/s); g_s(m/s) = $0.0224(T/T_f) \cdot (p_0/p) \cdot g_s$(mol/m^2/s); m is an empirical coefficient (= 9 for vegetation of class C_3, = 4 for C_4, = 6 for coniferous forests); b is an empirical coefficient ((mol/m^2/s or m/s), = 0.01 for C_3, = 0.04 for C_4); h_s is the relative moisture of the leaf surface, c_s is the CO_2 partial pressure in the stem (Pa); p is atmospheric pressure (Pa); p_0 is the normal atmospheric pressure (= 1.013×10^5 Pa); and T_f is absolute zero temperature (= 273.16 °K).

There is a relationship between partial pressure of CO_2 (c_i, c_s) and relative humidity h_s. This dependence can be obtained by analysis of the processes taking place on the "vegetation bud–atmosphere" border, the respective conductivities g_i and g_s, pure CO_2 flux, and water vapour flux E_{it} as a result of the process of photosynthesis.

This relationship is expressed as:

$$E_{it} = g_i(e_s - e_a)\frac{\rho c_p}{\lambda\gamma} = g_s(e_i - e_s)\frac{\rho c_p}{\lambda\gamma} \qquad h_s = \left[\frac{e_s}{e_i}\right]$$

$$A_n = \frac{(c_a - c_s)}{p}\frac{g_i}{1.4} = \frac{(c_s - c_i)}{p}\frac{g_s}{1.6}$$

where e_a, e_s, and e_i are partial pressures of water vapour in the open air, on the surface, and in the stem, respectively (Pa), c_p is specific heat capacity of air (Joule/kg·K), γ is a psychometric constant (Pa/K), g_l is the water vapour conductivity on one side of the leaf (mol/m^2/s or m/s, = $1/(2rl)$), E_{it} is respiration intensity (kg/m^2/s), c_a and c_s are partial pressures of carbon dioxide outside and inside the stem, respectively (Pa), and $e_i = e^*(T_c)$ (Pa).

Equations of the photosynthesis–conductivity model describe the processes taking place in a vegetation bud, and the next step is to integrate the relationships with respect to a leaf canopy to obtain total values of A_c and g_c. This was undertaken by Sellers *et al.* (1996a) with a supposition that the vertical distribution of vegetation ferments in the canopy is connected with the time-averaged distribution of PAR:

$$V_{\max_L} = V_{\max_0}\,\mathrm{e}^{-\bar{k}L}$$

where $V_{\max_L}$, and $V_{\max_0}$ are the $V_{\max}$ values on the leaf area index (LAI) at LAI = L and at the top of the canopy (LAI = 0), respectively (mol/m^2/s) and $\bar{k}$ is a time-averaged value of the coefficient of PAR attenuation.

An amount of PAR assimilated by the leaf canopy is:

$$F_\pi \cdot n \approx F_{\pi_0}\left[\frac{G(\mu)}{\mu}\right]\mathrm{e}^{-\bar{k}L}$$

where F_{π_0} is the value of the PAR flux caught by the leaf canopy (W/m^2) and $G(\mu)$ is a projection of the orientation of the canopy leaves towards the direction of the incoming radiation flux (μ). Sellers *et al.* (1996a) established that the use of the time-averaged $\bar{k}$ value instead of the instant k value leads to only negligible uncertainties in calculations.

Since now the w_c, w_e, and w_s parameters are connected (depending on the vertical coordinate) with LAI of the leaf canopy through an exponent, it is possible to use them to obtain an integral estimate of A_c. We have already discussed the difference of bushy and separately growing species ($V < 1$) and the notion of the photic share ($N < 1$). Therefore bearing in mind that in the case of dense thickets the photosynthesis intensity is reduced due to the lack of "green", rich-in-ferment vegetation material, and the integration is realized from 0 to L_T/V:

$$A_c = A_{n_0} \int_0^{L_T/V} VN\mathrm{e}^{-\bar{k}L}\,dL = A_{n_0}\Pi \tag{3.6}$$

where $A_{n_0} = A_n$ for leaves at the top of the canopy (mol/m^2/s):

$$A_{n_0} = f_c(V_{\max_0}, \ldots,)f_e(F_{\pi_0}, \ldots,)f_s(V_{\max_0}, \ldots,)$$

$$\Pi = \frac{VN(1 - \mathrm{e}^{-\bar{k}L_T/V})}{\bar{k}} \qquad \Pi \approx \mathrm{PAR}/\bar{k}$$

The A_c value is used to determine the respective conductivity of the leaf canopy using the modified expression:

$$g_c = m\frac{A_c}{c_S}h_S p + bL_T$$

where h_S and c_S are the volume analogues to values h_s and c_s relative to the leaf canopy.

The g_c parameter is used to estimate the respiration intensity λE_{ct}:

$$\lambda E_{ct} = \left[\frac{e^*(T_c) - e_a}{\dfrac{1}{g_c} + 2r_b}\right]\frac{\rho c_p}{\gamma}(1 - W_c)$$

where $e^*(T_c)$ is the saturated water vapour pressure at a temperature T_c (Pa), e_a is the water vapour pressure in an open atmosphere near the leaf canopy (Pa), ρ and c_p are the air density and specific heat capacity, respectively (kg/m^2, Joule/kg K), γ is a psychometric constant, and W_c is the share of the lead canopy wet area.

To complete a description of equations of the photosynthesis–conductivity model, the following expressions are given which estimate the ferment supply of vegetation cellulose V_m:

$$V_m = V_{\max}f_T(T_c)f_W(W_2)$$

$$K_c = 30f_T(T_c)$$

$$K_o = 30000f_T(T_c)$$

$$S = 2600 f_T(T_c), f_T(T_c) = \frac{2Q_t}{\{1 + \exp[s_1(T_c - s_2)]\}} \text{ for } C_3 \text{ and } V_m$$

$$f_T(T_c) = \frac{2Q_t}{\{1 + \exp[s_1(T_c - s_2)]\} \times \{1 + \exp[s_3(s_4 - T_c)]\}} \text{ for } C_4 \text{ and } V_m$$

$$f_T(T_c) = \frac{2Q_t}{\{1 + \exp[s_5(T_c - s_6)]\}} \text{ for } R_d \text{ and } V_m$$

$$f_T(T_c) = 2.1 Q_t \text{ for } K_c$$

$$f_T(T_c) = 1.2 Q_t \text{ for } K_o$$

$$f_T(T_c) = 0.57 Q_t \text{ for } S$$

$$Q_t = \frac{(T_c - 298)}{10}$$

$$f_W(W_2) = \frac{1}{\{1 + \exp[0.02(\psi_c - \psi_r)]\}}$$

where ψ_c is a critical water potential (m), ψ_s is the value of saturation potential for moisture content in the soil (m), B is an empirical parameter, s_l is the parameter of suppression of high and low temperatures, $\psi_r = \psi_s W_2^{-B}$ is the potential of moisture content in the root layer of the soil (m).

3.9.2 A dynamic photosynthesis–conductivity model

The physiological processes described by the A_c and g_c parameters do not exhibit an instant response to changes in the control derivatives. It takes about one minute to stabilize the photosynthesis process after changes in external conditions, but for conductivity the time step constitutes several minutes. Hence, for time steps longer than one minute the photosynthesis intensity A_c can be calculated using relationship (3.6), and a delay in changes of conductivity can be described reliably by the equation:

$$\frac{\partial g_c}{\partial t} = -k_g(g_c - g_{\text{inf}}) \tag{3.7}$$

where k_g is a time constant for the response of vegetation cellulose ($= 0.00113c^{-1}$), $g_{c_{\text{inf}}} = g_c$ for $t \to \infty$, and $g_{c_{\text{inf}}} = m\frac{A_c}{c_S} h_{SP} + b$.

The solution (3.6) is written as:

$$g_c = e^{-\bar{k}_g t} g_{c_0} + (1 - e^{-\bar{k}_g t}) g_{c_{\text{inf}}} \qquad g_{c_0} = g_c \text{ for } t = 0$$

and a change in the g_c during a time step Δt will be:

$$\Delta g_c = g_c - g_{c_0} = (1 - e^{-\bar{k}_g \Delta t})(g_{c_{\text{mf}}} - g_{c_0})$$

These correlations make it possible to calculate the energy fluxes in a multilayer structure of the "atmosphere–plant–soil" system. As a result, one can specify the elements of the climate models, and assess from the space-borne monitoring data the impact of surface vegetation on the synoptic situation. It is especially important in arid and over-humid zones where the state of vegetation cover depends strongly on small changes of temperature and other meteorological parameters. Another important conclusion from these correlations can be drawn for the study of the effect of vegetation cover on the processes of CO_2 assimilation from the atmosphere. Clearly, the use of such models increases an adequacy of global carbon cycle modelling.

3.10 MODELLING THE PRODUCTION PROCESSES IN THE CONIFEROUS FOREST

Coniferous forests play an important role in the global carbon cycle (Safonov *et al.*, 1999). Therefore in this section we shall consider a parameterization of the production process in the forest ecosystem with a pine forest stand of one age proposed and developed by Kirilenko (1990) for practical use in climate models. The model reflects correlations of mean annual and seasonal processes without consideration of the succession of vegetation communities. It describes the dynamics of carbon and water in the ecosystem using a system of five ordinary differential equations, one partial differential equation, and many algebraic equations reflecting the biotic and abiotic processes in the coniferous forest.

The model includes the following values: phase derivatives, input (external) and output (internal) derivatives, as well as parameters (constants, coefficients).

The phase derivatives include (kg/m^2 of soil):

- carbon of the structural mass of young pine needles m_1 and old ones m_2 and $m_L = m_1 + m_2$;
- carbon of the above ground parts of the grown trees (branches, boughs, trunks) and growing roots m_3;
- carbon of thin roots m_4; and
- fund of mobile carbon assimilants m_5.

The input variables characterize the environmental conditions:

- the incoming solar radiation intensity E;
- air temperature T and humidity H;
- length of the light period of day τ;
- daily amount of precipitation P; and
- concentration of carbon dioxide in the atmosphere C_o.

The output variables are photosynthesis, average height and diameter of trees, transpiration, respiration, increment of the tree's organs, and other characteristics of the coniferous forest ecosystem calculated in the model.

The biomass of the phase variables in the model is measured in carbon units, and the respective scheme of carbon fluxes is shown in Figure 3.7. Notations for dynamic

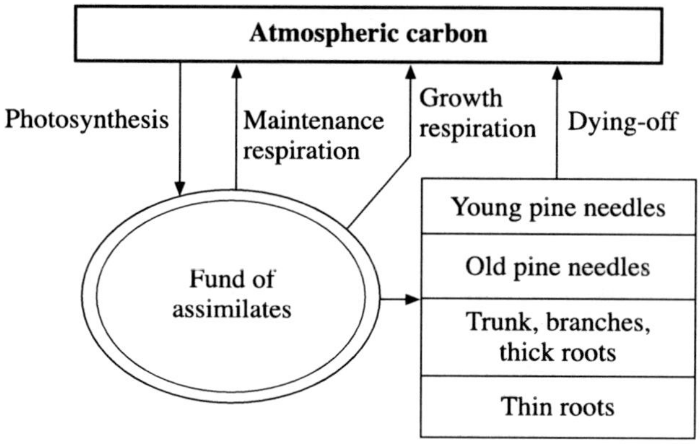

Figure 3.7. The conceptual scheme of the carbon cycle in the model of the production process in a coniferous forest, after Kirilenko (1990).

equations are:

- $s_3(m_3)$ is the area of the trunks per m^2 of soil;
- $\theta_i(\tau_B)$ is the function of distribution of assimilants, that is, the share of the fund of assimilants spent on the increment of the ith organ;
- τ_B is the biological time;
- d_i is the share of the dead biomass of the ith organ ($i = 1, \ldots, 4$);
- F is a photosynthesis value;
- r_i is the respiration intensity of the growth of the ith organ; and
- p_i is the respiration intensity of the maintenance of the ith organ.

3.10.1 Photosynthesis

To parameterize the process of photosynthesis we use the Chartier formula:

$$F = \frac{\sigma(T_L)\Omega(\tilde{\omega}_S)}{2(r - r_c)} \frac{S_L}{2} \tau \left(C_0 + r\alpha E - \sqrt{(C_0 - r\alpha E)^2 + 4r_c \alpha E C_0} \right)$$

where $\sigma(T_L)$ is the function of the dependence of photosynthesis intensity on the leaf temperature T_L, $S_L = \beta m_L$ is the area of pine needles, $\tilde{\omega}_S$ is the weighted-mean soil moisture, $\Omega(\tilde{\omega}_S)$ is the function of the impact of the over-moistened root-layer of the soil on photosynthesis, $r = r_a^c + r_{st}^c + r_c$, r_a^c and r_{st}^c are diffuse resistances for CO_2 molecules in the leaf boundary layer and in the leaf stoma, respectively, r_c is an efficient resistance of carboxylizing, τ is the duration of the light period of day, and α is an initial angle of inclination of the light curve of photosynthesis:

$$\tilde{\omega}_S = \int_0^{A_0} \omega_S(z)\, ds(z)$$

where $s(z)$ is the function of distribution of the root system along the vertical cross-section of the soil layer and A_0 is a maximum capacity of the root layer of the soil.

For conifers, $r_a^c \ll r_{st}^c$, and along the height of the tree the value of photosynthesis per unit mass of pine needles changes negligibly, and therefore we shall consider some point, averaged over the tree height, for which the intensity of the PAR is determined by the Beer–Lambert law:

$$E = \kappa_E E_e \exp(0,5 K_e b S_L H^{-1})$$

where E_e is the intensity of the incoming solar radiation over the forest, κ_E is the share of PAR in the total solar radiation, and K_e is the coefficient of extinction.

Assuming that the temperature of pine needles coincides with that of the atmosphere ($T_L = T$), we use the following approximation for the $\sigma(T)$ function:

$$\sigma(T) = \max\left\{0, \frac{T - T_1}{T_0 - T_1} \exp\left[l^{-1} - l^{-1}\left(\frac{T - T_1}{T_0 - T_1}\right)^l\right]\right\}$$

where the parameter $l = 1,8$, T_1 is a minimum temperature at which the photosynthesis still takes place, and T_0 is a temperature optimal for photosynthesis.

The Ω function characterizes the dependence of photosynthesis on soil moisture. With a short-term over-moistening of soil the photosynthesis does not decrease, and therefore $\Omega(\tilde{\omega}_S) = 1 \forall \tilde{\omega}_S \in [0, 1]$. In a general case we use the dependence:

$$\Omega(\tilde{\omega}_S) = \begin{cases} 1, & \tilde{\omega}_S \in (0, \omega_{HB}) \\ (1 - \tilde{\omega}_S)/(1 - \omega_{HB}), & \tilde{\omega}_S \in [\omega_{HB}, 1] \end{cases}$$

3.10.2 Distribution of enzymes

To describe the strategy of distribution of enzymes, let us introduce the function $\theta_i(\tau_B)$, where τ_B is the biological time. Determine the ratio of the sum of efficient temperatures T_l to the mean multi-year sum of efficient temperatures T_Σ accumulated during the vegetation period:

$$\tau_B(j) = \sum_{i=1}^{j} T_l / T_\Sigma$$

where j is the number of days.

The θ_i function is determined from the expression:

$$\theta_i = a_i^\theta \exp\left[-b_i^\theta (c_i^\theta - \tau_B)^2\right] \qquad (i = 1, \dots, 4)$$

where the a_i^θ coefficient is equal to a maximum of θ_i, c_i^θ is the biological time of a maximum of enzymes flux to the ith organ, the b_i^θ coefficient determines the steepness of the θ_i function.

The flux of enzymes at a time moment τ_B into the ith organ of the plant is equal to $m_5 \theta_i(\tau_B)$.

3.10.3 Respiration and dying-off

Assume that respiration connected with the growth of the ith organ of a tree is in proportion to the rate of its increment and is equal to $r_i\ dm_i/dt$, and the respiration for maintenance of leaves and roots draining moisture is in proportion to their mass and is equal to $p_i m_i\ (i=1, 2, 4)$ (the p_i coefficient depends on temperature):

$$p_i = a_i^p \exp\left[k_i^p(T - T^p)\right]$$

where a_i^p is the respiration for maintenance at a temperature T^p.

The respiration for maintenance of the trunk and thick roots is in proportion to the trunk area s_3 and is equal to $s_3 p_3 \sqrt{\beta_{st}}$, where β_{st} is the share of the trunk mass in m_3. The dying-off of the organs is considered to be in proportion to their masses with the coefficients $d_i\ (i = 1, \ldots, 4)$.

3.10.4 Water cycle

The water cycle in a tree is described by a system of equations that parameterize the water motion over the territory (Chapter 5), in soil, and in a tree and its elements. The water is one of the limiting factors of tree growth through a value of the stomatal resistance that affects the photosynthetic activity of plants. The water cycle on the forest territory is the sum of two basic parts – abiotic and biotic, which describe the processes of water motion in soil and its absorption by the root system, its motion in the plant, and transpiration. There are a lot of models of these processes. The difference between them is determined by different suppositions about the character of the soil structure and the location of the root system. Kirilenko (1990) proposed to describe the water motion in the soil with an equation that generalizes the Darsi classical model for the case when the soil is considered a homogeneous porous layer:

$$\frac{\partial \omega}{\partial t} = -\frac{\rho}{\rho_S}\left[\frac{K_{\Pi B}}{\Phi_{\Pi B}}\frac{\partial}{\partial z}\left(\mu\omega^n \frac{\partial \omega}{\partial z} + \omega^n\right) + \frac{f(z,t)}{\Phi_{\Pi B}}\right]$$

$$V(z,t) = K_{\Pi B}\left(\mu\omega^n \frac{\partial \omega}{\partial z} + \omega^m\right)$$

with the boundary conditions:

$$\Phi(z_g, t) = 1, \qquad V(0,t) = W(t) - L(t)$$

and the initial condition:

$$\omega(z_g, t_0) = \omega^0(z)$$

where $W(t)$ is precipitation intensity, $L(t)$ is the intensity of evaporation off the soil surface, ρ is water density, ρ_S is the volume density of soil, Φ is the relative soil moisture, $\Phi_{\Pi B}$ is the total water capacity of soil, V is the rate of moisture motion in soil, $\mu = n\phi_{\Pi B}/(\rho g), m = 1 + 2n$, where n is a constant, g is acceleration due to gravity, $\phi_{\Pi B} = \phi_S \omega^n$, where ϕ_S is the water potential of soil, $K_{\Pi B}$ is the conductivity corresponding to soil moisture $\Phi_{\Pi B}$, and z is the vertical coordinate. The f function describes the intensity of water absorption by the root system:

$$f(z,t) = \left[\xi_0 \bar{\phi} \varphi(z,t) + v(t)\right] dS/dz$$

where $\varphi(z,t) = \varphi_S(z,t) - \tilde{\varphi}_S(t)$ is the deviation of the dimensionless water potential from average value for the root layer of the soil, $\xi_0 = \xi_k \sigma_0$ is the conductivity of the entry of water into the root system, $\xi_k = 1/r_k$ is the specific conductivity of the entry of water to the root, r_k is the specific resistance of the entry of water into the root, and σ_0 is the total area of the sucking roots of the tree per unit soil area:

$$\varphi_S(z,t) = 1 - \left(\frac{\omega(z,t)}{\bar{\omega}}\right)^{-n}, \qquad \tilde{\varphi}_S(t) = 1 + \tilde{\phi}_S/\bar{\phi}$$

where $\tilde{\varphi}_S$ is an average dimensionless water potential for the root layer of the soil, $v = \xi_0(\tilde{\phi}_S - \phi_0)$ is the rate of water absorption by roots from the soil.

Water from the root system moves upward along the trunk, through branches into leaves and evaporates into the atmosphere. Assume that an amount of the transpiration moisture is equal to that assimilated by the root system, and the rate of motion is equal to v. The integral effect of the impact of the coniferous forest on the water cycle in its territory is the sum of the effects of each tree. The practical application of the developed model needs information about spatial heterogeneities and distributions of the characteristics of the forest stand. An application of the model to deciduous and mixed forest systems require additional units which would describe the processes of the transformation of water potentials as well as carbon cycles in leaves. Intensive studies are being carried out in this area (Peritunen, 1998; Peng, 1999; Papakyriakou and McCaughey, 1991; Kolbe, 1999; Ferm and Hultberg, 1999; Esteve, 1998).

3.11 MODELLING THE SUCCESSION PROCESSES IN THE TUNDRA–TAIGA SYSTEM

In connection with growing anthropogenic processes on the Earth and their propagation towards northern latitudes, some problems arise connected with prediction of the location of the forests northern boundary. Connections between the observed propagation of this boundary and forest fires, reindeer grazing, anthropogenic transformations, and climate change have been widely discussed in the scientific literature. Various hypotheses have been expressed about correlations between the natural–technogenic processes in the zone of this boundary. For instance, there is an opinion that forest fires can influence the northward propagation of the birch forests. This can lead to considerable changes in the native habitation and reindeer propagation. In the late 20th century some disagreements were observed between forestry and deer-breeding. So, in Finland the areas of reindeer propagation moved 400–500 km south from the boundary of the forests, that is, up to the boreal zone. In other words, a change of the forest boundary is followed by a change of the soil layer, and this, due to shifts in biogeocoenotic cycles, can direct the succession of the northern boundary of forests in an unpredicted direction. Moreover, this prediction, based on the pre-history of the development of this boundary in natural conditions, cannot be reliable. Therefore the use of the methods of eco-informatics with its models and technologies of monitoring is now a single adequate method for predicted estimates of the dynamics of the northern forest boundary. Here a lot of environmental problems appear including:

- combining various elements of the forest ecosystems in the zone of the northern boundary of the forests (reasons for the presence or absence of some type of trees);
- the role of various diseases of the northern forests and their sensitivity to climate changes and other external factors;
- the role of fires and their location and power in the change of the structure and configuration of the transfer boundary between forests and tundra;
- the role of insects, taking account of the dependence of their propagation and influence of the biomass on climate change in the dynamics at the boundary of the northern forests;
- patterns of scattering the seeds of various plants in the zone of the forests northern boundary and their effect on its dynamics; and
- the impact of the dynamics of the forests northern boundary on wild animal migration and habitat.

An expected climate warming can result in the substitution of some vegetation ecosystems for others (exogenic succession). It will be especially manifested in high latitudes where taiga transforms into tundra. Bogatyrev (1988) proposed a model, which parameterizes the transformation processes in the tundra–taiga system and makes it possible to reveal some important regularities. The phase variables are the biomass of moss (X_1), vegetation of the grass-bush level (X_2), conifers (X_3), forest cover (X_4), and dead organic matter of soil (X_5). The phase variables are measured in tonnes of carbon per unit area (t/ha), and their spatial distribution is assumed to be uniform. The carbon flux scheme is given in Figure 3.8.

The system of balance equations of the model of the transfer processes on the taiga–tundra border according to the scheme in Figure 3.8 is:

$$\frac{dX_i}{dt} = R_{0i} - R_{i4} \quad (i = 1, 2, 3)$$

$$\frac{dX_4}{dt} = \sum_{i=1}^{3} R_{i4} - R_{40} - R_{45} - R_4$$

$$\frac{dX_5}{dt} = R_{45} - R_{50} - R_5$$

where the R_{ij} functions depend on many parameters, and their determination is the principal stage of model formulation. Bogatyrev (1988) proposed the following

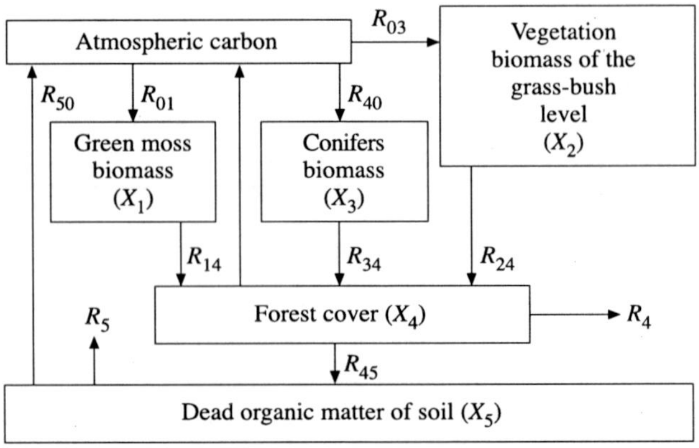

Figure 3.8. The scheme of the carbon cycle in the tundra–taiga system. After Bogatyrev (1988).

formulae for these functions:

$$R_{0i} = F_i\varphi_i(T_M)G_i(X_3)f_i(X_5)H_i(W) \qquad R_{i4} = k_iX_i \quad (i = 1, 2)$$
$$R_{03} = F_3\varphi_3(T_M)G_3(X_3)f_3(X_5)H_3(W)\chi(X_3)$$
$$R_4 = m_4X_4$$
$$R_5 = m_5X_5$$
$$R_{34} = k_3X_3[1 + \kappa(T_M - T_A)]^{-1}$$
$$R_{40} = \sigma k_4X_4\varphi_4(T_M)$$
$$R_{50} = k_5X_5\varphi_5(T_M)$$
$$R_{45} = (1 - \sigma)k_4X_4\varphi_4(T_M)$$

where the F_i coefficients express the dependence of the biomass increment on its type (the characteristic indicators of vegetation), the φ_i functions assign the laws of changes in the biomass increment as a function of atmospheric temperature beneath the canopy T_M, and the G_i, f_i and H_i functions describe deviations of the annual biomass increment from a maximum level at a given temperature T_M due to a deficit of illumination, an amount of mineral resource in the soil, and moisture, respectively. For trees, the annual production also depends on their biomass, and this dependence is represented by the χ function. For vegetation of the grass-bush level and moss, their annual increment is supposed to be independent of their biomass.

Determine the functions φ_i, G_i, f_i, H_i, and χ:

$$\varphi_i(T_M) = 1 + (\delta_i/100)(T_M - T_Y)(i = 1, 2)$$
$$\varphi_3(T_M) = 1 + (\delta_3/100)(T_M - T_A)$$
$$G_3(X_3) = \exp(-\alpha_iX_3)(i = 1, 2, 3)$$

$$f_1(X_5) \equiv 1$$

$$f_i(X_5) = \exp(-a_i/X_5)(i = 2, 3)$$

$$H_i(W) = \exp\left\{-b_i\left(W - W_A\frac{W_Y}{\sigma_i}\right)^2\right\}(i = 1, 2, 3)$$

$$\chi(X_3) = \left(\frac{X_3}{X_{3A}}\right)^\gamma$$

where T_Y and T_A are atmospheric temperatures in tundra and taiga, respectively, δ_i $(i = 1, 2, 3)$ are indicators of variability of the annual biomass increment with an increasing temperature beneath the canopy by 1°C, α_i, a_i, b_i, κ, and $\gamma \in (0, 1)$ are constant coefficients, W_A and W_Y are soil moisture in taiga and tundra, respectively, k_i $(i = 1, 2, 3)$ are proportions coefficients, and $\sigma \in (0, 1)$ is the share of the forest cover transforming into the dead organic matter of soil as a result of decomposition.

To parameterize the atmospheric temperature beneath the tree canopy, suppose that it cannot be below the environmental temperature T, increases with the mass of trees up to some level $X_3 = 0, 5\beta_2/\beta 1$, and then decreases:

$$T_M(X_3, T) = T \cdot \begin{cases} f(X_3), & 1 \leq f(X_3) \leq T_A/T \\ T_A/T, & f(X_3) > T_A/T \\ 1, & f(X_3) < 1 \end{cases}$$

where

$$f(X_3) = 1 - \beta_1 X_3^2 + \beta_2 X_3$$

and β_1 and β_2 are positive coefficients.

The model described can be used as an independent unit of the global model, providing a transformation of the spatial structure in Figure 3.4 and specifying the local elements of the model of the water balance on the territory described in Krapivin and Kondratyev (2002). For real application of the model, its parameters should be estimated with a concrete attachment to geographical latitude.

3.12 GLOBAL CARBON DIOXIDE MODEL TAKING INTO ACCOUNT THE SPATIAL DISTRIBUTION OF LAND BIOMASS

Assume that each site of land 4° by latitude and 5° by longitude belongs to one of the types of the soil–plant formations enumerated in Table 2.5, or without vegetation. On the ith site of land of the area S_i the soil–plant formation is characterized by an amount of carbon per unit area in the biomass of living plants B_i and in dead organic matter of soil D_i. Following Krapivin *et al.* (1982), let us write a system of balance equations:

$$\frac{\partial B_i}{\partial t} = F(T_i, P_i)[1 + (\delta/10)(C_a/C_a^0)] - K_i B_i$$
$$\frac{\partial D_i}{\partial t} = K_i B_i - E(T_i, P_i) D_i \tag{3.8}$$
$$\frac{\partial C_a}{\partial t} = \sum_{i=1}^{N} \{[F(T_i, P_i)(1 + (\delta/10)(C_a/C_a^0 - 1)) - E(T_i, P_i) D_i] S_i\} + V$$

where $F(T_i, P_i)$ is the annual production of plants per unit area (see Table 3.12), $E(T_i, P_i)$ is the specific rate of decomposition of dead organic matter of soil per unit area, T_i and P_i are the mean annual atmospheric temperature and the annual precipitation amount over the ith site of land, respectively, δ is the indicator of the annual production increase in % with a 10% increase of CO_2 content in the atmosphere, K_i is the specific intensity of the dying-off biomass of plants on the ith site of land, $C_a^0 = C_a(t_0)$, where t_0 is an initial time moment (in the global models it is usually assumed that $t_0 =$ the year 1850), and V is an anthropogenic input of carbon into the atmosphere.

Suppose that in the absence of anthropogenic impacts ($V = 0$) the amount of carbon in the APS system is constant, and in the early industrial period the system was in a state of equilibrium. Then, at the moment t_0 the following relationships are valid:

$$K_i = B_i^0 / F(T_i^0, P_i^0); \qquad E(T_i^0, P_i^0) = F(T_i^0, P_i^0) / D_i^0(T_i^0, P_i^0) (i = \overline{1, N})$$

As a result, the model makes it possible to calculate the dynamics of carbon in three reservoirs with their spatial distribution taken into account. Depending on the details of the types of vegetation cover (see Tables 1.12, 2.5, and 2.6), calculations of the concentration of carbon in the vegetation cover, soil, and atmosphere with formulae (3.8) require a knowledge of the functions F and E, as well as the δ coefficient.

4

Carbon cycle in the ocean

4.1 THE WORLD OCEAN AS A COMPLEX HIERARCHIC SYSTEM

To increase the reliability of assessing the role of the World Ocean in the global carbon cycle, a more detailed description is needed of the production processes in the oceans ecosystems (Orr *et al.*, 2001; Peterson *et al.*, 2002; Rau and Caldeira, 2002; Raupach, 2000; Raymond, 2002; Reid, 2001; Shindell, 2003; Stauffer *et al.*, 1998). Along with the physical and chemical processes of transformation and motion of carbon in the ocean medium, the biological processes play an important role. In particular, phytoplankton, like the nutrient elements, assimilates dissolved CO_2 from the salt water. As a result, an organic substance is formed that both partially goes to the nutrient chains of the trophic pyramid of a given basin of the World Ocean and partially descends to bottom sediments. A totality of all processes of carbon motion in the ocean medium creates a gradient of the CO_2 concentration between surface and deep waters. Therefore a study of the structure and functioning of the oceans ecosystems becomes one of the most important and rapidly developing directions of marine biology. Its various aspects are being developed in many countries within the framework of the International Biological Programme. In particular, the international programme JGOFS (US Joint Global Ocean Flux Study) is dedicated to a study of biochemical processes in the ocean to obtain a deeper knowledge of the response of the ocean to external forcings. One of the goals of the study is to attain the possibility of predicting the system's behaviour as a result of changes of some of its parameters. However due to the unique nature and broad spatial extent of the World Ocean, it is difficult to quantitatively estimate all the elements of the system at different moments of its development and in different regions of the World Ocean and, moreover, to assess the effect of their change on the functioning of the system as a whole.

The ocean covers 71% of the planet's surface yet produces 1% of the food products consumed by humans, the remaining 99% being obtained from cultivated land. At the same time, the total amount of organic matter produced in the ocean is approximately equal to that produced by land vegetation. By rough estimates, the total biomass of nekton constitutes $5.3 \cdot 10^9$ t. The catch of fish and other species from the World Ocean is estimated at $70 \cdot 10^6$ t/yr, which constitutes 20% of protein consumed by humans. The catch of traditional species is close to the maximum limit (~ 90–$100 \cdot 10^6$ t/yr). However, this has not been reached for the industrial ability of the ocean ecosystems in general, since the supplies of krill and other biological objects are still under exploited.

This disagreement between the role of the land and ocean ecosystems in food production is explained by the fact that on land, an intensive cultural economy is realized, which is weakly developed in the seas and oceans. Therefore possible ways of increasing the ocean bioproductivity have not been applied yet. First, humankind uses the final trophic chains of natural communities of the World Ocean – fish and whales. Each successive trophic level obtained about 0.1% share of energy accumulated in the previous level. On land, two levels of organisms (vegetation, herbivorous animals) are used, but in the ocean and in the seas, up to five levels. The direct use of non-fish species will make it possible to sharply increase the amount of proteins harvested from the ocean.

Second, a question arises about the transition from free fishing to cultural methods of economy in the World Ocean, that is, the question of an artificial increase of productivity of the biological communities of the ocean. To do this, it is necessary, first of all, to study the methods of controlling the output of the final produce in the biological systems of the World Ocean. To determine the rational ways of affecting the ocean communities, it is necessary to study their structure and functioning, to understand the production processes and transformations of matter and energy flux at different trophic levels of the ocean ecosystems. It is necessary to develop a theory of control in the biological systems of the coastal waters and the open ocean, which differ both in natural hydrophysical and biogeochemical parameters as well as in the extent of anthropogenic impacts.

Marine communities are complicated biological systems of populations of individual species. As a result of their interaction, communities are in dynamic development. Their spatial structure is mostly determined by the composition of numerous biotic and abiotic factors, which depend on a totality of the oceanic parameters. The latter are determined by the laws of general circulation of the ocean waters, including tides and ebbs, zones of convergence and divergence, wind, and thermohaline currents.

In the late 20th century an urgent problem arose of predicting the dynamics of the ocean systems in conditions of increasing anthropogenic impacts (chemical poisoning, mechanical removal of living organisms, environmental changes) as well as assessing their role in the dynamics of the whole biosphere. So, in recent studies of the climatic impact of greenhouse gases (GHGs) it has been shown that the role of the World Ocean in this process has been underestimated. In particular, data has been given in Kondratyev and Johannessen (1993) on the role of the arctic basins of the World Ocean in the formation of the global CO_2 cycle, from which it follows that previous assessments of this role were incorrect. This is connected with the fact that an account of biological and gravitational processes playing the combined role of a pump (that pumps carbon dioxide from the atmosphere to deep layers of the ocean), was inadequate in the earlier models of the global biogeochemical carbon cycle. Therefore specified models of the working regime for this pump with climatic zones taken into account may be of importance for predicting estimates of the greenhouse effect (Gorshkov, 1990); Krapivin, 2000; Anderson *et al.*, 1998; Aurela *et al.*, 2001; Bates *et al.*, 1998; Bousquet *et al.*, 2000; Caldeira and Duffy, 2000; Choudhury and Kukla, 1979; Doskey and Gaffney, 1992).

The impact of the ocean ecosystems on the biogeochemical cycles is manifested through the atmosphere–water border and is usually parameterized based on observational data. However, in this impact, of importance is the vertical structure of the processes taking place in the ocean medium. The nature of these processes depends significantly on external factors. For instance, according to Legendre and Legendre (1998), in the arctic zones of the World Ocean the patchy structure of the springtime blossoming of phytoplankton is determined by winter conditions of ice formation and subsequent ice melting. In other zones these external factors take the form of pollution of the atmosphere and ocean surface, changes in the phytoplankton living conditions, and functioning of the carbonate system.

Phytoplankton is positioned at one of the initial stages of the trophic hierarchy of the ocean system. As field observations have shown, the World Ocean has a patchy structure formed by a combination of non-uniform spatial distributions of insolation, temperature, salinity, concentration of nutrient elements, hydrodynamic characteristics, etc. The vertical structure of the phytoplankton distribution is less diverse and possesses rather universal properties. These properties are manifested through the existence of one-to-four vertical maxima of phytoplankton biomass.

Variability of the patchy topology and vertical structure is connected with seasonal cycles and has been well studied experimentally in many climatic zones of the World Ocean. The typical qualitative and quantitative indicators of this variability have been found and the combined distributions of abiotic, hydrological, and biotic components of the ocean ecosystems have been studied.

The complexity and mutual dependence of all the processes in the ocean substantially hinder the uncovering of the laws of formation of phytoplankton spots and the means to establish correlations between various factors regulating the trophic relationship intensity in the ocean ecosystems. For instance, many studies revealed a close relationship between primary production and amount of phytoplankton. At the same time, this relationship gets broken depending on a combination of the synoptic situation and insolation. It turns out that the extent to which the relationship is broken depends significantly on a combination of phytoplankton groups (Legendre and Legendre, 1998).

An analysis of the accumulated observation data on assessments of the produce of seas and oceans and an attempt of many authors to uncover the laws of the produce formation characteristic of various water basins have revealed various laws in local relationships between productivity and environmental parameters.

An efficient method of studying the vertical structure of ocean ecosystems is numerical modelling based on measurements of their characteristics (Kuck *et al.*, 2000). To derive the model, it is necessary to know the structure of trophic relationships in the ecosystems, specific features of hydrological conditions, and information about other characteristics of the environment. An experience of modelling has shown the possibility of an efficient prediction of the dynamics of the World Ocean communities. Examples of such models are the 3-D models of the ecosystem of the Peruvian Current (Krapivin, 1996) and the Okhotsk Sea (Aota *et al.*, 1993). In these models the unit of parameterization of the vertical structure of the ecosystem is central.

4.2 ATMOSPHERE–OCEAN EXCHANGE PROCESSES

Knowledge of exchange processes between the atmosphere and the ocean is very important in assessing the greenhouse properties of CO_2. In most of the known models of the global CO_2 cycle the relationship between partial pressure of CO_2 dissolved in water (p_c) and general concentration C_U of inorganic carbon in the surface waters is calculated based on the buffer coefficient:

$$\xi = (p_c - p_{c,o})C_{U,o}(C_U - C_{U,o})^{-1}/p_{c,o}$$

where the "*o*" index refers to the pre-industrial period. However, numerous observations show that the use of only this coefficient when calculating the fluxes H_2^C and H_3^C gives rough estimates, without the effect of many other factors being taken into account.

One of the first attempts to consider the spatial heterogeneity of the World Ocean and to simulate the impact of temperature gradients on CO_2 exchange between the upper layer of the ocean and the atmosphere was made by Bjorkstrom (1979). The idea to divide the World Ocean basins into two parts corresponding to warm and cold waters was later developed by numerous authors (Nefedova and Tarko, 1993). Pervaniuk (2001) developed a scheme dividing the World Ocean surface into 121 homogeneous basins. This discretization was based on combining basins $4^\circ \times 5^\circ$ in size. The adjacent cells of this grid with similar directions of vertical components of the water flow velocities were combined into one water basin and the average annual field of the water flow velocities was considered at a depth of 75 m. The size of water basins was determined with the dependence of CO_2 fluxes on the climatic factors taken into account and the maximum latitudinal size of the basin was assumed to be 8°. Above 80°N the World Ocean is presented as one site.

The most complete study of the physical mechanisms of the CO_2 transport for different conditions of the water–air interface was carried out by Alekseev *et al.* (1992). Here for the first time, detailed measurements of CO_2 fluxes were made, and their dependence on the parameters of the state of the atmosphere–ocean interface was analysed, with an account of the wind-driven mixing, the appearance of foam on the water surface, breaking waves, and pollution. Parametric descriptions of the processes of the ocean–atmosphere gas exchange have been proposed for each type of water–air interface condition. For instance, it was shown that with both increasing sea roughness and foam layer the intensity of gas exchange grows substantially. With the foam layer 5 cm thick the rate of gas exchange exceeds the gas exchange across a free surface by a factor of 2.4. This fact is important for the characteristic of the gas exchange in tidal zones and strong-storm regions, where foam stretches can be several cm thick. Also of interest is the result connected with availability of the active surface substances (ASS) in the upper layer of the ocean. With the ASS concentration about $7.8 \cdot 10^{-4}$% (by volume), all things being equal, the rate of gas exchange slows down to 60%. However, in the zones of foam formation this effect decreases considerably.

On the whole, over the World Ocean basins the estimates of the CO_2 flux between the atmosphere and the upper layer of the ocean vary between 16 and 1250 mol/m^2/yr). This variability means that the global model of the CO_2 biogeochemical cycle should accurately consider every feature mentioned.

In a stationary state the hydrosphere and atmosphere are in a certain equilibrium with respect to CO_2 exchange, being broken by fluctuations of temperature, ocean surface level, vertical circulation regime, etc. An amount of CO_2 assimilated and emitted in the process of exchange between the ocean and the atmosphere constitutes $55.6 \cdot 10^9$ t C/yr. Algae assimilates $16.7 \cdot 10^9$ t C/yr from the atmosphere. These values are non-uniformly distributed over the water surface and vary strongly with time. A characteristic place in the CO_2 exchange is occupied by the Arctic Ocean, whose low temperatures determine a high absolute content of CO_2 dissolved in the surface layer during any season. On average, in the summer, the arctic waters mainly assimilate CO_2 and emit O_2 to the atmosphere, and in winter, they emit CO_2 and assimilate O_2. This clearly expressed seasonal change is smoothed out in the basins of the North Atlantic seas where a decrease of ice cover, an increase of the period of photosynthetic activity, and the existing autumn–winter vertical convection leads to an intensification of the gas exchange with the atmosphere towards a prevailing of the H_3^C flux. Of the amount of CO_2 assimilated from the atmosphere, carbon constitutes $0.18 \cdot 10^9$ t C/yr. Compared to total assimilation of CO_2 by the World Ocean estimated at $5.2–6.6 \cdot 10^9$ t C/yr, the contribution of the Arctic water basins to the removal of excess carbonic acid from the atmosphere is small (Table 4.1). However, this assessment is rather underestimated, considering the results obtained by Kelley (1987). The average estimates show that the partial pressure of CO_2 in the atmosphere exceeds that in the Arctic seas by 110 ppm. The CO_2 deficit in the marine medium occurs mainly in the period of springtime blossoming and is estimated at ~450 g C/m^2. This means that the CO_2 flux from the atmosphere into the marine medium can vary from 1.5 to 4.0 g C/m^2/day. This assessment changes considerably as a function of longitude. For instance, the Norwegian Sea has a deficit of CO_2 partial pressure between 20 and 50 ppm, while for the Bering Sea $\Delta p_{CO_2} \cong 70$ ppm. Between the CO_2 content in the sea water and its temperature a clear linear correlation is observed with the proportion coefficient 10 ppm CO_2/°C.

Table 4.1. Annual budget of CO_2 exchange with the atmosphere for water bodies of the Arctic basins and North Altantic seas (10^6 t C/yr).

Region of the ocean	Summer			Winter			
	Assimilation	Emission	Balance	Assimilation	Emission	Balance	Year
Arctic basins	4.4	0.0	+4.4	–	–	–	+4.4
Arctic seas							
East Siberian Sea	1.0	2.1	−1.1	0.0	0.7	−0.7	−1.8
Kara Sea	11.4	2.4	+9.0	0.0	0.2	−0.2	+8.8
Laptev Sea	2.4	3.6	−1.2	0.0	2.9	−2.9	−4.1
Chuckchee Sea	11.2	0.0	+11.2	0.0	0.6	−0.6	+10.6
Total	*26.0*	*8.1*	*+17.9*	*0.0*	*4.4*	*−4.4*	*+13.5*
North Atlantic seas							
Barents Sea	71.9	0.0	+71.9	0.0	0.0	0.0	+71.9
Greenland Sea	19.1	0.0	+19.1	8.7	0.0	+8.7	+27.8
Norwegian Sea	64.8	0.0	+64.8	0.0	6.0	+6.0	+5.8
Total	*155.8*	*0.0*	*+155.8*	*8.7*	*6.0*	*+2.7*	*+158.5*
Arctic Ocean total	***186.2***	***8.1***	***+178.1***	***8.7***	***10.4***	***−1.7***	***+176.4***

All this suggests a conclusion that the Arctic waters pump away carbon dioxide from the atmosphere. Therefore a detailed study of this process is needed to specify the global model of carbon cycle and to raise the accuracy of estimation of the greenhouse effect due to emissions of anthropogenic CO_2.

On the whole, in the synthesis of the global model of the CO_2 biogeochemical cycle, the unit that simulates the ocean part of this cycle must describe the functioning of the ocean carbonate system. Alekseev *et al.* (1992), analysing the CO_2–HCO_3–CO_3^{2-} system and the distribution of pH values of the ocean waters, found out that more than 80% of the dissolved CO_2 is in the form of the hydrocarbonate ion of HCO_3. This means that in the synthesis of the model of the ocean carbonate system only a first stage of dissociation of carbonic acid can be reliably considered. As a result, the H_3^C flux of CO_2 dissolved in the upper layer of the ocean can be calculated with the formula:

$$H_3^C = \psi(T_L)\sqrt{P_a}/(1 + 0,5C_S)$$

where $\psi(T_L)$ is the function of the effect of temperature on CO_2 solubility and C_S is the water salinity. An expression of the ψ function has been studied inadequately. There are few experimental estimates (Alekseev *et al.*, 1992). Of course, a lot of simplified approaches to the determination of the direction of CO_2 fluxes on the atmosphere–ocean border prevail here. In the simplest case, the fluxes H_2^C and H_3^C can be considered as mutually exclusive. That is, a valve is supposed to exist on the atmosphere–ocean border. Its switching is determined by the pH parameter. A critical level of pH is about 8.11. At pH $\leq$ 8.11 the ocean assimilates CO_2, and at pH $>$ 8.11 the ocean emits CO_2.

On global scales, in the World Ocean regions with cold waters (northern latitudes, upwelling zones) the flux H_3^C prevails, and in warm waters flux H_2^C prevails. The dynamics of this regime is maintained by reactions in the hydrosphere between CO_2 and water resulting in the formation of carbonic acid:

$$HCO_3 + H_2O \Leftrightarrow H_2CO_3 + OH^-$$

$$CO_3^{2-} + 2H_2O \Leftrightarrow H_2CO_3 + 2OH^-$$

Dissolved carbon dioxide, non-dissociated molecules of H_2CO_3, anions of HCO_3, and CO_3^{2-} can be found in saltwater. They are all in equilibrium:

$$CO_2 + H_2O \Leftrightarrow H_2CO_3 \Leftrightarrow HCO_3^- + H^+ \Leftrightarrow CO_3^{2-} + 2H^+$$

An equilibrium state between carbonic acid and bicarbonate and carbonate ions establishes rapidly. The solution of CO_2 and formation of H_2CO_3 take place much more slowly. However, since many authors dispute the availability of H_2CO_3, the summarized concentration of CO_2 and H_2CO_3 should be considered, denoting it as $[CO_2]$. Let us introduce notations to characterize the total concentration of intermediate components:

$$\sum C = [CO_2] + [H_2CO_3] + [HCO_3^-] + [CO_3^{2-}] \tag{4.1}$$

The next characteristics of the hydrosphere are alkalinity:

$$A = [HCO_3^-] + 2[CO_3^{2-}] + [H_2CO_3] + [OH^-] - [H^+] \tag{4.2}$$

where $[Ca^{2+}]$ and $[H^+]$ are concentrations of ions of calcium and hydrogen, and $[HCO_3^-]$ and $[CO_3^{2-}]$ are concentrations of bicarbonate and carbonate ions.

Using this notation, let us formulate the equilibrium conditions:

$$K_0 p_c = [CO_2] \tag{4.3}$$

$$[HCO_3^-] = K_1[CO_2]/[H^+] \tag{4.4}$$

$$[CO_3^{2-}] = K_2[HCO_3^-]/[H^+] \tag{4.5}$$

$$[Ca^{2+}] = L_p[H^+]^2/(K_1 K_2[CO_2]) \tag{4.6}$$

where K_0 is the indicator of CO_2 solubility (=250 μmol/l atm at 30°C and 640 μmol/l atm at °C), L_p is the solubility indicator for $CaCO_3$, and K_1 and K_2 are first and second apparent coefficients of dissociation of carbonic acid (dependent on temperature and pressure).

From (4.1)–(4.6) we obtain:

$$\sum C = [CO_2](1 + K_1/[H^+] + K_1 K_2/[H^+]^2),$$

$$A = [CO_3^{2-}](K_1/[H^+] + 2K_1 K_2/[H^+]^2)$$

Eriksson (1963), using a differentiation operator δ, calculated relative changes of $[CO_2]$, $[Ca^{2+}]$, p_c, and A. Denote any of these components as U and taking $\Delta U = \delta U/U$, we obtain:

$$\Delta p_c = \Delta[CO_2] - \Delta K_0 \tag{4.7}$$

$$\Delta[Ca^{2+}] = \Delta L_p + 2\Delta[H^+] - \Delta K_1 - \Delta K_2 - \Delta[CO_2] \tag{4.8}$$

$$\Delta \sum C = \Delta[CO_2] + F_1\Delta K_1/F_0 + a_2\Delta K_2/F_0 - F_2\Delta[H^+]/F_0 \tag{4.9}$$

$$\Delta A = \Delta[CO_2] + \Delta K_1 + 2a_2\Delta K_2/F_2 - F_3\Delta[H^+]/F_2 \tag{4.10}$$

where

$$F_0 = 1 + a_1 + a_2 \qquad F_1 = a_1 + a_2 \qquad F_2 = a_1 + 2a_2$$

$$F_3 = a_1 + 4a_2 \qquad a_1 = K_1/[H^+] \qquad a_2 = K_1K_2/[H^+]^2$$

Based on relationships (4.7) through (4.10), and considering the experimental data, Eriksson (1963) showed that with a 1°C increase in the temperature of the upper layer of the marine medium the partial pressure of CO_2 in the atmosphere can grow by 4.2–5.8%, and with a 1% decrease of the water volume the partial pressure of CO_2 in the atmosphere grows by 3%, with 1% of carbon dioxide in the water medium precipitating as $CaCO_3$. With variations of the equilibrium state in deep layers of the hydrosphere a 1% output of CO_2 in the form of gas is followed by a 1% sedimentation of CO_2 in the form of $CaCO_3$. Moreover, variations in the CO_2 partial pressure correlate with changes of the concentration of phosphorus P, so that a decrease of P in deep layers by 1% leads to an increase of p_c by 5.6%.

Eriksson (1963) found out that a 1% increase of alkalinity causes a 2.26% decrease of the CO_2 partial pressure in the atmosphere and a 98% reduction of total CO_2 supply in the hydrosphere. In conditions lacking an external input of Ca into the hydrosphere, a 1% increase of alkalinity causes an increase in the rate of $CaCO_3$ deposition, that is, the total alkalinity decreases by 0.92%. Thus a 1% increase of water alkalinity is equivalent to an increase of pH by 0.5.

An equilibrium between various components of the hydrospheric carbonate system depends on temperature and pressure, a combination of which correlates with pH so that at given temperature and pressure the equilibrium is a function of only $pH = -\log[H^+]$. The effect of temperature on pH in the first approximation can be described by the Merrey law (Ivanov, 1978): $\Delta pH = -0.0111\,\Delta T$, valid at $pH \in [7.5, 8.4]$, $T \in [1 \div 30]$°C, and salinity from 10 to 40‰. The dependence of pH from pressure p_c follows the Buch–Grippenberg law: $pH = d\Delta p_c$, where, on average, $d = -0.0254$. A more accurate presentation of this law is given in Table 4.2.

The connection of the equilibrium condition by CO_2 exchange with pH on the atmosphere–ocean border is such that with the CO_2 pressure in the atmosphere of $330 \cdot 10^{-6}$ atm., an equilibrium occurs at 20°C for pH = 8.16 and at 0°C for pH = 8.11. With a lower pH value the ocean will assimilate CO_2, and with a higher pH it will emit CO_2. Hence, to describe the functions of the fluxes H_2^C and H_3^C on the atmosphere–ocean border, the structure of the ocean carbonate system

Table 4.2. Empirical dependence of pH on atmospheric pressure.

pH at atmospheric pressure	7.5	7.7	7.9	8.1	8.3
ΔpH with pressure increasing by 1000 dbar	−0.035	−0.028	−0.023	−0.021	−0.02

should be thoroughly studied. A simplified description of these fluxes is usually based on the comparison of the partial pressures of CO_2 in the atmosphere and in the ocean. According to the data in Alekseev *et al.* (1992), the fluxes H_2^C and H_3^C are well approximated by the function $H_i^C = k_i(p_a - p_c)^{1/2}$, where p_a and p_c are partial pressures of CO_2 in the atmosphere and in the ocean, respectively. The partial pressure of CO_2 in the atmosphere at the level of the ocean can be calculated with the formula:

$$p_a = 0.421542\ 10^{-18} M_C(273.15 + T)$$

where M_C is the CO_2 mass in tonnes, and T is the air temperature in °C.

According to Bjorkstrom (1979), the functional dependence of p_c on the parameters of the ocean carbonate system can be presented in the form $p_c = [CO_2]/K_0$. From the condition of chemical equilibrium, according to (4.4) and (4.5), it follows:

$$[CO_2] = [H^+] \sum C/a,$$

where $a = [H^+]^2 + [H^+]K_1 + K_1K_2$, and $\sum C = C_U/W_U$, where W_U is the volume of an elementary reservoir.

Denote:

$$K_W = [H^+][OH^-]$$
$$K_B = [H^+][B(OH)_4^-]/[B(OH)_3]$$
$$B_T = [B(OH)_3] + [B(OH)_4^-]$$

Then we obtain from (4.2):

$$A = K_W[H^+]^{-1} + K_B B_T (K_B + [H^+])^{-1} + C_U([H^+]K_1 + 2K_1K_2)/(aW_U) - [H^+]$$

or

$$A = [CO_2]\left(\frac{K_1}{[H^+]} + \frac{2K_1K_2}{[H^+]^2}\right) + \frac{B_T}{1 + [H^+]/K_B} + \frac{K_W}{[H^+]} - [H^+] \tag{4.11}$$

The solution of this equation relative to $[H^+]$ enables one to determine p_c as a function of C_U. From the estimates by Bjorkstrom (1979), $K_B = 2 \cdot 10^{-9}$, and $K_W = 10^{-14}$.

The right-hand part of equations (4.2) and (4.11) imply an account of all weak acids in the salt water. Ions of other compounds are taken into account through the dependence of the equilibrium constants on salinity or chlorine content. The characteristic parameter of equilibrium of the carbonate system is a variable:

$$X = \frac{(K_1 K_2)^{1/2}}{[H^+]} = \left[\frac{[CO_3^{2-}]}{[CO_2]}\right]^{1/2}$$

In the terms of this parameter we have:

$$\sum C = (1 + KX + X^2)[CO_2] \qquad (4.12)$$

$$A = (KX + 2X^2)[CO_2] + W(X) \qquad (4.13)$$

where $K = \sqrt{K_1/K_2}$

$$W(X) = B_T/\{1 + [K_1K_2]^{1/2}X^{-1}/K_B + K_W X[K_1K_2]^{-1/2} - [K_1K_2]^{1/2}X^{-1}$$

Exclude $[CO_2]$ from (4.12) and (4.13):

$$(2C - A')X^2 - K(A' - C)X - A' = 0 \qquad (4.14)$$

where $A' = A - W(X)$.

Since $W(X) \ll A$, to solve the equation (4.14) the following iterative procedure can be used beginning with an estimation of $X = X_1(\cong 1)$:

$$A_i' = A - W(X_i) \qquad X_i = 0.5[-b + (b^2 - 4ac)^{1/2}]/a \quad (i \geq 2)$$

where $a = 2C - A_{i-1}'$, and $b = K(A_{i-1}' - C)$.

The exchange processes taking place on the atmosphere–ocean border have been experimentally studied by Kiseleva (1990), Zaitsev (1988), and others. It has been shown in their works that at high wind speeds the rate of gas exchange sharply increases. This is connected with the mechanism of foam formation on wave crests as well as an intensive catching of air bubbles and their penetration beneath the water surface. The dependence of the amount of spray Q on the height over the water surface is well approximated by a linear function. For instance, for the wind speed $V = 11.1$ m/s this approximation is (Kiseleva, 1990):

$$Q = \begin{cases} Q' \exp\{-14,5z + 0,684\} & \text{without wave breaking} \\ Q' \exp\{-7,48z + 0,842\} & \text{with wave breaking} \end{cases}$$

where $Q'[1/(cm^2c)]$ is the dimensional multiplier.

Experimental estimates of the net gas transport due to the droplet mechanism have shown that the CO_2 flux between water and air can vary from 1.44 mg $CO_2/m^2/$hr in the absence of wave breaking to 26.6 mg $CO_2/m^2/$hr. The rate of the gas exchange on the atmosphere–ocean border is markedly affected by the foam formation on the ocean surface. In this case the rate of the gas exchange can increase up to 28%. All this suggests the conclusion that a detailed modelling of the mechanisms of the atmosphere–water CO_2 transport is needed. Satellite World Ocean monitoring allows us to identify the zones of foam formation and, hence, along with the modelling of the processes of foam formation, to perform direct measurements of the areas of the foam-covered basins.

The mechanism of the air bubbles behaviour in the upper layer of the ocean has been poorly studied. The available theoretical results are based, as a rule, on a number of suppositions, which in many cases can drastically distort the ideas about real processes of the gas exchange between the atmosphere and the ocean. Among these suppositions the following have been used most often:

- air bubbles in the upper layer of the ocean almost instantly acquire the temperature of the environment, since the molecular coefficient of heat conductivity exceeds that of gases by 2–3 orders of magnitude;
- air bubbles in water do not affect its dynamics; and
- air bubbles do not merge in water.

The size of air bubbles in water varies constantly because of their gradual solution. The gas flux through the bubbles is described by the relationship:

$$Q = 4\pi RD|\Delta p|S \cdot Nu$$

where $|\Delta p|$ is the difference of the partial pressures of gas in the water and in the atmosphere, R is the bubble radius and S is the area of the bubble surface:

$$Nu = \begin{cases} 1 + 0,5Pe + 0,5Pe^2 \ln(2Pe) & \text{for } R \leq 20\,\mu\text{m} \\ \sqrt{2/3\pi}Pe^{1/3} & \text{for } 20\,\mu\text{m} < R \leq 200\,\mu\text{m} \\ 0,45Pe^{1/3}Re^{1/3} & \text{for } 200\,\mu\text{m} < R \leq 400\,\mu\text{m} \end{cases}$$

Re is the Reynolds number ($Re = 2vR/D$), Pe is the Pekle number ($Pe = vR/\nu$), ν is water viscosity, v is the velocity of the bubble's motion, and D is the coefficient of the gas molecular diffusion.

Of course, the ocean surface condition substantially affects its gas exchange with the atmosphere. The size of the basins covered with either foam or white caps depends directly on a combination of the parameters such as wind speed, water temperature, and sea currents. An analysis of the statistical characteristics of the patchy pattern of the ocean surface made by many experts makes it possible to describe the percentage distribution of the areas covered with foam (S_f) and white caps (S_l) with the following binary functions of wind speed V (at a height of 10 m):

$$S_f = \begin{cases} 0 & \text{for } V < 5\,\text{m/s} \\ 0,65\{1 + 4,76 \cdot 10^{-2}(V - 5^2)\} & \text{for } V \geq 5\,\text{m/s} \end{cases}$$

$$S_l = \begin{cases} 0 & \text{for } V < 5\,\text{m/s} \\ 0,015\{1 + 2,2 \cdot 10^{-2}(V - 5^3)\} & \text{for } V \geq 5\,\text{m/s} \end{cases}$$

According to Kiseleva (1990), the relationship between S_f and S_l at $V \geq 5$ m/s obeys the following rule:

$$S_f/S_l = 50\text{–}3,4(V - 5)$$

It is clear that the structure of the atmosphere–ocean border can affect the gas exchange within widely varying values of fluxes H_2^C and H_3^C (see Figure 2.8). Unfortunately, the level of knowledge of the laws of changes of the atmosphere–ocean border structure as a function of synoptic situations does not permit a clear estimation of the limits of detailing the processes taking place on this border, to obtain the values of the fluxes H_2^C and H_3^C close to real ones. It means that at this stage of the synthesis of the global model of the CO_2 cycle some uncertainty still exists.

4.3 THE ZONAL MODEL OF THE CARBON CYCLE IN THE ATMOSPHERE–OCEAN SYSTEM

Consider one of the versions of parameterization of the carbon dioxide global dynamics in the atmosphere–ocean system proposed and studied by Nefedova (1994). The spatial heterogeneity of the World Ocean is approximated by the zonal scheme in accordance with the latitudinal dependence of the climatic processes as well as processes of mixing in the atmosphere and in the oceans. There are 14 latitudinal zones 10° in size. In the vertical, there are three layers in the World Ocean: the upper quasi-homogeneous layer (UQL) with a time-dependent thickness, thermocline, and deep ocean. As a result, the World Ocean is divided into 42 volume parts. An upwelling is assumed to exist between 40°N and 40°S along with a downwelling in high latitudes. In the UQL the water flows poleward from the equator, and in deep layers the water flows in the opposite direction (Figure 2.5). Over each water basin, the atmosphere is simulated by a point model. The carbon exchange between the zones of the atmosphere takes place due to advection H_i^a and turbulent diffusion H_i^d:

$$H_i = H_i^a + H_i^d, \qquad (i = \overline{1,14})$$

$$H_i^a = 2\pi R_0 h_a V_i^* (C_i^a - C_{i+1}^a) \cos\phi_i$$

$$H_i^d = [2\pi R_0 A_h h_a / \Delta\phi_i](C_i^a - C_{i+1}^a) \cos\phi_i$$

where $C_i^a = M_i^a / V_i^a$ is the concentration of carbon in the ith zone of the atmosphere, M_i^a is the carbon mass in the ith zone of the atmosphere, V_i^a is the volume of the ith zone of the atmosphere, R_0 is the Earth's radius, ϕ_i is the latitude of the southern boundary of the ith zone, h_a is the altitude of the atmosphere (10 km), V_i^* is the average velocity of the meridian transport of air masses in the atmosphere (0.2–1.0 m/s), and A_h is the coefficient (10^5 m^2/s).

The CO_2 exchange on the atmosphere–ocean border is described by the traditional law:

$$H_i^{a0} = k(u_i)(P_i^a - P_i^o)$$

where

$$P_i^a = k_a M_i^a R T_i^a S_i^{-1} \mu^{-1}$$

$k(u_i)$ is the proportion coefficient depending on wind speed, P_i^a and P_i^o are partial pressures of CO_2 in the ith zone of the atmosphere and the ocean, respectively, k_a is the share of the 100-m air column mass in the mass of a 10-km column (≈0,01602), R is the universal gas constant (8.31451 Joule/(mol · K) , T_i^a is the air temperature at the level of the ocean in the ith zone ($T_i^a = T_i^{as} + \Delta T_i^a$), T_i^{as} is the seasonal temperature component, and ΔT_i^a is the average annual change of air temperature caused by the increased CO_2 content in the atmosphere.

A sea surface temperature (SST) change is assumed to take place in phase with a change of air temperature by the same value:

$$T_i^o = T_i^{os} + \Delta T_i^a$$

The partial pressure of CO_2 dissolved in the surface waters is proportional to its concentration in the water and inversely proportional to its solubility. This dependence is established by solving the system of equations (4.12) and (4.13) that describe the functioning of the ocean carbonate system. For the quantitative solution of this system one can use, for instance, the methods of secants. As a result, we obtain $[CO_2]$ and P_i^o. Based on the data on the temperature dependence of the equilibrium constants for the respective chemical reactions, we find:

$$P_i^o = P_i^o(C_i^1, T_i^o)(i = \overline{1, 14})$$

The turbulent fluxes of carbon at the lower boundary of the UQL (H_{UPL}) and the upper boundary of thermocline (H_T) are described by the functions:

$$H_{UPL} = \begin{cases} (C_i^1 - C_i^2)(V_i + dh_i/dt) & \text{for } dh_i/dt + V_i > 0 \\ 0 & \text{for } dh_i/dt + V_i \leq 0 \end{cases}$$

$$H_T = \begin{cases} (C_i^1 - C_i^2)(V_i + dh_i/dt) & \text{for } dh_i/dt + V_i \leq 0 \\ 0 & \text{for } dh_i/dt + V_i > 0 \end{cases}$$

where C_i^1 and C_i^2 are concentrations of inorganic carbon in the UQL and thermocline, respectively, h_i is the UQL depth of the ith zone of the ocean (depends on the season), V_i is the water lifting (in upwelling) or lowering (in downwelling) (0.001–0.1 cm/s).

The turbulent fluxes of carbon on the thermocline–deep ocean border is considered to be proportional to the coefficient k_T of the difference of carbon concentrations in the bordering layers:

$$H_{TF} = k_T(C_i^3 - C_i^2)(i = \overline{1, 14})$$

where C_i^3 is the concentration of inorganic carbon in the deep ocean of the ith zone.

Production and circulation of organic carbon in the ocean has been studied in detail and parameterized by many authors (Krapivin, 1996; Nitu *et al.*, 2000 a,b; Krapivin and Kondratyev, 2002). In the model by Nefedova and Tarko (1993) the following approximations have been assumed:

$$B_i^p = B_i^o\{1 + (\delta_i - 1)/(\delta_i + 1)\sin(wt - \pi/2)\}$$
$$\text{for } \varphi \in [60°\text{N}, 90°\text{N}] \cup [60°\text{S}, 90°\text{S}] \quad (i = 1, 14)$$

$$B_i^p = B_i^o\{1 + (\delta_i - 1)/(\delta_i + 1)\sin(2wt - \pi/2)$$
$$\text{for } \varphi \in [10°\text{N}, 60°\text{N}] \cup [10°\text{S}, 60°\text{S}] \quad (i = 2\text{–}6, 9\text{–}13)$$

$$B_i^p = B_i^o \quad \text{for } \varphi \in [0°, 10°\text{N}] \cup [0°, 10°\text{S}] \quad (i = 7, 8)$$

where B_i^o is the average annual production of phytoplankton ($150–550 \cdot 10^9$ t/yr), and δ_i is the ratio of maximum and minimum values of the rate of organic matter production.

Organic matter produced in the UQL decomposes and descends to deeper layers. Let B_i^{1d}, B_i^{2d} and B_i^{3d} be the rates of organic matter decomposition in the UQL, thermocline, and deep ocean, respectively. Let γ be the ratio of the average annual amount of organic matter decomposing in the UQL to the average annual production of phytoplankton, let τ be the lag time of the organic matter decomposition, D the average depth of the ocean (3653–3795 m), and H the depth of the thermocline bottom (30–200 m). Then:

$$B_i^{1d} = \gamma B_i^p(t - \tau) \qquad (i = \overline{1, 14})$$

$$B_i^{2d} = (1 - \gamma)(H - h_i)/(D - h_i)\langle B_i^{1p}\rangle$$

$$B_i^{3d} = (1 - \gamma)(D - H)/(D - h_i)\langle B_i^{1p}\rangle$$

Here the angular brackets $\langle\,\rangle$ denote the calculation of the average value for the annual cycle.

So, the seasonal model of the global carbon cycle developed in the works by Nefedova and Tarko (1993) and Nefedova (1994) is simulated by a system of 56 ordinary differential equations with periodic coefficients ($i = 1, \ldots, 14$):

$$\frac{d}{dt}C_i^1 h_i - \left(\frac{dh_i}{dt} + V_i\right)C_i^1 = H_i^{ao} - H_{UPL} + L_i^1 - B_i^p + B_i^{1d}$$

$$\frac{d}{dt}C_i^2(H - h_i) + \left(\frac{dh_i}{dt} + V_i\right)C_i^2 = H_T + H_{TF} + B_i^{2d}$$

$$\frac{d}{dt}C_i^3(D - H) + V_i C_i^3 = L_i^3 - H_{TF} + B_i^{3d}$$

$$\frac{dM_i^a}{dt} = \gamma_i^a + V_i^a(H_i - H_{i-1}) - 0,012 S_i H_i^{ao}$$

where γ_i^a are the anthropogenic carbon emissions in the ith zone of the atmosphere.

4.4 THE 3-D MODEL OF THE CARBON CYCLE IN THE WORLD OCEAN

Along with the physicochemical processes of transformation of carbon compounds in the hydrosphere mentioned above, the general circulation in the World Ocean plays an important role in their long-range transport. According to the available estimates, about 80% (~1.7 Gt) of organic matter formed in the process of photosynthesis descends to deeper layers and oxidizes giving CO_2. This process is balanced by a slow lifting of water, and thereby a situation is formed for a stable CO_2 exchange. However, in local situations there are strong deviations from such stable conditions, which can only be described by the scheme of spatial–temporal structure of the ocean waters motion. The block scheme of this structure identifies the surface, intermediate, deep, and bottom waters as well as the currents topography on every horizon. In the scientific literature the simulation models of the World Ocean circulation vary widely, therefore the development of this part of the global model unit is not difficult.

Following Nitu *et al.* (2000b), by depth z the ocean can be divided into four basic layers: photic to well-heated depths ($\Omega_U = U[z_i, z_{i+1}], z_0 = 0; i = 0, 1, \ldots, m-1$); intermediate ($\Omega_P = U[z_i, z_{i+1}], i = m, \ldots, m+n-1$); deep ($\Omega_L = U[z_i, z_{i+1}]$, $i = m+n, \ldots, m+n+l-1$), and bottom Ω_F. Considering the hydro–physico–ecological characteristics the layer Ω_U as a function of latitude φ, longitude λ, and season t can be attributed to warm or cold waters, the layer Ω_P is photic but always with low water temperatures, and in layer Ω_L phytoplankton is not produced. Finally, the layer Ω_F plays the role of a boundary layer.

On average, the layer Ω_U has the area $360 \cdot 10^{12}\,\text{m}^2$ and depth $z_m \cong 75\,\text{m}$, the warm waters covering an area $240 \cdot 10^{12}\,\text{m}^2$. The layer Ω_P is located at depths z_m to $z_m + n \cong 200\,\text{m}$. In concrete model realizations given in Krapivin and Kondratyev (2002) the following assumptions have been made: $m = 10$, $n = 2$, and $l = 8$.

The vertical CO_2 transport in the ocean is determined by advective fluxes $H^C_{19,ij}$ and gravitational sedimentation of dead organic matter (the flux $H^C_{20,ij}$). An advective transport from the ith into the jth reservoir of the ocean is considered proportional to the concentration of carbon in the respective reservoirs: $H^C_{19,ij} = \lambda_{2,ij} C_{a,i} (a = U, P, L)$, where $\lambda_{2,ij} = V_{ij}/V_i$, V_{ij} is the water volume transported per unit time from the ith reservoir into the jth reservoir, and V_i is the volume of the ith reservoir.

The following algorithm is widely used for a parameterization of the process of carbon sedimentation. The flux under the unit area of the ocean is supposed to decrease exponentially with depth. If we denote the inflow of organic matter in the ith reservoir as g_i and the net outflow of organic matter from the water surface as H_{20T}, we obtain:

$$H^C_{20,1} = H_{20,T} \qquad H^C_{20,i} = g_{i-1}(\sigma_i/\sigma_{i-1}) \exp[-(z_i - z_{i-1})/D_s] \quad (i = 2, \ldots, m+n+l)$$

where σ_i is the area of the ith reservoir, and D_s is the characteristic time of the organic matter particles sedimentation before their decomposition. The rate of decomposition in each reservoir is equal to:

$$R_{D,i} = H^C_{20,i} - H^C_{20,i+1} \quad (i = 1, \ldots, m+n+l) \qquad R_{D,F} = H^C_{20,m+n+l} - H^C_{16}$$

However, if the time of the transition of the organic matter particles from one layer to another is short compared to D_s, then it is better to take $H^C_{20,i} = \lambda_1 C_{a,i}$ and $H^C_{16,i} = \lambda_4 C_{F,i}$. In addition to these fluxes one should take into account the fluxes of detritus decomposition, solution of bottom sediments, and carbon consumption in the process of photosynthesis:

$$H^C_{17,i} = \text{const} \qquad H^C_{18,i} = \lambda_3 D_{L,i} \qquad H^C_{22,i} = \lambda_3 D_{U,i} \qquad H^C_{21,i} = C_{31} R_{\phi,i}$$

The estimates for the modelled parameters of the particular oceanic processes of carbon cycle scatter widely. For instance, from the data of various authors the estimates of assimilation of carbon from the hydrosphere via the process of photosynthesis range from 10 to 155 Gt C/yr. The value 127.8 Gt C/yr is most widely used. However, because of large dispersions of these estimates the calculation of the C_{31} coefficient is fraught with great uncertainty and therefore to be specified, it requires numerical experiments with the use of other, more accurate data.

Finally, suppose that the surface layers of the ocean are filled with carbon due to its run-off from the land $H_{24,i}^C = C_7 W_{sio}$ and $H_{23,i}^C = C_8 W_{Hio}$, where W_{sio} and W_{Hio} are the river and underground run-offs into the World Ocean, respectively.

Considering the notations in Figure 2.8, the balance equations to describe the global carbon cycle are written as:

$$\frac{\partial C_A}{\partial t} = H_5^C - V_\varphi \frac{\partial C_A}{\partial \varphi} - V_\lambda \frac{\partial C_A}{\partial \lambda} + \begin{cases} -H_2^C + H_3^C, & (\varphi, \lambda) \in \Omega_O \\ H_1^C + H_4^C - H_6^C + \sum_{i=7}^{11} H_i^C & (\varphi, \lambda) \in \Omega \backslash \Omega_O \end{cases}$$

$$\frac{\partial C_{S1}}{\partial t} = H_6^C - H_7^C - H_8^C - H_{14}^C - H_{15}^C$$

$$\frac{\partial C_{S2}}{\partial t} = \sum_{i=1}^{4} H_{i+11}^C - H_4^C - H_9^C - H_{23}^C - H_{24}^C$$

$$\frac{\partial C_U}{\partial t} + \nu_\varphi^U \frac{\partial C_U}{\partial \varphi} + \nu_\lambda^U \frac{\partial C_U}{\partial \lambda} = Q_U + H_{22}^C + H_{25}^C - H_{21,U}^C - H_3^C + H_2^C + H_{19,U}^C - H_{20,U}^C$$

$$\frac{\partial C_P}{\partial t} + \nu_\varphi^P \frac{\partial C_P}{\partial \varphi} + \nu_\lambda^P \frac{\partial C_P}{\partial \lambda} = H_{19,P}^C - H_{19,U}^C + H_{20,U}^C - H_{20,P}^C - H_{21,P}^C + H_{22}^C + H_{25}^C$$

$$\frac{\partial C_L}{\partial t} + \nu_\varphi^L \frac{\partial C_L}{\partial \varphi} + \nu_\lambda^L \frac{\partial C_L}{\partial \lambda} = H_{18,L}^C + H_{19,L}^C - H_{19,P}^C + H_{20,P}^C - H_{20,L}^C$$

$$\frac{\partial C_F}{\partial t} + \nu_\varphi^F \frac{\partial C_F}{\partial \varphi} + \nu_\lambda^F \frac{\partial C_F}{\partial \lambda} = H_{17}^C - H_{16}^C + H_{18,F}^C - H_{19,L}^C + H_{20,L}^C$$

The flux Q_U is formed from H_{23}^C and H_{24}^C. Let $Q_U = 0$ for the pelagic regions Ω_{OP} of the World Ocean. Describe the formation of Q_U on the shelf by a simple algorithm with the supposed uniform distribution of the sink from the Kth region to the Mth water basin:

$$Q_U = \begin{cases} 0, & (\varphi, \lambda) \in \Omega_{OP} \\ (H_{23}^C + H_{24}^C)\sigma_{OP}/\sigma_L, & (\varphi, \lambda) \in \Omega_O \backslash \Omega_{OP} \end{cases}$$

where σ_{OP} and σ_L are the areas of the water basins Ω_{OP} and Ω_L, respectively.

4.5 MODELLING THE ORGANIC CARBON CYCLE IN THE WORLD OCEAN

4.5.1 General principles of deriving the models of the ocean ecosystems

Each element of the ocean ecosystem A can be described by some number of parameters, and the connection between the elements can be presented as the connection between the respective parameters. Then, on the whole, the ecosystem A can be described by N time-dependent parameters $x^*(t) = \{x_j(t), j = 1, \ldots, N\}$. The structure $|A(t)|$ and behaviour $\bar{A}$(t) of the ecosystem A which can be observed in more or less detail, are the functions of these parameters. Therefore the ecosystem itself $A(t) = \{|A(t)|, \bar{A}(t)\}$, as a collection of the structure and behaviour, is the function of these parameters:

$$A(t) = F(x^*(t)) \tag{4.15}$$

Hence, according to (4.15) with a change of the time t the ecosystem A will be characterized by the trajectory in the $(N+1)$-dimensional Eucledian space. The following abstract formation will be considered a model of the ecosystem $A(t)$:

$$A_M(t) = F_M(x^{*,M}) \tag{4.16}$$

that depends on $M \leq N$ components of the vector $x^*(t)(\{x^{*,M}(t)\} \in \{x^*(t)\})$ and considers all the connections existing between them.

The nearer M is to N and the more completely the connections between the vector $x^{*,M}(t)$ components are taken into account, the less is the disagreement of the trajectories of the ecosystem $A(t)$ and its model $A_M(t)$. The latter can be measured by any natural measure, for instance, by a maximum of absolute difference of all the respective coordinates of trajectories or by the integral of absolute difference of all respective coordinates for a final time period. In other words, introduce some goal functional:

$$V = Q(\{x_i(t)\}) \tag{4.17}$$

On the trajectory of the ecosystem $A(t)$. The form of the Q function is determined by the requirements made by the system A to the environment. The natural evolutionary process is considered to lead to an optimal system and therefore the system $A_{\text{opt},M}(t)$, which ensures an extremum of the functional (4.17), can be considered an optimal model of the ecosystem A.

The value of disagreement between the trajectories of the ecosystem A and an optimal model $A_{\text{opt},M}(t)$ is affected by the degree of correspondence of the goal functional V chosen by the relationship (4.17) to a real goal $\underline{A}$ of the ecosystem A. The set $\underline{G} = \underline{A}_1, \ldots, \underline{A}_r, \ldots, \underline{A}_m\}$ of possible reliable goals $\{\underline{A}_r\}$ of the ecosystem can be derived on the basis of the experience accumulated by oceanographers. Then, determining:

$$A_{\text{opt},M,r} = g_r(\underline{A}_r) \quad \underline{A}_r \in \underline{G}$$

we obtain a limited assembly of possible optimal systems $A_{\text{opt},M,r}(r = 1, \ldots, m)$, whose trajectories together with the trajectory of the ecosystem A, are within the

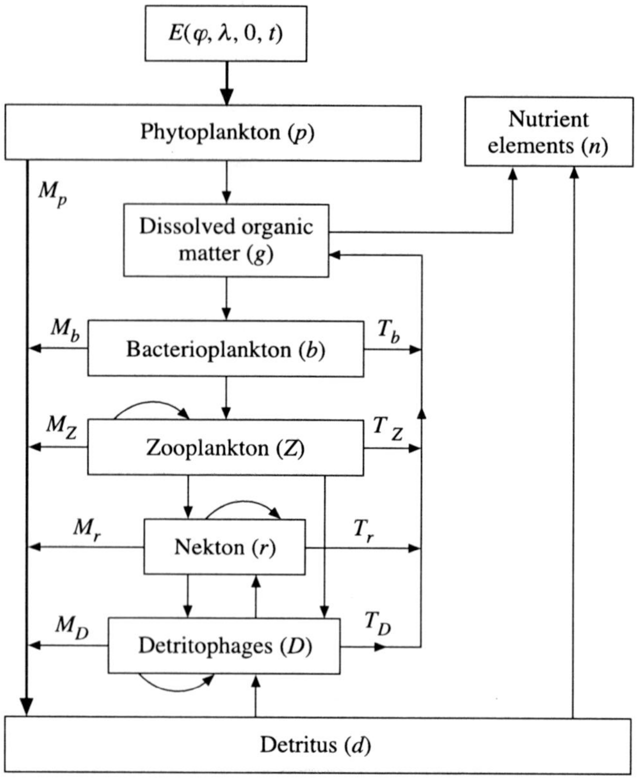

Figure 4.1. The standard block-scheme of the model of the ocean ecosystem. The arrows show the energy flux directions.

space of possible trajectories. If we find A_{opt,M,r_0}, whose trajectory has a minimum disagreement with $A(t)$, then, based on this, we can find the most reliable goal of the ecosystem:

$$A_{r_0} = (g_r)^{-1}(A_{\text{opt},M,r_0}(t))$$

According to the principles mentioned above, the derivation of a numerical model of the ocean ecosystem A (Figure 4.1) requires either a detailed description of its states or derivation of an adequate complex of numerical models of energy exchange between the trophic levels, taking place in A, as well as interactions of the biotic, abiotic, and hydrophysical factors. Of course, in this case an availability of a certain set of hypothesis is assumed concerning the character of the balance relationships in the ecosystem A.

The basic hypothesis is that in ecosystem A the only original source of energy and matter for all forms of life is the solar radiation energy (E). According to numerous theoretical and experimental studies, the sunlight penetration into deep layers of the ocean follows an exponential law:

$$E = uE_0 \exp\left[-\int_0^z \{\delta p(\varphi, \lambda, x, t) + \beta d(\varphi, \lambda, x, t) + \nu Z(\varphi, \lambda, x, t)\}\, dx - \alpha z\right]$$
$$+ (1-u)E_0 \exp(-\zeta z) \tag{4.18}$$

where $E_0 = E(\varphi, \lambda, 0, t)$ is the illumination of the ocean surface, α is the coefficient of light absorption by the filtered sea water, δ, β, and ν are the coefficients of light attenuation due to phytoplankton (p), detritus (d), and zooplankton (Z), respectively, and u and ζ are the parameters chosen in a concrete situation to bring $E(\varphi, \lambda, z, t)$ closer to a real pattern of illumination changing with depth. Note that here the impact of the biomass of other trophic levels on water transparency is considered to be negligibly small.

Illumination affects the rate of photosynthesis R_p. The R_p parameter as a function of E has a maximum at some optimal value of E_{max}, which drifts from this critical value when the illumination increases or decreases. A maximum of R_p at various latitudes φ is located at depths varying as a function of season (i.e., sun elevation). Thus in the tropical zones this variability with depth is most pronounced. On average, the photosynthesis maximum is located at depths of 10–30 m, and in the open water bodies it can be observed at depths below 30 m. Here $E_{max} = 65$–85 cal/cm^2/day. Beginning with the depths where $E = 20$–25 cal/cm^2/day, photosynthesis decreases in proportion to E. An apparent suppression of phytoplankton by light is observed at $E > 100$ cal/cm^2/day. These estimates are quite different in the northern latitudes, where the photosynthesis maximum is located, as a rule, on the surface.

The rate of photosynthesis at depth z depends on the water temperature T_W, the concentration of nutrient salts n, and the phytoplankton biomass p, as well as on other factors, which are not considered here. To express this dependence, various equations are used, which in any of its forms reflect the limiting role of the elements E, n, and p. Considering that $\partial p/p\partial z \to 0$ at $n \to 0$ and $\partial p/p\partial z \to$ const with increasing n, take the following function as the basic one to describe the photosynthesis intensity at a depth z:

$$R_p(\varphi, \lambda, z, t) = k_0(T_W)K_T f_2(p) f_3(n) \tag{4.19}$$

where

$$\left.\begin{array}{lll} K_T = Af_1(E) & A = kA_{max}/E_{max} & f_1(E) = E \cdot \exp[m(1 - E/E_{max})] \\ f_2(p) = [1 - \exp\{-\gamma_1 p\}] & f_3(n) = [1 - \exp\{-\gamma_2 n\}]^\theta & \end{array}\right\} \tag{4.20}$$

where k is the proportion coefficient, $k_0(T_W)$ is the function characterizing the dependence of the rate of photosynthesis on the water temperature T_W, A_{max} is an assimilation number in the region of maximum photosynthesis (an increment per unit weight of the phytoplankton organisms), and γ_1, γ_2, θ, and m are constants, the choice of which can determine the species characteristics of the phytoplankton elements. For A_{max} the following estimates are valid: $A_{max} = 5.94\, E_{max}$ in the region of the photosynthesis minimum and $2.69\, E_{max}$ for other regions. According to these estimates, an assimilation number of the tropical phytoplankton in the region of the maximum photosynthesis averages 11–12 mg C/hr. Thus for the Peru upwelling $A_{max} = 6.25$ mg C/hr. The light saturation of photosynthesis in the equatorial regions is reached at 9 cal/cm^2/day.

As for the dependence of $k_0(T_W)$, the specific intensity of the phytoplankton photosynthesis first increases with the temperature changing from low to high values, reaching at some interval of temperatures a maximum value, optimal for p, and then, with the temperature increasing further begins to decrease. Near the maximum the following approximation is often used:

$$k_0(T_W) = \exp\{(T_W - T_{W,\mathrm{opt}}) \ln(\theta_0)\} \quad 0 < \theta_0 \leq 2$$

The dependence of the rate of photosynthesis on the concentration of nutrient elements $n(\varphi, \lambda, z, t)$ (phosphorus, silicon, nitrogen, and others), expressed in the formula (4.19) by the exponential term, is, of course, more complicated. The nutrient element are one of the most important parts of the ecosystem, since it is they that regulate the energy flux in the ecosystem. The nutrient element supplies are spent in the process of photosynthesis at a rate R_n, usually approximated by the expression $R_n = \delta R_p$, where δ is the proportion coefficient. The nutrient elements supplies are replenished due to their sweeping out with the lifting of deep waters, where their supplies are formed as a result of the chemical processes of the dead organic matter decomposition. This process is controlled by several abiotic conditions characteristic of various climatic zones of the World Ocean. The vertical flux of nutrient elements is determined by conditions of water mixing. In the tropical zones, where the vertical structure of the water has a clear 3-layer configuration with a layer where the temperature leaps suddenly (the thermocline), the vertical motion of nutrient elements is confined to this layer. In the water bodies where the thermocline is located at depths 40–100 m, the upper layer is usually poor in nutrient elements, and their input into this layer takes place only in the zones of upwelling. In this case the average rate of vertical water lifting beneath the thermocline varies between 10^{-3} to 10^{-2} cm/s, and in the zones of upwelling (breakthrough of the thermocline) it can reach 0.1 cm/s.

4.5.2 Equations of the World Ocean ecosystems dynamics

The whole volume of the ocean waters is considered as a single biocoenosis, in which the flux of organic matter produced in the surface layers and then descending to maximum depths of the ocean is the main connecting factor. All the model parameters are assumed to be able to change as functions of place and time, and their parametric description is made by average characteristics, that is, deterministic models.

Suppose the food bonds between the trophic levels are described by the Ivlev model, that is, the consumption of various kinds of food by the ith trophic level is proportional to their efficient biomasses. Taking into account the diagram of food bonds assumed in Figure 4.1 and the structure of the trophic pyramid of the typical ocean ecosystem, let us consider each trophic level in detail.

Bacterioplankton b plays an important role in the trophic chains of the ocean. By available estimates, not less than 30% of the bacterioplankton mass are in natural aggregates exceeding 3–5 μm in size, therefore the filtrates can feed on this biomass. This fact must be taken into account when deriving a model of the ecosystem, since in many regions of the World Ocean the bacterioplankton production is comparable with the production of phytoplankton. Bacterioplankton, taking a special place in the trophic pyramid, differ by a variable exchange, strongly decreasing with the deficit of food, which is followed by the respective decrease of the rate of their growth. The food for bacterioplankton is mainly detritus d and dissolved organic matter g emitted by phytoplankton. As a result, the ration for bacterioplankton can be described by the Ivlev formula:

$$R_b = k_b b[1 - \exp(-k_{1,d} d - k_{1,g} g) \tag{4.21}$$

where k_b, $k_{1,d}$, and $k_{1,g}$ are the coefficients determined experimentally.

The equation describing the dynamics of the bacterioplankton biomass is written in the form:

$$\frac{\partial b}{\partial t} + V_\varphi \frac{\partial b}{\partial \varphi} + V_\lambda \frac{\partial b}{\partial \lambda} + V_z \frac{\partial b}{\partial z} = R_b - T_b - M_b - \sum_{s \in \Gamma_b} C_{bs} R_s + k_{2,\varphi} \frac{\partial^2 b}{\partial \varphi^2} + k_{2,\lambda} \frac{\partial^2 b}{\partial \lambda^2} + k_{2,z} \frac{\partial^2 b}{\partial z^2} \tag{4.22}$$

where T_b and M_b are losses of the bacterioplankton biomass due to energy exchange with the environment and dying-off, Γ_b is a set of the trophic subordination of bacterioplankton (in a typical case Γ_b is one element Z), and C_{bs} is the share of bacterioplankton in the food ration of the sth element of the ecosystem. The parameters T_b and M_b are described by the following relationships:

$$T_b = t_b b \tag{4.23}$$

$$M_b = \max\{0, \mu_b(b - \underline{B}_b)^\xi\} \tag{4.24}$$

where t_b is the specific expenditure on the exchange with the environment, μ_b is the rate of bacterioplankton dying-off, and $\underline{B}_b$ and ξ are the constants that determine the dependence of the intensity of the bacterioplankton dying-off on their concentration. The coefficient $k_2 = (k_{2,\varphi}, k_{2,\lambda}, k_{2,z})$ determines the process of the turbulent mixing of the ocean waters.

The dynamic equation for the phytoplankton biomass is:

$$\frac{\partial p}{\partial t} + V_\varphi \frac{\partial p}{\partial \varphi} + V_\lambda \frac{\partial p}{\partial \lambda} + V_z \frac{\partial p}{\partial z} = R_p - T_p - M_p - \sum_{s \in \Gamma_p} C_{ps} R_s + k_{2,\varphi} \frac{\partial^2 p}{\partial \varphi^2} + k_{2,\lambda} \frac{\partial^2 p}{\partial \lambda^2} + k_{2,z} \frac{\partial^2 p}{\partial z^2} \tag{4.25}$$

where Γ_p is the set of the trophic subordination of phytoplankton, C_{ps} is the share of the phytoplankton biomass in the food ration of the elements of the sth trophic level of the ecosystem, T_p is the expenditure on the energy exchange with the environment, and M_p is the dying-off of the phytoplankton cells. The latter parameters are determined by the following relationships:

$$M_p = \max\{0, \mu_p(p - \underline{p})^\theta\} \tag{4.26}$$

$$T_p = t_p p \tag{4.27}$$

where t_p is the specific expenditure on respiration of the phytoplankton cells, μ_p is the coefficient of the phytoplankton dying-off, and $\underline{p}$ and θ are the coefficients characterizing the dependence of the rate of the phytoplankton cells dying-off on their concentration.

Zooplankton is an important trophic element in the ocean ecosystem. In the scheme of Figure 4.1 it is presented by an integral level Z, which implies the presence of a large number of sub-levels with the intersecting trophic bonds. Zooplankton consumes phytoplankton and bacterioplankton. Zooplankton itself feeds a lot of animals (Figure 4.1), such as nekton r and detritophages D.

Let us describe the zooplankton ration by the Ivlev formula:

$$R_Z = k_Z(1 - \exp[-\nu\bar{B}]) \tag{4.28}$$

where $\bar{B}$ is the biomass of the accessible food ($\bar{B} = \max\{0, B - B_{\min}\}$), k_Z is a maximum of the ration with an excess of food, and ν is the coefficient characterizing the level of starvation. The ration maximum is assumed to be equal to the need for food, which, in turn, is determined by the exchange intensity T_1 and a maximum possible increment P_1 at a given intensity of exchange. Both latter parameters are related to the coefficient $q_2 = P_1/(P_1 + T_1)$, so that we obtain:

$$k_Z = T_1 u(1 - q_{2,\max})$$

where $1/u$ is the food assimilation, $q_{2,\max} = \max q_2$.

The formula (4.28) implies that with a small amount of food the ration of zooplankton grows in proportion to the amount of food, then, as the ration approaches a maximum of k_Z, its dependence on B diminishes. Since in fact one trophic level never completely consumes the other, there is a limitation in (4.28) that establishes $R_Z = 0$ at $B \leq B_{\min}$, where $B_{\min}$ is a minimum of the unconsumed food biomass. In (4.26) the $\underline{p}$ parameter plays the same role but in the process of the phytoplankton cells dying-off.

Thus a change of the zooplankton biomass follows the law described by the following differential equation:

$$\frac{\partial Z}{\partial t} + V_\varphi \frac{\partial Z}{\partial \varphi} + V_\lambda \frac{\partial Z}{\partial \lambda} + V_z \frac{\partial Z}{\partial z} = R_Z - T_Z - M_Z - H_Z - \sum_{s \in \Gamma_Z} C_Z R_s + k_{2,\varphi} \frac{\partial^2 Z}{\partial \varphi^2} + k_{2,\lambda} \frac{\partial^2 Z}{\partial \lambda^2} + k_{2,z} \frac{\partial^2 Z}{\partial z^2} \tag{4.29}$$

where Γ_Z is a set of the trophic subordination of zooplankton, $C_Z R_s$ is the share of the zooplankton biomass in the food ration of the sth trophic level, and H_Z, T_Z, and M_Z are the losses of the zooplankton biomass due to unconsumed food, and expenditures on respiration and dying-off, respectively. Let us describe the latter three parameters by the relationships:

$$H_Z = h_Z R_Z \qquad T_Z = t_Z Z \qquad M_Z = (\mu_Z + \mu_{Z,1} Z)Z \tag{4.30}$$

where the coefficients h_Z, t_Z, μ_Z, and $\mu_{Z,1}$ are determined empirically for a concrete species of zooplankton.

As seen from (4.29), zooplankton is considered as a passive element of the ecosystem subject to physical processes of transference in space with the motion of water masses. However, zooplankton is known to migrate mainly in the vertical direction. In the given model one can use a simple mechanism to simulate the process of the vertical migration of zooplankton. For this purpose, divide the whole water thickness into two layers: $0 \le z \le z_0$ and $z_0 < z \le H$. Suppose the zooplankton migration between these layers depends on food availability, that is, some part of zooplankton from the layer $[z_0, H]$ can satisfy its needs of food in the layer $[0, z_0]$. So that with $B_{\min}$ taken into account the whole vertical profile $B(\varphi, \lambda, z, t)$ is considered.

Determine the coefficients $C_{as}(a = p, Z)$ in the formulae (4.25) and (4.29) from the supposition that the consumption of various kinds of food in the sth trophic level is proportional to their efficient biomasses:

$$C_{as} = k_{sa}\bar{B}_a \left[\sum_{a \in S_s} k_{sa}\bar{B}_a\right]^{-1} \tag{4.31}$$

where $\bar{B}_a$ is an effective biomass of the ath food component, S_s is the set of trophic subordination of the sth component, and k_{sa} is the proportion coefficient that determines the significance of the sth constituent in the food ration of the ath element (for an estimation of these parameters see Section 4.5.6).

According to the scheme in Figure 4.1, the equations which describe the dynamics of the biomass of nekton, detritophages, detritus, dissolved organic matter, and nutrient salts will be:

$$\frac{\partial r}{\partial t} = R_r - H_r - T_r - M_r - \sum_{s \in \Gamma_r} C_{rs} R_s \tag{4.32}$$

$$\frac{\partial D}{\partial t} = R_D - H_D - T_D - M_D - \sum_{s \in \Gamma_D} C_{Ds} R_s \tag{4.33}$$

$$\begin{aligned}\frac{\partial d}{\partial t} + V_\varphi \frac{\partial d}{\partial \varphi} + V\lambda \frac{\partial d}{\partial \lambda} + V_z \frac{\partial d}{\partial z} &= M_b + M_D + M_r + M_p + M_Z \\ &\quad + H_Z + H_r + H_D - \mu_d d - C_{dD} R_D \\ &\quad + k_{2,\varphi} \frac{\partial^2 d}{\partial \varphi^2} + k_{2,\lambda} \frac{\partial^2 d}{\partial \lambda^2} + k_{2,z} \frac{\partial^2 d}{\partial z^2}\end{aligned} \tag{4.34}$$

$$\frac{\partial n}{\partial t} + V_\varphi \frac{\partial n}{\partial \varphi} + V_\lambda \frac{\partial n}{\partial \lambda} + V_z \frac{\partial n}{\partial z} = \mu_d d - \delta R_p + k_{2,\varphi} \frac{\partial^2 n}{\partial \varphi^2} + k_{2,\lambda} \frac{\partial^2 n}{\partial \lambda^2} + k_{2,z} \frac{\partial^2 n}{\partial z^2} \quad (4.35)$$

$$\frac{\partial g}{\partial t} + V_\varphi \frac{\partial g}{\partial \varphi} + V\lambda \frac{\partial g}{\partial \lambda} + V_z \frac{\partial g}{\partial z} = T_p + T_b + T_r + T_D + T_Z - C_{gb} R_b$$

$$+ k_{2,\varphi} \frac{\partial^2 g}{\partial \varphi^2} + k_{2,\lambda} \frac{\partial^2 g}{\partial \lambda^2} + k_{2,z} \frac{\partial^2 g}{\partial z^2} \quad (4.36)$$

Here $H_a = (1 - h_a)R_a$ is the unassimilated food of the ath element ($a = r, D$), $T_a = t_a$ where a is the expenditure on energy exchange, $M_a = (\mu_a + \mu_{a,1} a)$ where a is the dying-off, ρ_g is the indicator of the rate of replenishment of the supplies of nutrient elements due to decomposition of the dissolved organic matter, and δ is the coefficient of consumption of nutrient elements in the process of photosynthesis.

As follows from (4.32) and (4.33), a supposition is made in the model that the elements of nekton and detritophages do not move in space with the motion of the water masses. These elements are assumed to migrate independently of the hydrophysical conditions of their environment. Consider two possible versions of modelling the processes of migration. The first version is connected with an addition to the right-hand parts of equations (4.32) and (4.33) of the terms describing the turbulent mixing with the coefficients $k_2^* > k_2$. In other words, the process of migration is identified with the process of an intensified turbulent mixing, that is, it is accidental. However, the process of fish migration manifests some expediency in the choice of the direction of their movement. According to the biological principle of adaptation, the migration of fish is subject to the principle of the complex maximization of the efficient food ration, with the environmental parameters remaining within the conditions of their habitat. Hence, the motion of fish in space with characteristic velocities ensures their location in the regions where at this time the food and other abiotic conditions (temperature, salinity, dissolved oxygen, chemicals concentration) are most favourable. This means that fish migrate in the direction of a maximum gradient of the efficient food with the preserved limitations of the environmental parameters.

4.5.3 Analysis of the vertical structure of the ocean ecosystem

To qualitatively assess the role of the World Ocean ecosystems in the global carbon cycle, it is useful to consider some aspects of the laws of formation of the vertical distributions of organic matter. It follows from the equation (4.25) that the number s of maxima in its solution, $p(\varphi, \lambda, z, t)$ at a fixed time at any point (φ, λ) is determined by the number l of changes of the sign of the derivative $\partial p/\partial z$, with $s \leq 0.5(1 + l)$. It is important to know the location of the phytoplankton biomass maxima as a function of the depth in order to understand the whole range of the biogeocoenotic processes in the ocean ecosystem. Of course, the amount and the location of the p function maxima are determined by many characteristics of the water medium, among which the hydrological heterogeneity between various layers of the water space is one of the most important ones. The results of numerous field studies show that there can be four maxima of phytoplankton and by depth these maxima can locate anywhere from the surface to depths of 500 m. Of course, in most cases there are one to two maxima down to 50–100 m.

A theoretical analysis of equation (4.25) is possible after its simplification:

$$\frac{dp}{dz} = F(z)p - \nu N(z) - w(z) \tag{4.37}$$

where F, N, and w are the rate of photosynthesis, the zooplankton density, and the rate of sedimentation, respectively, at a depth z. The concentration of nutrient elements $n(z)$ is supposed to be taken into account in the form of the function $F(z)$.

The equation (4.37) is linear relative to p, and therefore it will not change if it is multiplied on both sides by the value:

$$\exp\left\{-\int_0^z [F(\xi) - \nu N(\xi)]\right\}$$

We have:

$$\exp\left\{-\int_0^z [F(\xi) - \nu N(\xi)]\right\}\left\{\frac{dp}{dz} - [F(z) - \nu N(z)]p\right\}$$
$$= w(z)\exp\left\{-\int_0^z [F(\xi) - \nu N(\xi)]\right\} \tag{4.38}$$

Note that:

$$\exp\left\{-\int_0^z [F(\xi) - \nu N(\xi)]\right\}\left\{\frac{dp}{dz} - [F(z) - \nu N(z)]p\right\}$$
$$= \frac{d}{dz}\left[p\exp\left\{-\int_0^z [F(\xi) - \nu N(\xi)]\right\}\right] \tag{4.39}$$

Substituting (4.39) into (4.38) and integrating both parts with respect to z, we have the simplifications:

$$p(z)\exp\left\{-\int_0^z [F(\xi) - \nu N(\xi)]\right\} = -\int_0^z w(y)\left\{\exp\left[-\int_0^y [F(x) - \nu N(x)]\,dx\right]\right\}dy + C \tag{4.40}$$

Substituting $z = 0$ into (4.40), we obtain $C = p(0)$. Finally, from (4.40) we obtain the formula of the vertical distribution of the phytoplankton biomass:

$$p(z) = \left\{p(0) - \int_0^z w(y)\exp\left\{-\int_0^y [F(x) - \nu N(x)]\,dx\right\}dy\right\}$$
$$\times \exp\left\{\int_0^z [F(\xi) - \nu N(\xi)]\,d\xi\right\} \tag{4.41}$$

Based on the dependence (4.41) let us investigate the character of the $p(z)$ distribution from the viewpoint of the number of maxima and their location. This is important for specification of the role of various depths in the functioning of the ocean carbonic system. Denote:

$$g(h) = \int_0^h N(z)\,dz \qquad \psi(h) = \int_0^h F(z)\,dz$$

then the expression (4.41) is rewritten as:

$$p(h) = \left\{p(0) - \int_0^h w(y)\exp\left[\nu g(y) - \psi(y)\right] dy\right\} \cdot \exp\left[\psi(h) - \nu g(h)\right] \tag{4.42}$$

Thus $p(h)$ is the functional of $g(h)$ and $y(h)$. The g function has a simple interpretation if $h_2 > h_1$, then the difference $g(h_2) - g(h_1)$ represents the zooplankton biomass located in the layer $h_1 \le z \le h_2$. The functions $g(h)$ and $\psi(h)$ are monotonically nondecreasing, since, $T(h)$ and $F(h)$ are not negative.

Before proceeding to a qualitative analysis of the function $p(h)$, let us obtain some auxiliary results. Let us introduce the following sets:

$$H^+ = \{h \ge 0\colon F(h)/N(h) = \nu/F'(h)/N(h) > \nu\}$$

$$H^- = \{h \ge 0\colon F(h)/N(h) = \nu/F'(h)/N'(h) < \nu\}$$

and the function:

$$\varphi(h) = \nu g(h) - \psi(h)$$

Lemma 4.1. *The $\varphi(h)$ function is non-decreasing at $\forall h \in \Omega_1^+ \cup \Omega_2^+$ where:*

$$\Omega_1^+ = \{h \ge 0\colon h \in [0, h_1^+]/h_1^+ = \inf H^+, F(0)/N(0) < \nu\}$$

$$\Omega_2^+ = \{h \ge 0\colon h \in [h_i^-, h_j^+]/h_i^- \in H^-, h_j^+ \in H^+, (h_i^-, h_j^+) \cap H^- = \varnothing\}$$

Lemma 4.2. *The $\varphi(h)$ function is non-increasing at $\forall h \in \Omega_1^- \cup \Omega_2^-$ where:*

$$\Omega_1^- = \{h \ge 0\colon h \in [0, h_1^-]/h_1^- = \inf H^-, F(0)/N(0) < \nu\}$$

$$\Omega_2^- = \{h \ge 0\colon h \in [h_i^+, h_j^-]/h_i^+ \in H^+, h_j^- \in H^-, (h_i^+, h_j^-) \cap H^+ = \varnothing\}$$

The validity of Lemmas 4.1 and 4.2 follows from the elementary properties of the function $\varphi(h)$. In particular, we have: $\varphi(0) = 0, \varphi'(0) = \nu N(0) - F(0), \ldots,$ $\varphi^{(i)}(0) = \nu N^{(i-1)}(h) - F^{(i-1)}(h)$.

Note 1. The conclusion of Lemma 4.1 also remains valid at $F(0)/N(0) = \nu$, provided $(\exists s)[F^{(s)}(0)/N^{(s)}(0) < \nu, F^{(j)}(0)/N^{(j)}(0) = \nu, j = 0, 1, \ldots, s1]$.

Note 2. The conclusion of Lemma 4.2 also remains valid at $F(0)/N(0) = \nu$, provided $(\exists s)[F^{(s)}(0)/N^{(s)}(0) > \nu, F^{(j)}(0)/N^{(j)}(0) = \nu, j = 0, 1, \ldots, s1]$.

According to (4.42) the function $p(h)$ is the function of $\varphi(h)$:

$$p(h) = \left\{p(0) - \int_0^h w(y) \exp[\varphi(y)]\, dy\right\} \exp[-\varphi(h)] \tag{4.43}$$

Based on the results of Lemmas 4.1 and 4.2, from (4.43) one can easily formulate the monotonicity properties of the function $p(h)$. Since $\varphi(h)$ does not decrease in the set $\Omega^+ = \Omega_1^+ \cup \Omega_2^+$, the function $\exp[-\varphi(h)]$ will be non-increasing at $\forall h \in \Omega^+$. At $w(h) \geq 0$ the function $p(0) - \int w(y) \exp[\varphi(y)]\, dy$ is non-increasing at any values of $h \geq 0$. Therefore $p(h)$ is the non-increasing function of $\forall h \in \Omega^+$.

In the set $\Omega^- = \Omega_1^- \cup \Omega_2^-$ $p(h)$ is the product of the non-decreasing and non-increasing functions. Hence, in Ω^- $p(h)$ is either non-decreasing, non-increasing, or has a maximum. Since for $\forall h \in \Omega^- F(h)/N(h) > \nu$, then $p(h)$ will be the non-increasing function at $\forall h \in H_0 = \{h: h \geq h_0\}$, where:

$$p(0) = \int_0^{h_0} w(y) \exp[\varphi(y)]\, dy$$

Denote the multitude of values $h \geq 0$ by H_1, for which the following condition is valid:

$$w(h)[F(h) - \nu N(h)]^{-1} \exp[\varphi(h) + \int w(y) \exp[\varphi(y)] dy > p(0)$$

Then $p(h)$ will be the non-increasing function at $\forall h \in H_1 \cap \Omega^-$.

Formulate the obtained results as the following theorem.

Theorem *The function $p(h)$ is non-increasing at $\forall h \in H^+ \cup (H_0 \cap H^-) \cup (H_1 \cap H^-)$ and non-decreasing at $\forall h \in H^- \cap (E \backslash H_0) \cap (E \backslash H_1)$, where $E = \{h: h \geq 0\}$.*

In a more general case $\partial p/\partial$ can be presented in the form of the sum:

$$\frac{\partial p}{\partial z} = \left(\frac{\partial p}{\partial z}\right)_{EB} + \left(\frac{\partial p}{\partial z}\right)_O + \left(\frac{\partial p}{\partial z}\right)_\tau + \left(\frac{\partial p}{\partial z}\right)_M + \left(\frac{\partial p}{\partial z}\right)_Z + \left(\frac{\partial p}{\partial z}\right)_T \tag{4.44}$$

where each of the terms on the right-hand side reflects a change of the phytoplankton biomass due to changes in illumination and concentration of nutrient elements (EB), sedimentation (O), temperature changes (τ), dying-off (M), consumption by zooplankton (Z), and turbulent mixing (T). According to (4.43), it can be supposed that:

$$\left(\frac{\partial p}{\partial z}\right)_{EB} \approx F(t, p, z) = Af_1(E)f_2(n)[1 - 10^{-\gamma p}]$$

At $\gamma p \ll 1$ we obtain:

$$p^{-1}\left(\frac{\partial p}{\partial z}\right)_{EB} \approx Af_1(E)f_2(n)$$

With an increasing depth in other stationary conditions the water temperature is known to monotonically decrease, dropping sharply in the layer of the thermocline, that is, $(\partial p/\partial z)t \leq 0$. The similar term $(\partial p/\partial z)_M$ has the form of the sectionallycontinuous non-positive function. However, due to the physical processes of mixing a situation can appear when this monotonicity gets broken.

The values of the constituent $(\partial p/\partial z)_Z$ depend on the distribution of $N(z)$. The term $(\partial p/\partial z)_T$ is totally determined by the hydrological conditions through the water thickness, and its behaviour can change arbitrarily. Apparently, in the layer above the thermocline, $(\partial p/\partial z)_O \leq 0$.

Studies of all components in totality will make it possible to determine the number of changes of sign of the derivative $(\partial p/\partial z)$, which uniquely determines the number of the phytoplankton maxima. In particular, if $\partial p/\partial z$ changes sign seven times in the layers under study, $(\partial p/\partial z)_{z=0} > 0$ and $(\partial p/\partial z)_{z\to\infty} < 0$, then the function $p(z)$ has four maxima.

Finally, assuming that zooplankton uniformly consume phytoplankton through the depth of the layer considered, and neglecting in (4.43) all constituents apart from the first one, based on (4.41) we obtain the following expression for the specific rate of the change of the phytoplankton biomass:

$$p^{-1}(\partial p/\partial t) \approx b - ap \tag{4.45}$$

where

$$a = (k\gamma^2/2)E \cdot 10^J \qquad b = k\gamma E \cdot 10^J \qquad m = (E_{\max}/E_0) \cdot 10^Y$$
$$J = m(1 - E/E_{\max}) \qquad Y = \alpha H_{\max}$$

The equation (4.45) has the solution:

$$p(t,h) = [\gamma/2 + \exp\{-tk\gamma E \cdot 10^S\}]$$

where

$$S = (E_{\max}/E_0) \cdot 10^c \qquad c = \alpha H_{\max}(1 - E/E_{\max})$$

Since at $h \to \infty, E \to 0$, then $p(t,h) \approx [\gamma/2 + K\exp\{-tk\gamma E\}]^{-1}$, where $K = \exp\{-10^{-m}\}$.

4.5.4 The organic carbon cycle in the ecosystem of the Peruvian current

Each region of the World Ocean has characteristic hydrophysical and ecological properties of its own. Therefore a specification of the model of the carbon cycle in the World Ocean, with these properties taken into account, will raise the reliability of the model assessment of the role of the ocean systems in the assimilation of excess CO_2 from the atmosphere. With this aim in view, consider the region of the Peruvian current characterized by high concentrations of nutrient elements (μg-at/l): nitrites (0.6–0.8); nitrates (5.5–14.9); phosphates (0.6–1.31); silicic acids (3–45); and large values of primary production (cal/m^2/day): 17.9–69.3 for phytoplankton; 3.26 for bacteria; 0.12–0.73 for zooplankton, and 0.7–3 for nekton (mainly, anchovy). Near the Peruvian coastline there is the most powerful upwelling of deep waters that feeds the photic layer of the bordering region of the ocean with nutrient elements. Here the main catch of industrial fishing is the anchovy whose average production is estimated at $100 \cdot 10^8$ kg/yr. From the available data, anchovies constitute above 95% of the catch, and therefore the derivation of the model of the functioning of the Peruvian ecosystem is also important for the economy of the bordering states. Of course, the real limits of the interaction of the natural–anthropogenic system of the concrete state of South America with the global environment are impossible to assess without taking account of the role of the ecosystem of the Peruvian current.

The Peruvian current refers to the regions of the World Ocean where complicated interconnected and interdependent physical, biological, and chemical processes take place in a relatively small area. Near the coastlines of Peru and Chili the production of organic substance is so large that the available oxygen is insufficient for it to be oxidized, and as a result, a layer with an oxygen deficit appears at depths 100–800 m. This is accompanied by underfeeding of fauna.

In the hierarchic structure of the trophic relationships of the Peruvian current community, the following components can be selected: B_1 – bacterioplankton; B_2 – phytoplankton; B_3 – phytophages; B_4 – predatory zooplankton; B_5 – commercial anchovy; B_6 – anchovy larvae; B_7 – predatory fish; B_9 – young anchovy; B_{10} – detritus; B_{13} – dissolved organic matter (cal/m^3); B_8 – birds (birds/m^2); B_{12} – nutrient salts (mg/m^3); B_{14} – dissolved oxygen (ml/l); and B_{11} – the biomass of fished anchovy (t/km^2). The characteristic feature of the trophic graph of the system is the availability of connections between the components of the ocean biocoenoses (B_5, B_7, B_9) and the land biocoenoses (B_8). An introduction of this connection into the model has turned out to be essential, since the maritime birds (cormorants, pelicans, etc) in the regions of the Peruvian current markedly affect the dynamics of the whole system. From the available estimates, anchovy constitutes between 80 to 96% of the food spectrum for birds. Therefore the characteristic feature of the developed model is an account of the element of interaction of the land and ocean biocoenoses expressed by the introduction of trophic bonds between birds and anchovies. The trophic pressure on anchovies from other fish is expressed by a single connection with the component B_7 – which is mainly Chili pelamyd (*Sarda chiliensis*), though anchovy is also in the food spectrum for other animals.

The block-scheme of the modelled ecosystem of the Peruvian current (MEPC) reflects the elements of atmosphere–ocean interaction reduced to a mutual exchange with oxygen, with the given rates of invasion and evasion.

The whole system receives the solar radiation energy E whose intensity is estimated by the formula (4.18) at $\alpha = 0.013$ and $\delta = \beta = \nu = 5 \cdot 10^{-5}$. A change of the biomass of the ith component ($i = 1$–$7, 9$) is described by the balance equation similar to (4.29):

$$\frac{\partial B_i}{\partial t} + V_\varphi \frac{\partial B_i}{\partial \varphi} + V_\lambda \frac{\partial B_i}{\partial \lambda} + V_z \frac{\partial B_i}{\partial z} = R_i - T_i - M_i - \sum_{j \in \Gamma_i} C_{ij} R_j + k_{2,\varphi} \frac{\partial^2 B_i}{\partial \varphi^2} + k_{2,\lambda} \frac{\partial^2 B_i}{\partial \lambda^2} + k_{2,z} \frac{\partial^2 B_i}{\partial z^2} - \nu_i \frac{\partial B_i}{\partial z} + Q_i - H_i \quad (4.46)$$

where Q_i is the characteristic of the age transitions between the stages of the anchovy development ($i = 5, 6, 9$), $Q_i = 0$ for $i = 1$–4, 7, R_i is the produce, T_i is the expenditure on energy exchange, M_i is an amount of the died-off biomass, H_i is the unassimilated food (cal/m^3/day),

$$C_{ij} = k_{ij}\bar{B}_i \Big/ \sum_{m \in S_j} k_{jm}\bar{B}_m,$$

Γ_i is the set of the trophic subordination of the ith component, S_j is the food spectrum of the jth component, k_{ab} is the indicator of the degree of satisfaction of the needs of food of the ath component at the expense of the biomass of the bth component, $\bar{B}_i = \max\{0, B_i - B_{i,\min}\}$, where $B_{i,\min}$ is the minimum biomass of the ith component consumed by other trophic levels, $k_2 = (k_{2,\varphi}, k_{2,\lambda}, k_{2,z})$ are the constituents of the coefficient of the turbulent diffusion (with the supposed isotropy of the eddy diffusion in the horizontal plane $k_{2,\varphi} = k_{2,\lambda} = \nu_H$), and ν_i is the rate of the gravitational sedimentation of inactive elements.

The rate of photosynthesis as the function of illumination, nutrient elements, temperature, and pollutions θ is modelled by the relationship:

$$R_2 = \min\{\rho_2 B_2, R_{22}\}$$

where

$$R_{22} = 0.07B_2[1 - \exp(-k_{2,12}B_{12})]E\exp[0.2(1 - E/E_{\max})]e^{-10.7\theta} \cdot (1 + 0.02\,\Delta t^o)$$

The produce R_i on the remaining trophic levels is calculated by the formula:

$$R_i = R_{i,\max}\left[1 - \exp\left(-\sum_{j \in S_i} k_{ij}\bar{B}_j\right)\right](1 + 0,02\,\Delta t^o)B_i$$

The function $T_i (i = 2\text{–}9)$ is postulated as the dependence $T_i = \tau_i B_i^{\omega_i}$, where $\omega_i = 0.7\text{–}0.95$. The coefficients τ_i are functions of temperature and oxygen concentration:

$$\tau_i(t^o, B_{14}) = \begin{cases} t_i^*(t^o) & \text{for } B_{14} > B_{14,i}^* \\ t_i^*(t^o)B_{14}/B_{14,i}^* & \text{for } 0 \le B_{14} \le B_{14,i}^* \end{cases}$$

$$t_i^*(t_o) = \begin{cases} \tau_i^* & \text{for } t^o \in [t_{1,i}^o, t_{2,i}^o] \\ \tau_i^*[a_i\exp(-b_i t^o)]^s & \text{for } t^o \notin [t_{1,i}^o, t_{2,i}^o] \end{cases}$$

where $s = (t^o - t_i^{oo})/m_i$, $t_{1,i}^o$, and $t_{2,i}^o$ are the lower and upper boundaries of the temperature zone of adaptation of the ith trophic level, $t_i^{oo} = 0.5(t_{1,i}^o + t_{2,i}^o)$, and a_i, b_i, and m_i are constant values, selected from the condition of the measurement data approximation.

In contrast to other plankton elements, bacteria are characterized by a substantial dependence of gas exchange on the food concentration. Therefore to describe T_1, assume the following relationship:

$$T_1 = t_1 B^{\omega_1}[1 - \exp(-t_1'\bar{B}_{10} - t_1''\bar{B}_{13})]$$

where t_1 is the expenditure of bacteria on energy exchange per unit biomass, ω_1 is the indicator of the dependence of expenditures on energy exchange on the amount of bacterioplankton biomass; and t_1' and t_1'' are indicators of the dependence of expenditures on the energy exchange of the efficient concentrations of detritus and dissolved organic matter, respectively.

For phytoplankton, the losses due to respiration are characterized by reduced assimilation of carbon in photosynthesis. For the Peruvian current these losses average 74.6% of assimilated oxygen, which constitutes 1.45 g/cm^2/day. Determine the dynamic indicator of mortality M_i based on the assumptions made above in the form of the dependence $M_i = (\mu_i + \mu_B' i)B_i$, where $\mu_i = 0.01$–0.02 and $\mu_i' = 0.1\mu_i$. The H_i parameter is calculated by the formula $H_i = h_i R_i$.

Among living components, anchovy takes a special place in the model, and for its description the following three stages have been chosen: larvae, young anchovy, and commercial anchovy. It is assumed that the rate of the transition from the ith stage to the jth stage is constant and it takes one year for these three stages to occur.

The inert components in the model are described by the following equations:

$$\left.\begin{aligned}
\frac{\partial B_{10}}{\partial t} + (V\cdot\nabla)B_{10} &= \sum_{s=1}^{9}(M_s + H_s) - \rho_{10}B_{10} - \nu_{10}\frac{\partial B_{10}}{\partial z}\\
&\quad - \bar{B}_{10}(k_{1,10}R_1/P_1 + k_{3,10}R_3/P_3 + k_{4,10}R_4/P_4)\\
&\quad + (k_2\nabla^2)B_{10}\\
\frac{\partial B_{12}}{\partial t} + (V\cdot\nabla)B_{12} &= \rho_{10}B_{10} + \rho_1\left(T_9 + \sum_{s=1}^{7}T_s\right) - \delta_{12}R_2\\
&\quad + (k_2\nabla^2)B_{12} + \beta_{12}\frac{\partial B_{12}}{\partial z}\\
\frac{\partial B_{13}}{\partial t} + (V\cdot\nabla)B_{13} &= k_{13,0}R_2 - \bar{B}_{13}k_{1,13}R_1\bar{B}_{13}k_{1,13}R_1/P_1\\
\frac{\partial B_{14}}{\partial t} + (V\cdot\nabla)B_{14} &= \delta_{14}R_2 - \sum_{s=1}^{7}\xi_s T_s - \xi_9 T_9 + (k_2\nabla^2)B_{14}\\
&\quad - \delta_{10}\rho_{10}B_{10} + \max\{0, z_{14} - z)/|z - z_{14}|\}w_{14}
\end{aligned}\right\}\qquad (4.47)$$

The equation to describe changes in the size of the bird population is written as:

$$\frac{\partial B_8}{\partial t} = (1 - h_8)R_8 - M_8$$

where

$$R_8 = \max\{0, (\Delta^* z - z)/|z - \Delta^* z|\}k_{8,0}B_8(1 - \exp[-V_8])$$

$$V_8 = \sum_{s=5}^{7}k_{8,s}\bar{B}_s + k_{8,9}\bar{B}_9 + \gamma_8' \qquad M_8 = \mu_8 B_8$$

γ_8' is the share of the non-fish food in the ration of birds.

The Q_i function of the age transitions is presented only for the anchovy elements and written as a linear approximation:

$$Q_6 = \lambda_{56}B_5 - \lambda_{69}B_6 \qquad Q_5 = \lambda_{95}B_9 \qquad Q_9 = \lambda_{69}B_6 - \lambda_{95}B_9$$

4.5.5 The organic carbon cycle in the ecosystem of upwelling

The zones of the deep waters lifting in the World Ocean called upwellings are known to exhibit a high productivity. An upwelling results from the surface waters wind-driven from the shore, dispersing currents (or water retreat from the shore for other reasons). The rates of water lifting and the upwelling stability are determined by several synoptic parameters. A most characteristic value of the vertical velocity of water lifting in the zone of upwelling is $0.77 \cdot 10^{-3}$ cm/s. The depths from which the water starts lifting vary widely, but depths around 200 m prevail.

Consider the ocean ecosystem in the zone of upwelling proceeding from the concept of the successive development of a community from the moment of its formation in the region of the deep waters lifting to the climax state in the oligotrophic region of convergence, and derive a model of the upwelling ecosystem (MUE). Between these moments the system developed in time and, respectively, moves together with the water flow. In this case changes take place both in the total energy supply in the community and in its structure (spatial, trophic, specific). The basic moments of these changes have been generally studied in numerous field investigations. Therefore the correspondence of the model calculations to these ideas can be one of the criteria of the model adequacy.

Suppose the ecosystem from the zone of upwelling develops uniformly horizontally, so that changes are observed both through the depth z with a step $\Delta z = 10$ m and in time with a step $\Delta t = 1$ day. The horizontal velocity of water flow from the zone of upwelling is assumed to be constant and equal to $V = V_\varphi = V_\lambda$, so that the distance of a unit volume of water from the zone of upwelling is equal to $\Delta r = (\Delta\varphi^2 + \Delta\lambda^2) = V \cdot \Delta t$.

The state of the ecosystem on each horizon $z =$ const. is determined by illumination $E(z,t)$, concentration of nutrient elements $n(z,t)$ and detritus $d(z,t)$, biomass of phytoplankton $p(z,t)$, bacteria $b(z,t)$, protozoa $Z_1(z,t)$, microzooplankton $Z_2(z,t)$, small filtraters $Z_3(z,t)$, large filtraters $Z_4(z,t)$, small predatory cyclopoids $Z_5(z,t)$, predatory calamoids $Z_6(z,t)$, and large predatory chaetognaths and polychaete $Z_7(z,t)$. Protozoa include infusorian and radiolarian, microzooplankton includes naupliuses of all copepods. Based on studies of the food for tropical plankton animals, small filtraters (<1 mm in size) are plankton animals such as *Oikopleura*, copepods *Clausocalanus*, *Paracalanus*, *Acartia*, and *Lucicutia*, and small ostracode *Conchoecia*. Large filtraters include copepods larger than 1 mm, such as *Undinula*, *Pleuromamma*, *Centropages*, *Temora*, *Scolecythrix*, baby euphasiids, pteropods, etc. In the trophic pyramid the impact of nekton elements is considered negligibly small. The food bonds between the enumerated elements are shown in the scheme in Figure 4.2.

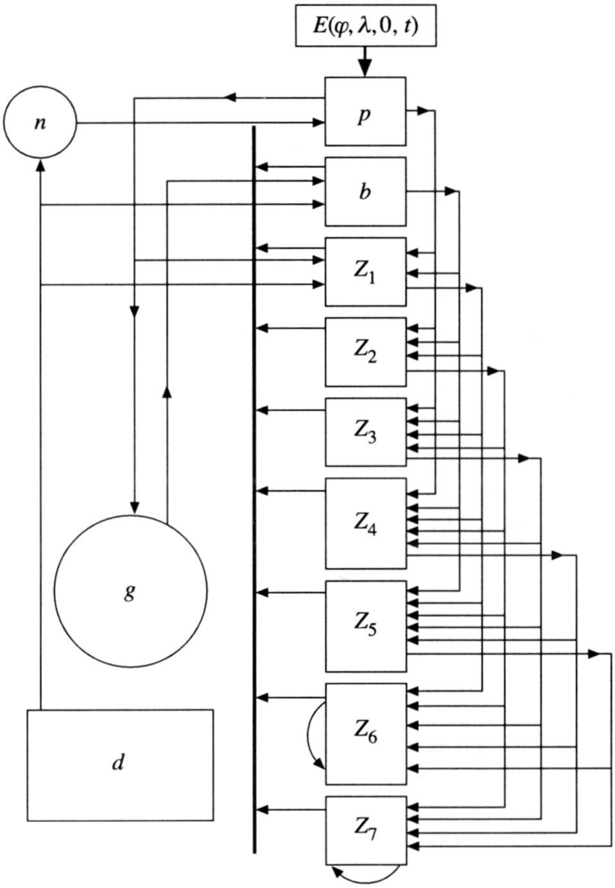

Figure 4.2. The scheme of energy fluxes (cal/m^3/day) through the community of pelagic organisms residing in the 200-m upper surface layer in the oligotrophic tropical region of the World Ocean.

It is assumed that 30% of the mass of bacteria are in natural aggregates >3–5 μm in size, which can be eaten and assimilated by filtraters (Z_3, Z_4); nauplii (Z_2), protozoa (Z_1), and small filtraters (Z_3), which can also assimilate non-aggregated bacterioplankton.

The trophic bonds between the elements are described based on the energy principle. Biomass, rates of production and exchange, and rations are expressed in energy units (cal/m^3 or cal/m^2).

Like in the general model of the ocean ecosystem, the original source of energy and substance in the community is the primary production of phytoplankton (R_p). From outside come the solar radiation energy (E) and nutrient elements (n). The vertical structure of the water medium is described by a "3-layer model": a layer above the thermocline; a layer of sharp gradients; and a layer of lower gradients beneath the thermocline. The upper (z_b) and lower (z_l) boundaries of the layer with sharp gradients are assumed to be located (in metres of depth):

$$z_b = \begin{cases} 10 + 2.2t & \text{for } 0 \le t \le 50 \text{ days} \\ 120 + 0.6(t - 50) & \text{for } t > 50 \text{ days} \end{cases}$$

$$z_l = \begin{cases} 30 + 2.4t & \text{for } 0 \le t \le 50 \text{ days} \\ 150 + 1.4(t - 50) & \text{for } t > 50 \text{ days} \end{cases}$$

Thus the thermocline at an initial time moment $t = 0$ is located in the layer 10–30 m, descending gradually to 120–150 m on the 50th day and to 150–190 m on the 100th day.

The supply of nutrient elements in the layer 0–200 m is replenished at the expense of detritus decomposition and emission of unassimilated food by living elements of the community, as well as at the expense of the input from deep layers ($z > 200$ m) due to turbulent mixing (coefficient k_2) and vertical advection (velocity V_Z). The content of nutrients in organic matter is constant and constitutes 10%.

Taking into account the assumptions mentioned equation (4.35) becomes:

$$R_p = k_T[1–10^{-0.25p\gamma(t)}][1 - 10^{-0.1n}]^{0.6}$$

where the function $\gamma(t)$ characterizes the time dependence of the produce/biomass ratio (P/B-coefficient) for phytoplankton. A maximum of P/B for phytoplankton at the point of upwelling ($t = 0$) is assumed to be 5. The P/B-coefficient decreases and on the 15th day it reaches 1, after which it remains constant. Hence the equation (4.25) is rewritten:

$$\frac{\partial p}{\partial t} = R_p - t_p p - \mu_p p - \sum_{j \in \omega_0 \setminus p} C_{pj} R_j + k_2 \frac{\partial^2 p}{\partial z^2} + (V_z - w_p)\frac{\partial p}{\partial z}$$

where the coefficient w_p (cm/day) describes the process of gravitational sedimentation and is estimated at:

$$w_p = \begin{cases} 50 & \text{for } 0 \le z \le z_b \\ 10 & \text{for } z_b < z < z_l \\ 30 & \text{for } z \ge z_l \end{cases}$$

Bacteria feed on detritus and dissolved organic matter (DOM) emitted by phytoplankton. The allochthonous DOM is neglected. In equation (4.36) an approximation is considered:

$$T_p + T_b + T_Z + T_r + T_D = 0.3R_p$$

It is also assumed that $\mu_b = 0.01$, $t_b = 0,75[1–10^{\varsigma}]$, $R_p = 3b[1–10^{\varsigma}]$, $\varsigma = -0,2d - 0,3R_p$.

A limitation is established – during a 24 hr period bacterioplankton can consume not more than 10% of all detritus located in the same water layer.

With these assumptions the equation (4.22) is written as:

$$\frac{\partial n}{\partial t} = -0,1R_p + 0,1d + 0,05\sum_{i\in\omega_0} t_iB_i + k_2\frac{\partial^2 n}{\partial z^2} + V_z\frac{\partial n}{\partial z}$$

where $\omega_0 = \{p, b, Z_j(j = 1\text{–}7)\}$ and B_i is the biomass of the ith element.

The coefficients k_2 (cm^2/day) and V_z (cm/day) are assumed to be:

$$V_z = \begin{cases} 5 & \text{for } 0 \le z \le z_b \\ 0,1 & \text{for } z_b < z < z_l \\ 1 & \text{for } z \ge z_l \end{cases} \qquad k_2 = \begin{cases} 200 & \text{for } 0 \le z \le z_b \\ 50 & \text{for } z_b < z < z_l \\ 150 & \text{for } z \ge z_l \end{cases}$$

The formula (4.18) for the case considered is simplified:

$$E(z_n, t) = E_0 \cdot 10^{-\alpha z_n}$$

where z_n is the depth of the nth layer (m):

$$\alpha = 0,01 + 0,001\sum_{k=0}^{n}[p(z_k, t) + d(z_k, t)]/z_n$$

A marked suppression of phytoplankton is observed at $E > 100\,\text{cal/cm}^2\text{/day}$. Hence, the formula (4.20) is rewritten:

$$K_T = 0,041E \cdot 10^{0,25[1-E/E_{\max}]}$$

where $E_{\max} \approx 70\,\text{cal/cm}^2\text{/day}$.

For this case the formula (4.19) is reduced to the form:

$$\frac{\partial b}{\partial t} = R_b - \mu_b b - t_b b - \sum_{j\in\omega_0\backslash(p,b)} C_{bj}R_j + k_2\frac{\partial^2 b}{\partial z^2} + (V_z - w_p)\frac{\partial b}{\partial z}$$

The rate of the change of the biomass of each element of zooplankton $\partial Z_i/\partial t$ ($i = 1\text{–}7$) like in equation (4.29) is determined by the level of consumption (ration) $R_{Z,i}$, food assimilation $1/u_{Z,i}$, expenditures of energy exchange $T_{Z,i} = t_{Z,i}Z_i$, rate of dying-off $\mu_{Z,i} = 0.01Z_i$, consumption of the ith element by the jth element with the coefficient C_{ij}, and the age transition of nauplii into copepod of filtraters and predators. Thus:

$$\frac{\partial Z_i}{\partial t} = \Omega_{m,i} + (1 - h_{Z,i})R_{Z,i} - (t_{Z,i} + \mu_{Z,i})Z_i - \sum_{j\ge i}^{7} C_{ij}R_{Z,j} \quad (i = \overline{1,7})$$

where $h_{Z,i} = 1 - 1/u_{Z,i}$ and $\Omega_{m,i}$ characterizes the age transition of nauplii to another category of zooplankton. The transition of nauplii Z_2 to copepod of filtraters (Z_3, Z_4) takes place with an intensity:

$$\Omega_{m,3} = Z_2Z_3\{15(Z_3 + Z_4 + Z_5 + Z_6)\} - 1$$

$$\Omega_{m,4} = Z_2Z_4\{20(Z_3 + Z_4 + Z_5 + Z_6)\}^{-1}$$

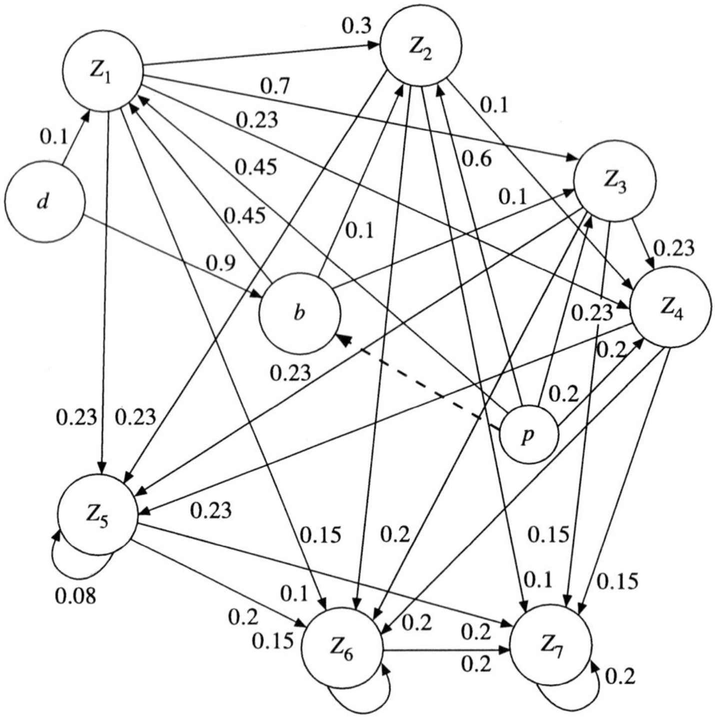

Figure 4.3. The scheme of trophic loads between the elements of the upwelling ecosystem. The letter identifiers correspond to Figure 4.2. The arrows indicate the values of the coefficients C_{ij}.

and for predatory copepods (Z_5, Z_6) with an intensity:

$$\Omega_{m,4} = Z_2 Z_5 \{15(Z_3 + Z_4 + Z_5 + Z_6)\}^{-1}$$

$$\Omega_{m,6} = Z_2 Z_6 \{20(Z_3 + Z_4 + Z_5 + Z_6)\}^{-1}$$

To calculate the ration needed by zooplankton by the formula (4.28), the coefficient ν is determined as $\nu = 0.01/B_{\min}$. When calculating the coefficients C_{ij}, it is supposed that food requirements are related to fodder in proportion to their biomass, taking into account their nutritional needs. The coefficients C_{ij} for all elements of the ecosystem are given in Figure 4.3.

The daily migrations of zooplankton are simulated by an addition to the daily ration of the elements $Z_j (j = 4\text{–}7)$, located in the layer 0–50 m, and of some share (k_Z) of total food needs of the same elements of the community but in the layer 50–200 m. The coefficient k_Z is considered to be time-dependent:

$$k_Z = \begin{cases} 0,02 + 0,0016t & \text{for } t \leq 50 \text{ days} \\ 0,1 & \text{for } t > 50 \text{ days} \end{cases}$$

Various scenarios are possible here both for migration and nutrition of the elements of zooplankton. It is only known for certain that some share of the zooplankton elements from deep layers satisfy their needs for food in the upper layers of the ocean.

In this case the equation (4.34) is written as:

$$\frac{\partial d}{\partial t} = \sum_{i \in \omega_0} (H_{Z,i} + M_{Z,i}) - \sum_{j \in \omega_0 \backslash p} C_{dj} R_j + k_2 \frac{\partial^2 d}{\partial z^2} - \mu_d d + (V_z - w_d) \frac{\partial d}{\partial z}$$

where the coefficient of the rate of the gravitational sedimentation of detritus is taken in the form (cm/day):

$$w_d = \begin{cases} 25 & \text{for } z_b \leq z \leq z_l \\ 50 & \text{for } z < z_b, z > z_l \end{cases}$$

4.5.6 The organic carbon cycle in conditions of freezing seas

The formation of phytoplankton productivity in the northern seas, most of the time covered with ice, is connected with the composition of complicated synoptic, hydrodynamic, geochemical, ecological, and energy variations of the parameters of the marine medium. The complexity of this problem has been mentioned in Legendre and Krapivin (1992), and the problem of the formalized description of the kinetic processes of the formation of the vertical distributions of phytoplankton biomass has been formulated for the first time in the work by Legendre and Legendre (1988), where a stationary oceanographic criterion has been derived to estimate a new produce of phytoplankton.

The ice–water ergocline plays an important role in the formation of the biological productivity of the northern seas as well as the community of algae that get into snow and ice. Therefore in deriving a model of the formation of the vertical structure of the marine ecological community it is necessary to take into account the mechanisms of the snow cover formation, ice growth, and transformation of hydrodynamic processes. In other words, the dynamic formalization of the marine ecosystem in the northern latitudes, covering the season of the ice cover formation, should be of a system character and take into account the variability of the synoptic parameters. The latter, apparently, cause the variability of other parameters in the system.

Here a complex model is suggested of the marine ecosystem dynamics in conditions of the northern climate. The model takes into account the seasonal variations of temperature and precipitation, either prescribed or determined by the respective models. The climatic situations in the region considered were parameterized on the basis of time series of temperature and precipitation. The growth of ice thickness is described by the Stefan model. The formation of the ecological situation in the marine medium is described by the relationships that make it possible to calculate the intensity of light at any depth and to determine the dynamic characteristics of the vertical flux of nutrient elements.

To derive a general model and to obtain a forecast, it is also important to determine the temporal and spatial scales of the processes under study as well as to select the characteristic and special features of the ecosystem's functioning in conditions of the northern climate. Most reliable studies of the stability and characteristic scales have been carried out when studying the atmosphere. Theoretically it has been proven that the time of predictability of actual weather conditions (temperature, wind speed, cloudiness, precipitation) is confined to 1–3 days, with the accuracy of the forecast rapidly decreasing when the term of the forecast increases. At the same time, the experiments reveal an increase of predictability of the same parameters but averaged over time and space, and on this basis a weather forecast is possible for a period of 10–15 days. Longer-term (seasonal and annual) predictions use the data averaged over a decade or a month. That is, the ratio of the time of prediction to the time of averaging can be estimated as ≤ 10. The spatial scales are closely connected with the timescales and characterize various phenomena: motions of fronts, cyclones and anticyclones, low-frequency seasonal oscillations, etc. For instance, the daily resolution in time enables one to describe the motion of cyclones and has a characteristic spatial scale ~ 1000 km.

The characteristic temporal and spatial scales of the ocean and sea water flows have been adequately studied, too, which has made it possible to easily estimate them. The horizontal scales vary from 100 km for synoptic formations to 1000 km for large-scale currents. Oceanographers have found out a substantial connection between primary synoptic hydrophysical processes and elementary circulation mechanisms in the atmosphere with an average duration of 5 days. The seasonal hydrophysical cycles described mainly by mixing of isolines in the meridional circulation have been well estimated. In the interannual variability of the hydrophysical fields the interaction of currents, upwellings, and large-scale atmospheric processes plays an important role. Here the characteristic timescales constitute several months.

The ice component of this system under the influence of atmospheric and water heat inflow as well as snow fall varies in the periods of the formation and melting with a characteristic time step from 1 to 15 days, and in the remaining period varies more slowly.

The phytoplankton concentration should be predicted considering the characteristic rate of growth and dying-off which constitutes between a few hours to 24 hours. This means that the time resolution for the model of the growth of phytoplankton in conditions of possible freezing should not exceed 24 hours in the periods of ice formation and melting and about a decade in the interval between these events. The characteristic spatial scales should be chosen based on the compartment principle of division of the water body with the structure of the currents taken into account.

Figure 4.4 shows a conceptual scheme of energy fluxes of solar radiation and nutrient elements from deep layers of the sea where their supplies are supposed to be infinite. The nutrient background at a depth z is formed due to dead organic matter decomposition d, and the rate of vertical advection v. The nutrient elements are absorbed by phytoplankton at a rate inherent to a given kind of algae at given temperature and illumination. The formation of the phytoplankton biomass is

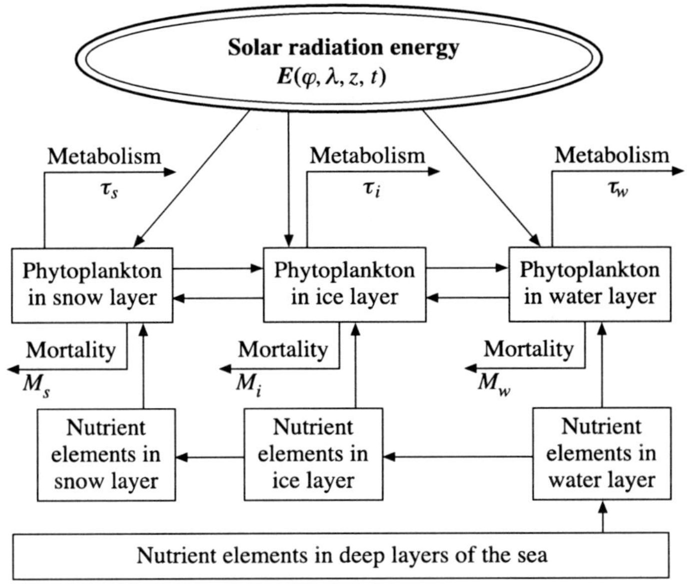

Figure 4.4. The conceptual structure of the model of the organic matter cycle in the ice-covered sea.

affected by solar radiation intensity, concentration of nutrient elements, dying-off, and expenditures on energy exchange with the media. In principle, to close the trophic relationships, it is worthwhile to consider an impact on phytoplankton of the process of its consumption by zooplankton and other organisms of other trophic levels. In this case, when the time intervals with low temperatures are considered, this effect can be neglected. Therefore the final realization of all dynamic equations, that describe the interaction of ecological factors in the marine medium, leads to a set of model units shown in Figure 4.4.

The time and space steps also determine the means of parameterization of the processes taken into account. Thus the atmospheric part of the model is of statistical character and realizes a daily simulation and inertial prediction of the following meteorological parameters: average daily temperature T_a, humidity deficit in the surface air layer D, short-wave solar radiation E, and daily total of precipitation x.

Consider an application of this approach to model a random vector $Y = \{T_a, E, D\}$. Let $Y_i = Y_i + y_i$, $(i = 1, 2, 3)$, where Y is the climatic norm, y_i is a deviation from the climatic norm caused by external accidental forcings. Then $dy_i/dt = A_{ij}y_j + F_i$, where F_i is an accidental forcing and A_{ij} are the coefficients of relaxation.

Let us substitute a differentiation procedure by a difference operator following the simplest scheme $dy_i/dt = [y_i(t_{k+1}) - y_i(t_k)]/\tau$, where $\tau = t_{k+1} - t_k = \text{const}$ and k is the number of the grid knots of time quantization. Taking account of this scheme and multiplying the linear equation given above by y_l, we obtain, after statistical averaging: $\langle y_i(t_{k+1})y_l(t_k)\rangle - \langle y_i(t_k)y_l(t_k)\rangle = \tau A_{ij}\langle y_j(t_k)y_l(t_k)\rangle$. The angular brackets denote the statistical averaging over an ensemble of possible weather scenarios. Hence, if the covariance matrix $\langle y_j(t_k)y_l(t_k)\rangle$ is non-singular, the coefficients A_{ij} are easily calculated:

$$A_{ij} = (\langle y_i(t_{k+1})y_l(t_k)\rangle - \langle y_i(t_k)y_l(t_k)\rangle)(\langle y_j(t_k)y_l(t_k)\rangle)^{-1}/\tau$$

Then the covariance of random forcings are calculated:

$$\langle F_iF_j\rangle = \langle (y_i(t_{k+1}) - y_i(t_k) - \tau A_{il}y_l(t_k))(y_j(t_{k+1}) - y_j(t_k) - \tau A_{jm}y_m(t_k))\rangle$$

Usually the sums of precipitation are assumed to follow the gamma-distribution:

$$P(x) = x^{\kappa-1}\exp(-\kappa/\beta_0)/[(\beta_0)^{\kappa}\Gamma(\kappa)]$$

where κ and β_0 are empirical parameters for the form and scale of distribution. Then the daily sums of precipitation obtained are distributed by the dates of the highest probability in accordance with oscillations of other meteorological elements and the value of the probability coefficient:

$$\lambda_P = \{+N_2[Y_i - \bar{Y}_{i,2}]C_{ij}[Y_j - \bar{Y}_{j,2}]/(1+N_2)\}$$
$$\times \{+N_1[Y_i - \bar{Y}_{i,1}]C_{ij}[Y_j - \bar{Y}_{j,1}]/(1+N_1)\}$$

where N_1 and N_2 are the number of days with and without precipitation respectively, $\bar{Y}_1$ and $\bar{Y}_2$ are vectors of the averaged meteorological parameters for the days with and without precipitation respectively, and C_{ij} is the correlation matrix of meteorological parameters.

The input data for the model can be either obtained directly from the meteorological data for the territory considered or calculated with the use of traditional algorithms. For instance, in the absence of short-wave radiation measurements one of the empirical connections with cloudiness can be used:

- *The Kimbell formula* – $E = E_y(1 - 0.71n_o)$, where n_o is the average point of cloudiness and E_y is the solar radiation in clear skies;
- *The Budyko formula* – $E = E_y(1 - 0.68n_o)$;
- *The Berliand formula* – $E = E_y(1 - [m_o + 0.38n_o]n_o)$ where m_o is the correction coefficient to take latitude into account.
- *The Matauike formula* – $E = Ey(1 - 0.52[n_o]^{1.3})$.

The results of calculations with these formulae scatter most in conditions of continuous clouds. On average, the error of calculations is determined by variations of the parameter E/E_y, which constitute 0.25–0.35 for continuous cloudiness with stratus and cumulostratus clouds and 0.65–0.85 for stratus and cirrostratus clouds.

The ice cover thickness $z_s(t)$ can be described by statistical data with indicated dispersive characteristics $z_s(t) = G(t) + g(t)$, where the function $G(t)$ characterizes an average thickness of snow cover during a chosen time period, and the function $g(t)$ determines its average variation during the same time period. There are numerous models that describe the process of the snow cover formation and melting. Within the global model it is expedient to use one of the simplest parameterizations, such as the Sellers model: $S_F = x(27.5 - 0.1T_a)$, $T_a \in [265, 275]$K, $S_A = S_M + Q_0$, $Q_0 = \rho_a C_a U_a[r^*(T_a) - r_a]$, $S_M = 0.045(T_a - 273.16)$, where S_F is the share of the snowfall at air temperatures close to zero, S_A is the snow ablation (evaporation + melting), S_M is the amount of snow melted per day (water equivalent in cm), r^* is the saturating specific humidity, r is specific humidity, C_a is the resistance coefficient, U_a is the surface wind speed, and ρ_a is the air density.

As a result, the snow cover thickness is calculated by the formula:

$$z_s(t + \Delta t) = z_s(t) + [S_F - S_A]\Delta t$$

We will describe the process of ice formation on the water surface by the Stefan model. Let the ice thickness be described by the function $R(t) = z_I - z_s$, with the part $R_1(t)$ being below the level $z = 0$. Denote by $\rho G, \rho_R$, and ρ_W as the densities of snow, ice, and water, respectively. Then $R_1 = (z_s\rho_G + R\rho_R)/\rho_W$. The process of ice formation will be subject to the regularity following from the system of equations:

$$\frac{\partial Ta}{\partial t} = a^2(z)\frac{\partial 2T_a}{\partial z^2}$$

$$\lambda_1 \frac{\partial T_a}{\partial z}\bigg|_{z=R_1} = q\rho_R \frac{dR}{dt} + \lambda_2 \frac{\partial T_a}{\partial z}\bigg|_{z=R_1}$$

with boundary and initial conditions: $T_a(R_1 - R) = f(t) \leq 0°\text{C}$, $T_a(R_1, t) = 0°\text{C}$, $T_a(\infty, t) = P(t)$, $R(0) = 0$ and $T_a(z, 0) = \psi(z)$, where λ_1 and λ_2 are coefficients of the ice and water heat conductivity, respectively, q is the specific heat of ice melting, f is the function of variability of the surface air temperature in the absence of snow either prescribed as a scenario or calculated by the climate model, or snow temperature, P is the prescribed temperature of deep water layers, and $a^2(z)$ is the coefficient of the medium heat conductivity:

$$a^2(z) = \begin{cases} a_1^2(z) & \text{for } R_1 - R \leq z \leq R_1 \\ a_2^2(z) & \text{for } z > R_1 \end{cases}$$

Now proceed to a description of the ecological part of the model. The formation of the nutrient background in the case considered, compared to ice-free conditions, has certain specific features manifested through the characteristic process of their motion in the environment indicated in the scheme in Figure 4.5. Select three types of nutrient elements: $n_{AN}(z, t)$ – nitrogen, $n_{AP}(z, t)$ – phosphorus, and $n_{ASi}(z, t)$ – silicon. Symbols $A = (s$ – snow, i – ice, and w – water) and $X = \{N, P, Si\}$ are used to make the formulae shorter. The dynamic equations to describe the kinetics of nutrient elements will be:

$$\frac{\partial n_{AX}}{\partial t} = k_1 \frac{\partial n_{AX}}{\partial z} + k_2 \frac{\partial^2 n_{AX}}{\partial z^2} + Q_{AX} - \delta R_{Ap}$$

where k_1 and k_2 are the coefficients of advection and turbulent diffusion, respectively, R_{Ap} is the rate of photosynthesis in the medium A, Q_{AX} is the inflow of nutrient elements into the medium A from other sources, and δ is the consumption of

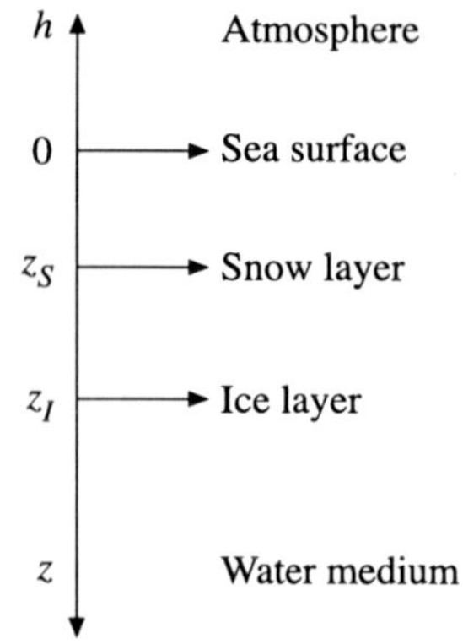

Figure 4.5. The vertical section of the model medium of an arbitrary sea basin in conditions of freezing.

nutrient elements by unit biomass of pure produce of phytoplankton per unit time. In the simplest case the flux Q_{AX} is identified with the input of nutrient elements due to decomposition of dead organic matter and phytoplankton respiration $\tau_X : Q_{AX} = \delta_X R_d + \lambda_p \tau_X$, where the coefficient δ_X characterizes the content of the nutrient element X in detritus d, and λ_p is the indicator of the rate of replenishing of the nutrient elements supplies due to respiration of the phytoplankton cells.

To calculate the level of illumination $E(z, t)$, apply an exponential law that reflects the impact of the optical properties of the medium A:

$$E(z,t) = \begin{cases} E_0 & \text{for } z = 0 \\ E_1(z,t) = E_0 \exp\{-\alpha_1 z\} & \text{for } z \in (0, z_s] \\ E_2(z,t) = E_1(z_s,t)\exp\{-\alpha_2 z\} & \text{for } z \in (z_s, z_I] \\ E_3(z,t) = E_2(z_I,t)\exp\{-\alpha_3 z\} & \text{for } z > z_I \end{cases}$$

Let us approximate the rate of photosynthesis in the medium A using the relationship:

$$R_{Ap} = k_p p_A E \exp(1 - E/E_{\text{opt}})(1 + a \cdot \Delta T_A)[1 - \exp(-b \cdot \min\{n_{sX}, n_{iX}, n_{wX})]$$

where k_p is a maximum of phytoplankton productivity at a control temperature T_c, $\Delta T_A = T_A - T_c$, E_{opt} is an illumination optimal for photosynthesis, a is the indicator of a change of the rate of photosynthesis when the temperature of the medium deviates from the control one, and b is the degree of limitation of photosynthesis by a nutrient salt following the Libich principle.

The dynamic equations for the elements of the sea ecosystem in conditions of freezing are written by analogy to the balance equation given above in other paragraphs of this chapter.

4.5.7 The organic carbon cycle in the Okhotsk Sea

The Okhotsk Sea is most characteristic of the northern latitudes. Numerous investigations have concentrated on the Okhotsk Sea as a unique ecosystem (Vasiliev, 1991; Terziev *et al.*, 1993). This is connected with the desire to understand the climatic and ecological processes taking place here and to establish their correlations with the global processes, especially in the Arctic basin. For Russia and Japan this is also important for economic reasons. Therefore an urgent problem is to derive a model of the ecosystem of the Okhotsk Sea (MEOS) that would make it possible to predict the state of the ecosystem as a function of global and regional changes in the environment.

The MEOS structure and the scheme of energy fluxes in the ecosystem of the Okhotsk Sea are shown in Figures 4.6 and 4.7, respectively. The equations of the trophic interconnections are written by analogy to the equations (4.21) through (4.36). Let us approximate a change of the oxygen regime by solving the equations:

$$\frac{\partial O}{\partial t} + V_\varphi \frac{\partial O}{\partial \varphi} + V_\lambda \frac{\partial O}{\partial \lambda} + Vz \frac{\partial O}{\partial z} = \xi_p R_p - \xi_Z T_Z - \xi_r T_r - \xi_D T_D - \xi_d R_d + O_A$$

where O_A is the function describing the oxygen exchange on the atmosphere–sea border, ξ is the indicator of the oxygen production through photosynthesis,

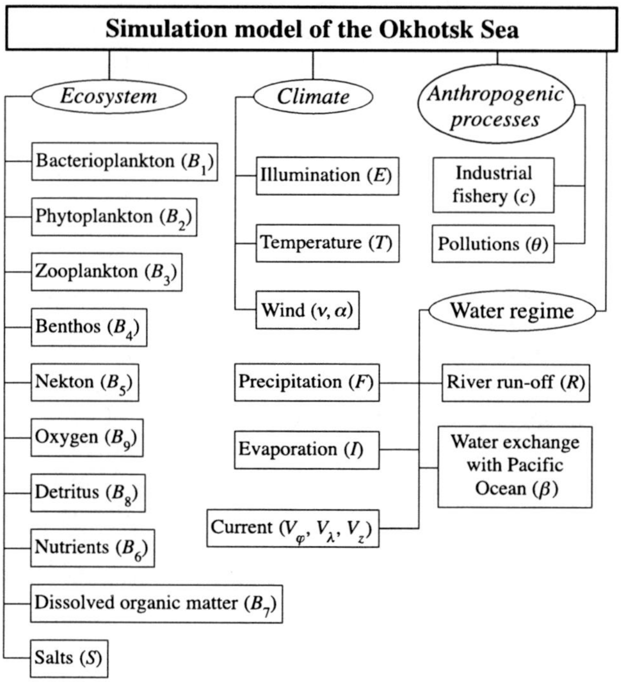

Figure 4.6. The MEOS structure.

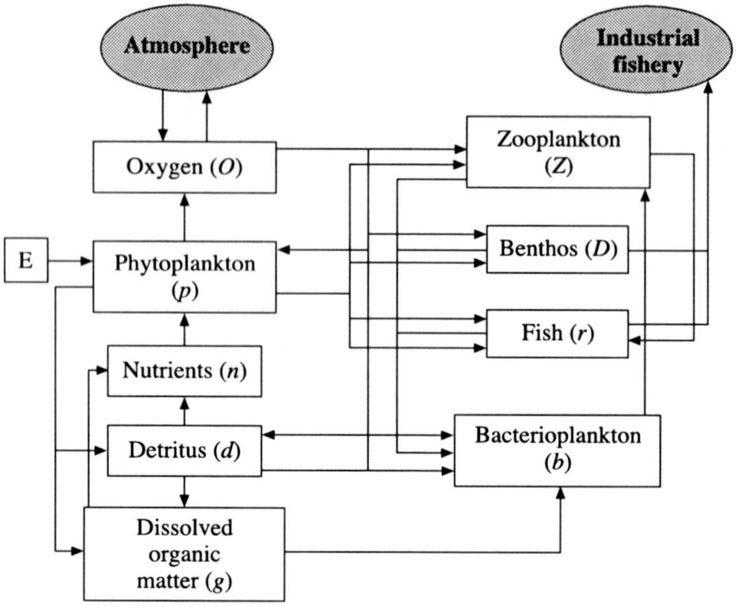

Figure 4.7. Energy fluxes in MEOS.

ξ_i $(i = Z, r, D)$ is the coefficient of the oxygen expenditure on respiration of plankton i, and ξ_d is the indicator of the oxygen expenditure on the decomposition of detritus.

Describe the spatial and seasonal dynamics of salinity by the balance relationships:

$$V_i \frac{dS_i}{dt} = \sum_{\substack{j=1 \\ j \neq i}}^{N} [\alpha_{ij} W_j S_j + \mu_{ij}(S_j - S_i)] + (1 - \beta_i) W_i S_i + f_i \quad (i = \overline{1, N})$$

where N is the number of segments of the volume $V_i = \Delta\varphi \cdot \Delta\lambda \cdot \Delta z_i$, appearing after division of the water body of the Okhotsk Sea in the process of quantization of latitude φ, longitude λ, and depth z; μ_{ij} is the volume coefficient of diffusion equal to $\delta_{ij}\sigma_{ij}/L_i$ (L_i is the average length of the bordering boundary of the ith segment, σ_{ij} is the interface of the segments i and j, δ_{ij} is the coefficient of turbulent diffusion), f_i is the function of the source describing the external input of salts to the ith segment (with precipitation F, river run-off R, from the Pacific Ocean β, from the bottom sediments with water uplifting); W_j is the total water flow from the jth segment, α_{ij} is its share directed to the ith segment ($\alpha_{ij} = 0$ for non-bordering segments), and β_i is the share of the flux W_i, directed outside.

The temperature regime of the Okhotsk Sea is presented by the heat balance equation:

$$\rho c V_i \frac{dT_i}{dt} = \sum_j W_{ij,A} + \sum_j W_{ij,D} - W_{i,S} + \Omega_{i,1} + \Omega_{i,2}$$

where T_i is the average temperature of water in the ith segment (°K), $W_{ij,A}$ is the input of heat with the water flow from the jth segment into the ith one, $W_{ij,D}$ is the heat exchange between the ith and jth segments due to turbulent motion of water masses, $W_{i,S}$ is the total heat output from the jth segment; $\Omega_{i,1}$ is the heat input with water fluxes across the external boundary, $\Omega_{i,2}$ is the heat exchange across the external boundaries not connected with water fluxes, ρ is the sea water density (g/cm^3), and c is heat capacity (cal/g/deg).

The scheme to calculate the temperature regime is in fact an approximate simulation. This semi-empirical approach is acceptable for the climatic zone of the Okhotsk Sea and does not require complicated calculations. The exchange processes with water masses between the segments V_i ($i = 1, \dots, N$) are simulated with account of information on directions and velocities of the currents contained in the database.

The functioning of the Okhotsk Sea ecosystem is determined by severe climatic conditions. During 6–7 months in a year the sea is almost completely covered with ice 80–100 cm-thick. Therefore in the MEOS the vertical structure of the Okhotsk Sea is presented as a vertical section shown in Figure 4.5. With account of this scheme, in the zones of the frozen sea the energy fluxes in Figure 4.7 get complicated due to partition of the component "phytoplankton" into three possible media. This is shown schematically in Figure 4.4.

5

Modelling the interactive cycles of carbon and other chemicals

5.1 INTRODUCTION

A discussion of the problem of the greenhouse effect cannot be constructive without a thorough consideration of feedbacks of the carbon dioxide cycle with the biogeochemical processes in the presence of other elements, such as nitrogen, sulphur, phosphorus, methane, ozone, and others (Chagas *et al.*, 2002; Coconea, 2000; Hadley and Toumi, 2002; Johnson and Ulyatt, 2000; Liping *et al.*, 2000; Lowry *et al.*, 1998; Mészàros, 2002; Novozhilov, 2002; Peng, 1999; Pennise *et al.*, 2001; Potosnak *et al.*, 1999; Raschke *et al.*, 2001). The processes of CO_2 assimilation from the atmosphere are affected by numerous natural and anthropogenic factors manifested through a long chain of cause and effect bonds. For instance, the acid rains affect the state of the vegetation cover and the latter affects the CO_2 exchange on the atmosphere–land border. The use of nitrogen and phosphorus fertilizers in agriculture changes the role of cultural plants in the assimilation of atmospheric CO_2 and affects the rates of decomposition of the soil organic matter. In totality, many chemicals, especially greenhouse gases (GHGs) that have got into the environment from anthropogenic sources, become the subject of not only biogeochemical analysis but also economic consideration. Such a diverse analysis was carried out, for instance, in connection with the CH_4 dynamics problem at the Second International Conference in Novosibirsk in 2000 (Bazhin, 2000; Byakola, 2000). Relevant connections should be thoroughly systematized and parameterized. Without it, the reliability of the estimate of the role of the biosphere in the assimilation of excess CO_2 from the atmosphere raises doubts. Complex studies in this direction are being carried out (i.e., in several laboratories in the USA). Measurements of the spatial and temporal distributions of gases involved in the global CO_2 cycle are made at surface observatories, with the use of flying laboratories and satellite remote sounding. An accumulation of such information will make it possible to reveal dependences needed for global modelling (Francey *et al.*, 2000; Frantz *et al.*, 2000; Fung, 2000; Gasse and

Lin, 1998; Gillon and Yakir, 2001; Godderis *et al.*, 2001; Kennedy, 2001, 2003; Kennedy *et al.*, 2002; Kerr, 2001).

All this confirms an apparent fact that fragmentary studies of the global carbon cycle not based on a constantly complicating global model, of the type described in Krapivin and Kondratyev (2002), will always raise doubts. For global conclusions like the Kyoto Protocol recommendations these doubts become dangerous with unpredicted global consequences. Nevertheless, such conclusions and assessments are necessary. Unfortunately, most of the international programmes on the subject considered are not aimed at the development of the global modelling technology and do not concentrate the efforts of specialists on deriving numerical models of the nature–society system.

The existing global models are simple and inadequately provided with databases. Three directions in global modelling have been described in the works of Kondratyev *et al.* (2002b), Sellers *et al.* (1996 a–c), Boysen (2000), and Krupchatnikov (1998). In each of them one or several components are absent, but on the whole, conceptually they are mutually additive. This makes it possible to combine them and, hence, to derive a global model which would take into account the most important processes in the nature–society system.

The photochemical system of the atmosphere is one of the most poorly studied systems and is practically neglected in global models. Knowledge is needed of the laws of changes of the incoming solar radiation intensity in connection with the effects of its absorption by gases and aerosols. First of all, of importance here is the role of molecular nitrogen, ozone, water vapour, nitrogen oxide, sulphur dioxide, nitrogen dioxide, CO_2, and other gases. Therefore in this chapter some models are suggested which make it possible to parameterize part of the elements indicated.

5.2 GENERAL CHARACTERISTICS OF GLOBAL BIOGEOCHEMICAL CYCLES

An interaction of abiotic factors of the environment and living organisms of the biosphere is followed by continuous cycles of matter in nature. Various species of living organisms assimilate substances needed for their growth and life support, emitting into the environment the products of metabolism and other complex mineral and organic compounds of chemical elements in the form of unassimilated food or dead biomass. The biospheric evolution has resulted in a stable chain of global biogeochemical cycles, whose breaking in the second half of the 20th century has posed many principal problems, such as unpredictable climate changes due to the possible greenhouse effect, a decrease of biodiversity, progressing desertification, and many other problems. The question as to what happens with the Earth's climate and what are the prospects for the ozone layer depletion remain unanswered despite huge economic expenditures on their study. Now it is clear that these and other nature-protection questions cannot be answered without developing an efficient global system of monitoring based on the global model of the nature–society system, one of the basic units of which simulates the biogeochemical cycles of the main chemicals in the biosphere. This approach will enable one to assess the anthropogenic fluxes of pollutants and to estimate the permissible emissions of hydrogen, chlorine, fluorine, methane, and other chemicals into the environment (Reichle *et al.*, 1999; Romashkin *et al.*, 1999; Schafer and Victor, 1999; Sedjo, 2001; Sellers *et al.*, 1992, 1997; Shaw *et al.*, 2002; Siani *et al.*, 2001; Stephen *et al.*, 2000; Streets *et al.*, 2001; Strich, 2002; Tenn and Davidesin, 1998; Woodbury *et al.*, 2002).

The global biogeochemical cycle of CO_2, described in this book, has been the centre of attention for scientists. Specialists from many countries try to answer the questions: (1) What are the concentrations of CO_2 that can be expected in the future with the existing or predicted rates of organic fuel burning? (2) What climatic changes can result from the increasing concentrations of CO_2? (3) What are the consequences of such climatic changes for the biosphere? (4) What processes can mankind undertake in order to either reduce negative consequences of climate change or even prevent them?

Clearly, now, according to rough model estimates, the industrial civilization should search for new sources of energy, which reduce the rates of the organic fuel burning and, hence, reduce the external impacts on natural biogeochemical cycles.

During the last decade the notion of the "greenhouse effect" has been used in numerous publications on the problems of global climate change on the Earth. This notion of the "greenhouse effect" refers to an explanation of changes in the thermal regime of the atmosphere caused by the effect of some gases on the process of solar radiation absorption as well as longwave radiation transfer. Many gases are characterized by a high stability and long residence in the atmosphere. Carbon dioxide is one of them. Its time of residence in the atmosphere is estimated at 2–3 years.

Numerous long-term observations in various latitudinal belts show a high level of correlation between temperature and atmospheric CO_2 content (Bazzaz, 1986; Bratus *et al.*, 2000, 2001; Carlslaw *et al.*, 2002; Charette *et al.*, 2001; Gruber *et al.*, 2002; Hansell and Carlson, 2002; Hood, 1981; Jones and Cox, 2001; Kamall, 2000; Khotuntsev, 2001; Kuznetsov and Bolshakov, 2002; Lal *et al.*, 1998). The atmosphere–ocean interaction contributes most into this dependence. Though the atmosphere and the ocean are in equilibrium with respect to CO_2 exchange, this equilibrium still gets regularly broken. The most serious causes of this breaking of equilibrium include: (1) SST variations; (2) changes in the ocean volume; and (3) changes in the regime of the vertical circulation of the ocean. In general, an efficiency of these causes can be characterized by the following ratio of forcing on CO_2 concentrations in the atmosphere. The first cause contributes about 65% towards the change of CO_2 partial pressure in the atmosphere (p_a). The remaining 35% results from contributions from the second and third causes. Quantitatively, this relationship is characterized by a 6% increase of the atmospheric CO_2 partial pressure per 1°C increase of the temperature of the upper layer of the ocean. Also, a 1% decrease of the ocean volume increases p_a by 3%.

An assessment of the greenhouse effect requires a complex consideration of the interaction of all processes of the energy transformation on the Earth. However, in the diversity of the processes (from astronomical to biological) that affect the climatic system on various timescales, there exists an hierarchy of their significance. But this hierarchy cannot be constant, since the role of some processes significantly varies for climatic variations. In fact, the impact of the greenhouse effect is determined by an exceeding of surface temperature T_L over the efficient temperature T_e. The Earth's surface temperature T_L is a function of surface emissivity κ. The effective temperature T_e is a function of emissivity α of the atmosphere–land–ocean system. In a general case, the parameters κ and α depend on many factors, in particular, on the CO_2 concentration in the atmosphere. There are a lot of simple and complicated numerical models where attempts have been made to parameterize these dependences. Unfortunately, there is not a single model that would meet the requirements of adequacy and reliably describe the pre-history of the climatic trends on the Earth. Nevertheless, one can state that the greenhouse effect depends nonlinearly on the difference $T_L - T_e$, that is, on atmospheric turbidity, especially in the long-wave region. The more CO_2 in the atmosphere, the stronger is the atmospheric turbidity. The strongest effect of CO_2 on the atmospheric turbidity is in the long-wave region 12–18 μm. This effect is weaker in the wavelength intervals 7–8, 9–10, 2.0, 2.7, and 4.3 μm. It is clear that with the increasing partial pressure of CO_2 in the atmosphere the role of various bands of CO_2 will grow, and this means that with the intensified CO_2 absorption bands the upward long-wave radiation flux will decrease. At the same time, the downward long-wave radiation flux on the Earth surface will increase. From the available estimates, a reduction of the global mean upward flux and an increase of the downward flux is estimated at 2.5 and 1.3 W/m^2, respectively.

Thus to estimate the level of the greenhouse effect due to CO_2 and other gases, it is necessary to know how to predict their concentration in the atmosphere with account of all feedbacks in their global biogeochemical cycles (Watson *et al.*, 2000; Grant *et al.*, 2001; Grogan *et al.*, 2001; Guaduong and Masao, 2001; Grossman, 2001). This problem touches upon several spheres of science – biogeochemistry, geochemistry, soil science, ecology, agrochemistry, geology, oceanography, physiology, and radiochemistry. The present methods of global eco-informatics enable us to combine knowledge accumulated in these fields (Kondratyev and Krapivin, 2002).

Of course, the global cycle of chemicals should be studied not only to be able to assess the climatic consequences of the anthropogenic activity but also to understand the prospects of the environmental dynamics from the viewpoint of its quality and life support. Since the cycles of chemicals in nature are closely connected with living organism activity, one can single out the geological, biogenic, and biological cycles of this activity. The biogenic cycle includes sub-cycles, such as biogeochemical, biogeocoenotic, and geochemical (Haan *et al.*, 2001; Heans, 2001; Hensen *et al.*, 1998; Korner, 1998).

5.3 MODEL OF THE GLOBAL SULPHUR CYCLE

The global model of the biosphere can be improved by increasing the number of biogeochemical cycles taken into account. The necessity to include a unit describing the sulphur fluxes in natural systems is dictated by the dependence of biotic processes on the content of sulphur in the compartments of the biosphere. The available data on supplies and fluxes of sulphur compounds in the atmosphere, soils, vegetation cover, and hydrosphere enable one to formulate numerical relationships to simulate the global cycle of sulphur (the MGSC unit).

The global sulphur cycle consists of the mosaic structure of local fluxes of its compounds with other elements formed due to water migration and atmospheric processes (Hadley and Toumi, 2002). The conceptual schemes of the global and regional cycles of sulphur have been described in detail by many authors (Nitu *et al.*, 2000b; Xu and Carmichael, 1999; Stein and Lamb, 2000; Howarth *et al.*, 1992). However, the existing models have been developed for autonomous functioning and usage, which makes it difficult to include them in the global model without substantial changes of their parametric space. The model of the global sulphur cycle proposed here has been derived in the form of a unit with inputs and outputs, which enables us to match it with other units of the global model via their inputs and outputs.

In contrast to hydrogen, sulphur compounds cannot be attributed to long-lived elements of the biosphere. Therefore in the unit of sulphur the spatial digitization of its natural and anthropogenic reservoirs should be planned to reflect the local distributions of sulphur in the vicinity of its sources and to enable us to estimate the intensities of the inter-regional fluxes of sulphur compounds. The version of the sulphur unit proposed here, in contrast to the known hydrodynamic models of the long-distance transport, takes into account the fluxes of sulphur compounds between the hydrosphere, atmosphere, soil, and biota. The model does not consider the vertical stratification of the atmosphere. The characteristics of sulphur fluxes averaged vertically are calculated for both the land and ocean. The spatial digitization of the biosphere and the World Ocean corresponds to the criterion inherent to the global model. The block-scheme of the model of the biogeochemical cycle of sulphur is shown in Figure 5.1, a description of the fluxes of sulphur compounds is given in Table 5.1. This scheme is realized in every cell Ω_{ij} of the Earth's surface and in every compartment Ω_{ijk} of the World Ocean. The interaction between the cells and the compartments is organized through the climate unit of the global model. Therefore the equations of the sulphur unit lack the terms reflecting the dynamic pattern of the spatial transformation of the sulphur reservoirs. With account of notations assumed in Figure 5.1 and in Tables 5.1 and 5.2, the equations describing the balance relationships between the reservoirs of sulphur compounds will be written in the form:

$$\frac{dAH2SL}{dt} = C_1 + C_2 + C_3 - C_4 + C_{21} \tag{5.1}$$

$$\frac{dASO2L}{dt} = C_4 + C_5 + C_6 - C_7 - C_8 - C_9 \tag{5.2}$$

$$\frac{dASO4L}{dt} = C_9 + C_3 + C_{20} - C_{11} - C_{12} \tag{5.3}$$

$$\frac{dS}{dt} = C_{17} - C_{16} - C_{19} \tag{5.4}$$

$$\frac{dSO4L}{dt} = C_{10} + C_{11} + C_{12} + C_{16} - C_3 - C_{13} - C_{14} \tag{5.5}$$

$$\frac{dFIX}{dt} = C_7 + C_{15} - C_{17} + C_{22} \tag{5.6}$$

$$\frac{dH2SO4L}{dt} = C_8 - C_{18} - C_{21} - C_{22} \tag{5.7}$$

$$\frac{dAH2SO}{dt} = H_1 + H_3 + H_4 + H_{26} - H_2 \tag{5.8}$$

$$\frac{dASO2O}{dt} = H_2 + H_5 + H_6 - H_7 - H_8 - H_{24} \tag{5.9}$$

$$\frac{dASO4O}{dt} = H_8 + H_9 + H_{12} - H_{10} - H_{11} \tag{5.10}$$

$$\begin{aligned}\frac{\partial SO4OU}{\partial t} + v_z\frac{\partial SO4OU}{\partial z} + k_z\frac{\partial^2 SO4OU}{\partial z^2} &= H_7 + H_{10} + H_{11} + H_{20} + H_{22} \\ &\quad + H_{27} + C_{14} - H_{12} - H_{13}\end{aligned} \tag{5.11}$$

$$\frac{\partial H2SOU}{\partial t} + v_z\frac{\partial H2SOU}{\partial z} + k_z\frac{\partial^2 H2SOU}{\partial z^2} = H_{21} + H_{23} - H_4 - H_{22} \tag{5.12}$$

$$\frac{\partial H2SOD}{\partial t} + v_z\frac{\partial H2SOD}{\partial z} + k_z\frac{\partial^2 H2SOD}{\partial z^2} = H_{17} - H_{18} - H_{21} \tag{5.13}$$

$$\frac{\partial SO4OD}{\partial t} + v_z\frac{\partial SO4OD}{\partial z} + k_z\frac{\partial^2 SO4OD}{\partial z^2} = H_{18} - H_{19} - H_{20} \tag{5.14}$$

$$\frac{\partial DU}{\partial t} + v_z\frac{\partial DU}{\partial z} + k_z\frac{\partial^2 DU}{\partial z^2} = H_{14} - H_{15} - H_{23} \tag{5.15}$$

$$\frac{\partial DD}{\partial t} + v_z\frac{\partial DD}{\partial z} + k_z\frac{\partial^2 DD}{\partial z^2} = H_{15} - H_{16} - H_{17} \tag{5.16}$$

$$\frac{\partial FI}{\partial t} + v_z\frac{\partial FI}{\partial z} + k_z\frac{\partial^2 FI}{\partial z^2} = H_{13} - H_{14} \tag{5.17}$$

$$\frac{\partial BOT}{\partial t} = H_{16} + H_{19} \tag{5.18}$$

where v_z is the velocity of the vertical water motion in the ocean (m/day) and k_z is the coefficient of the turbulent mixing (m^2/day).

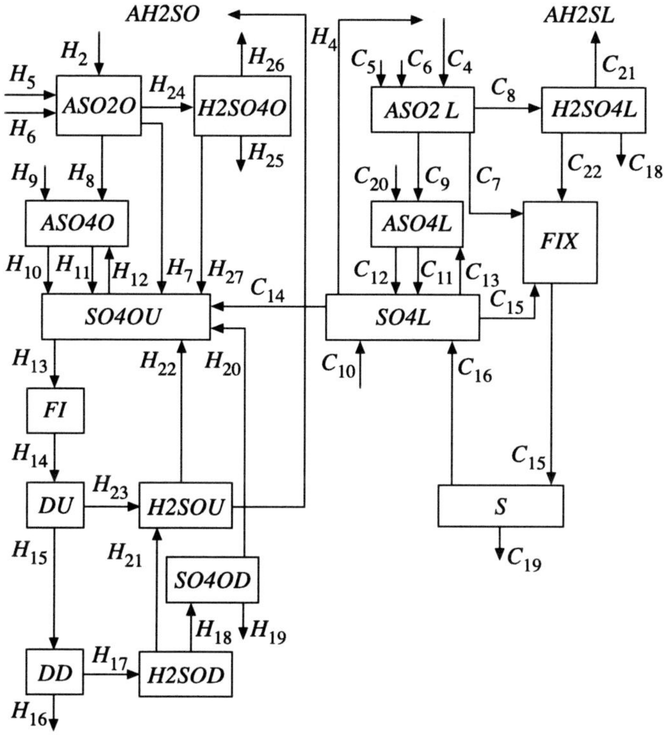

Figure 5.1. The scheme of sulphur fluxes in the environment. Notations are given in Tables 5.1 and 5.2.

The equations (5.1) through (5.18) in each cell of the spatial division of the ocean surface are supplemented with initial conditions (Table 5.2). The boundary conditions for the equations (5.11) through (5.18) are zero. The calculation procedure to estimate the sulphur concentration consists of two stages. At each time moment t_i, first for all cells Ω_{ij}, the equations (5.1)–(5.18) are solved by the method of quasi-linearization, and all reservoirs of sulphur are estimated for $t_{i+1} = t_i + \Delta t$, where a time step Δt is chosen from the condition of convergence of the calculation procedure. Then at the moment t_{i+1} with the use of the climate unit of the global model these estimates are specified with account of the atmospheric transport and ocean currents over the time Δt.

The sulphur supplies in the reservoirs are measured in mg S/m^3, the sulphur fluxes have the dimensionality mg S/m^3/day. The sulphur supplies in the water medium are calculated with the volumes of the compartments Ω_{ijk} taken into account. To estimate the sulphur supplies in the atmosphere, it is assumed that an effective thickness of the atmosphere h is an input parameter either introduced into the model by the user, prescribed as the constants from Table 5.2, or received from the climate unit of the global model. The quantitative estimates of the fluxes in the right-hand parts of the equations (5.1) through (5.18) are obtained in different units

Table 5.1. The characteristics of the land and hydrosphere fluxes of sulphur in the structure in Figure 5.1. Numerical estimates of the fluxes (mg/m^3/day) are obtained by averaging over the respective territories (Krapivin and Nazarian, 1997).

	Land		Hydrosphere	
Sulfur flux	Identifier	Estimate	Identifier	Estimate
Volcanic invasions				
H_2S	C_1	0.018	H_3	0.0068
SO_2	C_5	0.036	H_5	0.0073
SO_4^{2-}	C_{20}	0.035	H_9	0.0074
Anthropogenic emissions				
H_2S	C_2	0.072	H_1	0.00076
SO_2	C_6	0.92	H_6	0.038
SO_4^{2-}	C_{10}	0.47		
Oxidation of H_2S to SO_2	C_4	1.13	H_2	0.3
Oxidation of SO_2 to SO_4^{2-}	C_9	1.35	H_8	0.16
Dry sedimentation of SO_4^{2-}	C_{12}	0.37	H_{11}	0.11
Fall-out of SO_4^{2-} in the rain	C_{11}	1.26	H_{10}	0.38
Biological decomposition and emission of H_2S into the atmosphere	C_3	1.03	H_4	0.31
Assimilation of SO_4^{2-} by biota	C_{15}	0.41	H_{13}	1.09
Biological decomposition and formation of SO_4^{2-}	C_{16}	1.13	H_{17}	0.43
			H_{23}	0.12
Sedimentation and deposits	C_{18}	0.22	H_{15}	0.98
	C_{19}	0.11	H_{16}	0.55
			H_{19}	0.0076
			H_{25}	0.036
Wind-driven return to the atmosphere	C_{13}	0.25	H_{12}	0.33
Replenishing sulphur supplies due to dead biomass	C_{17}	0.86	H_{14}	1.1
Assimilation of atmospheric SO_2	C_7	0.46	H_7	0.18
Washing-out of SO_2 from the atmosphere	C_8	0.27	H_{24}	0.061
River run-off of SO_4^{2-} to the ocean	C_{14}	1.17		
Transition of gas phase H_2SO_4 to H_2S	C_{21}	0.018	H_{26}	0.0076
Assimilation of the washed-out part of atmospheric SO_2 by biota	C_{22}	0.036	H_{27}	0.015
Oxidation of H_2S to SO_2 in water medium			H_{18}	0.045
			H_{22}	0.19
Advection of SO_2			H_{20}	0.38
Advection of H_2S			H_{21}	0.37

of the global model. The anthropogenic fluxes of sulphur H_1, H_6, C_2, C_6, and C_{10} are simulated in the unit of scenarios. The fluxes H_3, H_5, H_9, C_1, C_5, and C_{20} are prescribed either by the climate unit or formed in the unit of scenarios. The accuracy of different functional presentations of the fluxes in the equations (5.1) through (5.18) corresponds to the accuracy of similar fluxes of the biogeochemical

Table 5.2. Some estimates of the sulphur reservoirs that can be used as initial data.

Reservoir	Identifier in Equations (5.1)–(5.18)	Quantitative estimate of the sulphur reservoir (mg/m^2)
The atmosphere over the oceans		
H_2S	*AH2SO*	10
SO_2	*ASO2O*	5.3
SO_4^{2-}	*ASO40*	2
The atmosphere over land		
H_2S	*AH2SL*	36.9
SO_2	*ASO2L*	17.9
SO_4^{2-}	*ASO4L*	12.9
Land		
SO_4^{2-}	*SO4L*	11.2
Biomass	*FIX*	600
Soil	*S*	5,000
Photic layer of the World Ocean		
H_2S	*H2SOU*	1.9
SO_4^{2-}	*SO4OU*	$19 \cdot 10^7$
Biomass	*FI*	66.5
DOM	*DU*	730
Deep layers of the World Ocean		
H_2S	*H2SOD*	$2 \cdot 10^6$
SO_4^2	*SO4OD*	$3.4 \cdot 10^9$
DOM	*DD*	13,120

cycles of hydrogen, phosphorus, and nitrogen. The rate of emission of H_2S into the atmosphere at humus decomposition is described by the linear function $C_3 = \mu_1(\text{pH}) \cdot SO4L \cdot T_L$, where μ_1 is the proportion coefficient depending on soil acidity ($\text{day}^{-1} \cdot \text{K}^{-1}$) and T_L is the soil temperature (°K). The initial value of *SO4L* in equation (5.5) is estimated proceeding from the humus supply considering that the content of sulphur in humus is prescribed by the parameter a_g (%). According to the available observations of the input of H_2S into the atmosphere from the ocean, the flux H_4 varies widely from low values to high values at the transition from stagnant waters to zones of upwelling. The flux H_4 is assumed to be the function of the ratio of the rates of H_2S oxidation in the photic layer to the rate of the vertical uplifting of water. Therefore to describe the H_4 flux, use the parameter t_{H2SU}, which reflects the lifetime of H_2S in the water medium: $H_4 = H2SU/t_{H2SU}$. Determine the value of t_{H2SU} as a function of the rate of the vertical advection v_z and concentration of oxygen in the upper layer Z_{H2S} thick: $t_{H2SU} = H2SOU \cdot v_z(\theta_2 + O_2)/[O_2(\theta_1 + v_z)]$, where the constants θ_1 and θ_2 are determined empirically, the value of O_2 is either calculated by the oxygen unit of the global model or prescribed from the global database (Keeling *et al.*, 1998).

The reaction of oxidation of H_2S to SO_2 in the atmosphere, on land, and over the water surface is characterized by a rapid process of the reaction of hydrogen sulphide with atomic and molecular oxygen. At the same time, the reaction of H_2S with O_3 in the gas phase is slow. It is impossible to describe within the global model the diversity of the situations appearing here. However, an inclusion of the fluxes H_2 and C_4 to the unit of sulphur enabled us to take into account the correlation between the cycles of sulphur and oxygen. These fluxes are parameterized with the use of the indicator t_{H2SA} of the lifetime of H_2S in the atmosphere: $C_4 = AH2SL/t_{H2SA}$ and $H_2 = AH2SO/t_{H2SA}$. The mechanism to remove SO_2 from the atmosphere is described by the fluxes H_7, H_8, H_{27}, C_7, C_8, and C_9. Schematically this mechanism consists of a set of interconnected reactions of SO_2 with atomic oxygen under the influence of various catalysts. A study of the succession of the reaction enables us to estimate the lifetime of SO_2 for oxidation over land t_{SO2L} and the water surface t_{SO2A1}. This makes it possible to assume the following parameterizations of the fluxes H_8 and C_9: $H_8 = ASO2O/t_{SO2A1}$ and $C_9 = ASO2L/t_{SO2L}$.

Sulphur dioxide is assimilated from the atmosphere by rocks, vegetation, and other Earth surfaces. Over the water surface this assimilation is connected with the intensity of turbulent gas fluxes and surface roughness. We describe a dry deposition of SO_2 over the vegetation by the model $C_7 = q_2 RX$, where $q_2 = q_2' \cdot ASO2L/(r_{tl} + r_s)$, r_{tl} is the atmospheric resistance to the SO_2 transport over the vegetation of type l (day/m), r_s is the surface resistance of the type s to the SO_2 transport (day/m), RX is the produce of vegetation of type X (mg/m^2/day) (calculated by the biogeocoenotic unit of the global model), and q_2' is the proportion coefficient. The parameters r_{tl} and r_s are functions of the types of the soil-vegetation formations and are estimated, respectively, at 0.05 and 4.5 for the forests, 0.9 and 3 for grass cover, 0.5 and 2 for bushes, 0.8 and 1 for bare soils , 1.9 and 0 for water surfaces, and 2 and 10 for snow cover.

The process of washing-out of SO_2 from the atmosphere with changing phase to H_2SO_4 and a subsequent neutralization on the surface of the type l is described by the function $C_8 = q_{1l} W \cdot ASO2L$ with the Langmuir coefficient q_{1l} and precipitation intensity $W(\varphi, \lambda, t)$. The interaction between acid rain and Earth surface elements is reflected in the scheme in Figure 5.1 by the fluxes C_{18}, C_{21}, and C_{22} for land and H_{25}, H_{26}, and H_{27} for water surfaces. To parameterize these fluxes, assume the hypothesis that the reservoirs of $H2SO4L$ and $H2SO4O$ are spent in proportion to the outfluxes, and the coefficients of this proportion are the controlling parameters of the numerical experiments: $C_{18} = h_1 \cdot H2SO4L$, $C_{22} = h_2 \cdot RX \cdot H2SO4L$, $C_{21} = h_3 T_a \cdot H2SO4L$, $H_{25} = h_6 \cdot H2SO4O$, $H_{26} = h_4 T_a \cdot H2SO4O$, $H_{27} = h_5 \cdot RFI \cdot H2SO4O$, $h_1 + h_2 \cdot RX + h_3 T_a = 1$, and $h_4 T_a + h_5 \cdot RFI + h_6 = 1$, where $T_a(\varphi, \lambda, t)$ is the air surface temperature. Parameterize the fluxes H_7 and H_{24} by the relationships: $H_7 = ASO2O/t_{SO2A2}$, $H_{24} = q_{1l} W \cdot ASO2O$, where t_{SO2A2} is the lifetime of SO_2 over the water surface.

Sulphates interacting with the ecosystems and establishing an interaction of the sulphur cycle with other biogeochemical processes are one of the most important elements in the global cycle of sulphur. Numerous complicated transformations of sulphates in the environment are described by a set of fluxes H_7, H_8, H_{10}, H_{11}, H_{12}, C_9, C_{11}, C_{12} and C_{13} for the atmospheric reservoir and the fluxes H_{13}, H_{18}, H_{19}, H_{20}, H_{22}, C_3, C_{14}, C_{15} and C_{16} for land and the World Ocean.

Physical mechanisms for the transport of sulphates from the atmosphere to the soil and water medium are connected with dry and wet sedimentation. An efficient model of the wet removal of particles and gases from the atmosphere was proposed by Langmann (2000). A substitution of the mechanism of the aerosols and gases by a simplified binary model enables one to match it with other units of the global model: $H_{10} = \mu W \cdot ASO4O$, $H_{11} = \rho \nu_O \cdot ASO4O$, $C_{11} = b_3 W \cdot ASO4L$, $C_{12} = d_1 \nu_a \cdot ASO4L$, where ν_O and ν_a are the rates of the aerosols dry deposition over the water surface and land, respectively; μ, b_3, ρ, and d_1 are the constants.

The return of sulphates from the soil and water medium to the atmosphere is connected with rock weathering and spray above the rough water surface: $C_{13} = d_2 \cdot RATE \cdot SO4L$, $H_{12} = \theta \cdot RATE \cdot SO4U$, where $RATE(\varphi, \lambda, t)$ is the wind speed over the surface (m/s) and d_2 and θ are the empirical coefficients.

The flux C_{14} relates the surface and water reservoirs of sulphur. Let σ be the share of the river system area on land and d_3 the proportion coefficient, then $C_{14} = d_3 W \cdot SO4L + (C_{11} + C_{12})\sigma$.

The surface part of the sulphur cycle is connected with the functioning of the atmosphere–vegetation–soil system. Plants adsorb sulphur from the atmosphere in the form of SO_2 (fluxes C_7 and C_{22}) and assimilate sulphur from the soil in the form of SO_4^{2-} (flux C_{15}). In the hierarchy of the soil processes two levels can be selected defining the sulphur reservoirs as "dead organics" and "SO_4^{2-} in soil". The transitions between them are described by the flux $C_{16} = b_2 ST_L$, where the coefficient $b_2 = b_{2,1} b_{2,2}$ reflects the rate $b_{2,1}$ of transition of sulphur contained in dead organics into the form assimilated by vegetation. The coefficient $b_{2,2}$ indicates the content of sulphur in dead plants.

The fluxes of sulphur in the water medium according to studies by Bodenbender *et al.* (1999) depend on the biological processes in the water bodies and constitute an isolated part of the global cycle of sulphur that contains only the fluxes that connect it with the atmospheric and surface cycles. Rough estimates show that the rates of the sulphur cycle in the water of the seas and oceans do not play a substantial role for the remaining parts of its global cycle. And though this specific feature does exist, for the purity of the numerical experiment, in the proposed model the internal hydrospheric fluxes of sulphur compounds are separated in space and parameterized with the same details as other fluxes of sulphur in the atmosphere and on land. This excessiveness is important for other units of the global model as well. In particular, it is important for the parameterization of photosynthesis whose rate *RFI* affects the closure of other biogeochemical cycles. Finally, assume: $H_{13} = \gamma \cdot RFI$, $H_{14} = b \cdot MFI$, $H_{15} = f \cdot DU$, $H_{16} = p \cdot DD$, $H_{17} = q \cdot DD$, $H_{18} = H2SOD/t_{H2SOD}$, $H_{19} = u \cdot SO4D$, $H_{20} = a_1 \nu_D \cdot SO4D$, $H_{21} = b_1 \nu_D \cdot H2SOD$, $H_{22} = H2SOU/t_{H2SOU}$, $H_{23} = g \cdot DU$, where MFI is the mass of dead phytoplankton, t_{H2SOU} and t_{H2SOD} is the time of complete oxidation of H_2S in the sea water in the photic and deep layers, respectively; γ, b, f, p, q, u, a_1, b_1, and g are the constants.

The anthropogenic input to the sulphur unit is realized through the prescribed fluxes C_2, C_6, C_{10}, H_1, and H_6 in the form of the functions of the spatial coordinates and time.

5.4 MODEL OF THE GLOBAL PHOSPHORUS CYCLE

In contrast to nitrogen, the main reservoir of phosphorus in the biosphere is not the atmosphere but the rocks and other deposits formed in the past geological epochs, which, being subject to erosion, emit phosphates. Besides these, there are other mechanisms of the return of phosphorus to the biospheric cycle, but, as a rule, they are not that efficient. One of these mechanisms is fish catching, returning to land from the hydrosphere about $60 \cdot 10^3$ t P/yr, as well as an extraction of phosphorus-containing rocks estimated at $1\text{–}2 \cdot 10^6$ t P/yr. The present cycle of phosphorus is closed due to its flux to the bottom deposits in the World Ocean where it is transported along with sewage by coastal and river run-off.

The flux diagram of the model of the global phosphorus cycle (MGPC) is presented in the scheme in Figure 5.2, according to which the balance system of equations will be:

$$\frac{\partial P_A}{\partial t} + V_\varphi \frac{\partial P_A}{\partial \varphi} + V_\lambda \frac{\partial P_A}{\partial \lambda} = H_1^P + H_{19}^P + \begin{cases} -H_{16}^P & (\varphi, \lambda) \in \Omega_O \\ H_7^P - H_8^P & (\varphi, \lambda) \in \Omega/\Omega_O \end{cases}$$

$$\frac{\partial P_U}{\partial t} + v_\varphi \frac{\partial P_U}{\partial \varphi} + v_\lambda \frac{\partial P_U}{\partial \lambda} + v_z \frac{\partial P_U}{\partial z} = H_{11}^P + H_{15}^P + H_{16}^P - H_9^P - H_{10}^P$$

$$\frac{\partial P_L}{\partial t} + v_\varphi \frac{\partial P_L}{\partial \varphi} + v_\lambda \frac{\partial P_L}{\partial \lambda} + v_z \frac{\partial P_L}{\partial z} = H_{12}^P + H_{14}^P - H_{13}^P - H_{15}^P$$

$$\frac{\partial P_S}{\partial t} = H_2^P + H_8^P + H_9^P + H_{10}^P - H_6^P - H_7^P - H_{11}^P$$

where $P_U = P_{U1} + P_{U2} + P_{U3}$, and $P_S = P_{S1} + P_{S2}$. With this detailing we have:

$$\frac{\partial P_{U1}}{\partial t} + v_\varphi \frac{\partial P_{U1}}{\partial \varphi} + v_\lambda \frac{\partial P_{U1}}{\partial \lambda} + v_z \frac{\partial P_{U1}}{\partial z} = H_{17}^P - H_9^P - H_{10}^P - H_{20}^P$$

$$\frac{\partial P_{U2}}{\partial t} + v_\varphi \frac{\partial P_{U2}}{\partial \varphi} + v_\lambda \frac{\partial P_{U2}}{\partial \lambda} + v_z \frac{\partial P_{U2}}{\partial z} = H_{20}^P - H_{18}^P$$

$$\frac{\partial P_{U3}}{\partial t} + v_\varphi \frac{\partial P_{U3}}{\partial \varphi} + v_\lambda \frac{\partial P_{U3}}{\partial \lambda} + v_z \frac{\partial P_{U3}}{\partial z} = H_{11}^P + H_{18}^P - H_{17}^P$$

$$\frac{\partial P_{S1}}{\partial t} = H_3^P - H_4^P - H_5^P$$

$$\frac{\partial P_{S2}}{\partial t} = H_2^P + H_4^P + H_5^P + H_8^P - H_3^P - H_6^P - H_7^P$$

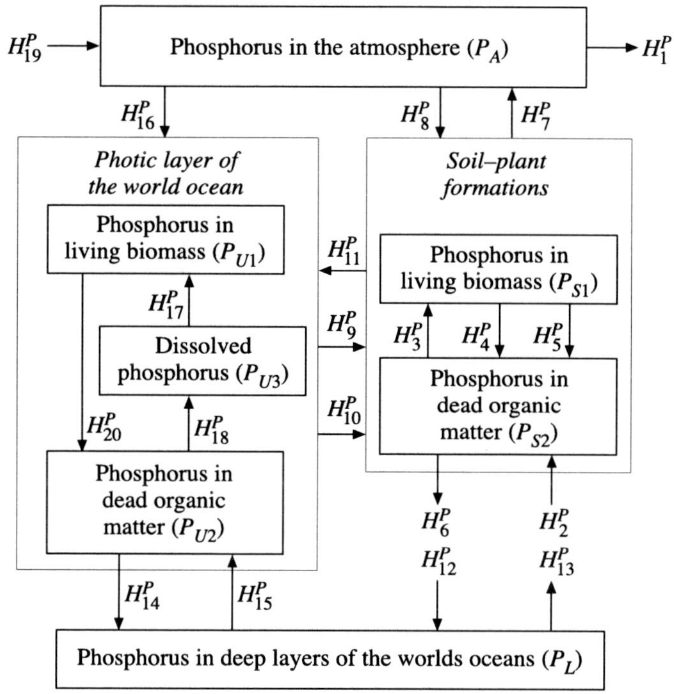

Figure 5.2. The scheme of phosphorus fluxes in the biosphere. Notations are given in Table 5.3.

Determine now the functional and dynamic characteristics of the fluxes of phosphorus (Table 5.3) based on analysis of the existing ideas about their nature. The atmospheric cycle is governed by rock weathering, volcanic eruptions, and by washing-out of phosphorus by precipitation. From the available estimates, the phosphorus content in the lithosphere constitutes 0.093%, and the processes of weathering delivers annually into the atmosphere about $0.2 \cdot 10^6$ t P, on average. Since these processes are of a complicated stochastic nature and their models are absent, in the first approximation the fluxes H_1^P and H_{19}^P will be considered constant.

The continental cycle of phosphorus is determined by ten fluxes (Figure 5.2) closed by a single component P_S indicating the phosphorus supplies on land in the soil–vegetation formations and in animals. The supplies of phosphorus in soils are replenished due to the fluxes H_l^P ($l = 2, 4, 5, 8, 9, 10$). The loss of phosphorus from the soil is determined by the fluxes H_j^P ($j = 3, 6, 7, 11$). With a more complicated detailing of the surface reservoirs of phosphorus and consideration of more ingenious effects in the interaction of these reservoirs, the classification of the surface fluxes of phosphorus becomes more complicated. Within the assumed detailing of the surface reservoirs of phosphorus the following functional presentations of the fluxes $\{H_i^P\}$ can be considered:

Table 5.3. The characteristics of fluxes (10^6 t/yr) and reservoirs (10^6 t) of phosphorus in the biosphere (Krapivin, 2000b).

Reservoirs and fluxes of phosphorus	Identifier of the MGPC unit	Estimate
Phosphus supplies		
In the atmosphere	P_A	3
On land	P_S	1,546
In the photic layer of the World Ocean	P_U	$2 \cdot 10^4$
In deep layers of the World Ocean	P_L	$12 \cdot 10^4$
Volcanic emissions	H_1^P	0–2
Fertilization	H_2^P	19
Assimilation by plants	H_3^P	45.34
Input with dead plants	H_4^P	39.34
Input due to organisms functioning		
On land	H_5^P	5
In the World Ocean	H_{20}^P	81.5
Transition into the unassimilable form	H_6^P	2.9
Weathering	H_7^P	5
Falling-out with precipitation		
On land	H_8^P	1.8
On the oceans	H_{16}^P	2
Removal with fish catch	H_9^P	0.06
Removal by birds	H_{10}^P	0.04
Washing-out and sink into the World Ocean	H_{11}^P	4–14
Input due to detritus lysis in the oceans		550
Photic layer	H_{18}^P	159
Deep layers	H_{12}^P	
Exchange between photic and deep layers of the ocean		
Lifting	H_{15}^P	96.1
Descending	H_{14}^P	22
Precipitation	H_{13}^P	13–83.9
Rock weathering	H_{19}^P	1
Photosynthesis	H_{17}^P	630–1,300

$$H_2^P = p_1 GMG_0^{-1} M_0^{-1}$$

$$H_3^P = p_V R_V$$

$$H_4^P = p_4 M_L$$

$$H_5^P = p_5 H_F$$

$$H_6^P = p_6 P_{S2} / P_{S2,0}$$

$$H_9^P = p_3 I$$

$$H_7^P = p_7 \theta^{\Delta T} P_{S2}/P_{S2,0}$$

$$H_{11}^P = p_2 P_{S2}[1 - \exp(-k_{su} W_{so})]/P_{S2,0}$$

where p_V is the content of phosphorus in the living biomass of plants, p_4 and p_5 is the content of phosphorus in organic matter of the vegetable and animal origin, respectively, I is the amount of produce extracted from the ocean, G is the population, M is mineral resources, R_V is the produce of the vegetation photosynthesis, H_F is the biomass of the unassimilated food of animals, $M_L = \mu_V L$, μ_V is the rate of vegetation dying-off, L is the vegetation biomass, θ is the temperature coefficient of the rate of the dead organic matter decomposition on land, ΔT is the saturated air temperature (SAT) variation with respect to the control value, W_{so} is the volume of the river run-off into the oceans, and $p_i (i = 1, \ldots, 7)$ are the constants. The index "0" in G_0, M_0, and $P_{S2,0}$ attributes these parameters to some control time moment t_0, when all parameters of $MGPC$ are known.

Describe the hydrochemical cycle of phosphorus by a totality of fluxes $H_k^P (k = 9\text{–}18, 20)$: $H_{12}^P = p_{14} R_{DL}$, $H_{13}^P = p_8 P_L/P_{L,0}$, $H_{14}^P = p_9 P_U/P_{U,0}$, $H_{15}^P = p_{10} P_L/P_{L,0}$, $H_{16}^P = p_{12} R_{WO} P_A/P_{A,0}$, $H_{17}^P = p_{13} R_\Phi$, $H_{18}^P = p_{15} R_{DU}$, $H_{20}^P = p_{16} M_\Phi$, where R_D is the rate of the dead organic matter decomposition, R_{WO} is precipitation over the ocean, R_Φ is the produce of phytoplankton and other living organisms in the ocean, M_Φ is the rate of the living biomass dying-off, and p_i $(i = 8\text{–}16)$ are the proportion constants.

5.5 MODEL OF THE GLOBAL NITROGEN CYCLE (MGNC)

The MGNC unit simulating the fluxes of nitrogen in the environment is necessary in the global biospheric model for several indisputable reasons: nitrogen compounds can affect the environmental conditions, change the food quality, affect the climate, and transform the hydrospheric parameters. An abundant use of nitrates leads to water pollution and deteriorates the quality of food products. It is well known that an intensive exploitation of soils without taking into account the consequences of the misuse of nitrogen fertilizers breaks the stability of the agroecosystems and human health. Besides, nitrous oxide (N_2O), nitrogen dioxide (NO_2), and nitrogen oxide (NO), being minor gas components of the atmosphere, substantially affect the formation of the processes of absorption of optical radiation in the atmosphere. Small deviations in their concentrations can cause significant climatic variations near the Earth's surface (Kondratyev, 1999a–c; Gorshkov and Makaryeva, 1999; Sellers *et al.*, 1995; Stockwell *et al.*, 1999).

The nitrogen cycle is closely connected with the fluxes of hydrogen, sulphur, and other chemicals (Smith *et al.*, 1998; Dimitroulopoulou and Marsh, 1997).

5.5.1 Conceptual schemes of the nitrogen cycle in nature

The global cycle of nitrogen as one of the nutrient elements is a mosaic structure of local processes of its compounds formed due to water migration and atmospheric processes. The present-day nitrogen cycle is strongly subject to anthropogenic forcings manifested through interference into the nitrogen cycle both directly and via the influence of related processes. Therefore the construction of an adequate model of the nitrogen cycle in its present nature should be based on the description of the whole complex of natural processes and those initiated by humans.

The natural sources of nitrogen oxides are connected with the vital functions of bacteria, volcanic eruptions, as well as several atmospheric phenomena (e.g., lightening discharges). The biogeochemical cycle of nitrogen includes the processes of fixation, mineralization, nitrification, assimilation, and dissimilation. The structural schemes of these processes have been described in detail by many authors. Their complexity level is determined by the goal of studies, availability of data on the rates of transformation of the nitrogen-containing compounds and their supplies, by the level of detailing, etc.

Nitrogen moves in the biosphere by a complicated meandering structure of fluxes consisting of a hierarchy of cycles at various levels of life organization. From the atmosphere, nitrogen enters the cells of micro-organisms, from which it enters the soil and then progresses to higher plants, animals, and humans. Dying-off of living organisms results in the return of nitrogen into the soil, from which it either again goes to plants and living organisms or is emitted to the atmosphere. A similar scheme of nitrogen oxide cycling is inherent to the hydrosphere. The characteristic feature of these cycles is their openness connected with the available processes of removal of nitrogen from the biospheric balance into rocks, from which it returns much slower. Taking into account the nature of the nitrogen cycle in the biosphere and its reservoir structure enables one to formulate a global scheme of nitrogen fluxes.

5.5.2 The block-scheme of the global nitrogen cycle in the biosphere

A comparative analysis of the model schemes of the flux diagrams of nitrogen compounds in nature proposed by various experts makes it possible to construct a block-scheme presented in Figure 5.3. Here the atmosphere, soil, lithosphere, and hydrosphere are considered as nitrogen reservoirs. The first three reservoirs are described by 2-D models, and the hydrosphere is described by a 3-D multilayer model. The characteristics of nitrogen fluxes between these reservoirs are given in Table 5.4. The equations of the model are written as:

$$\frac{\partial N_A}{\partial t} + V_\varphi \frac{\partial N_A}{\partial \varphi} + V_\lambda \frac{\partial N_A}{\partial \lambda} = H_1^N + \begin{cases} H_{20}^N - H_{16}^N & (\varphi, \lambda) \in \Omega_O \\ H_7^N + H_{19}^N - H_8^N - H_9^N + H_{22}^N - H_2^N - H_{10}^N & (\varphi, \lambda) \in \Omega/\Omega_O \end{cases} \quad (5.19)$$

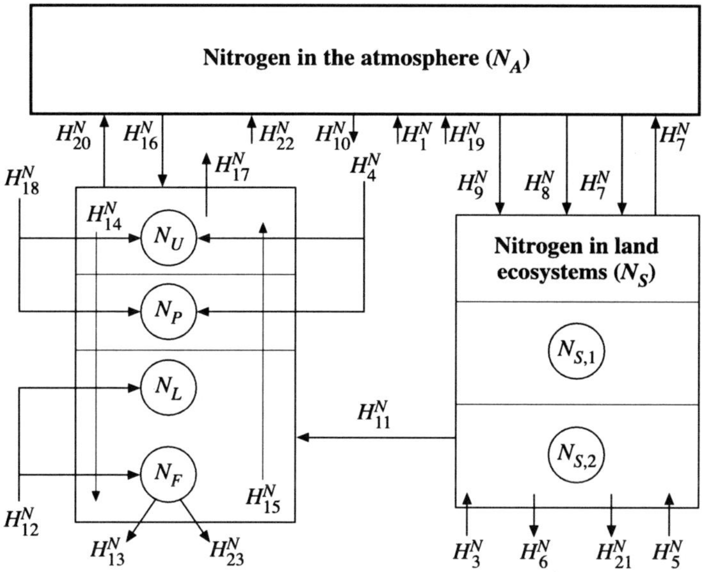

Figure 5.3. The scheme of nitrogen fluxes in nature (see Table 5.4).

$$\frac{\partial N_{S1}}{\partial t} = H_8^N + H_6^N - H_3^H \tag{5.20}$$

$$\frac{\partial N_{S2}}{\partial t} = H_2^N + H_3^N + H_5^N + H_9^N - H_6^N - H_7^N - H_{11}^N - H_{21}^N \tag{5.21}$$

$$\frac{\partial N_U}{\partial t} + v_\varphi \frac{\partial N_U}{\partial \varphi} + v_\lambda \frac{\partial N_U}{\partial \lambda} = H_{16}^N + H_{4,U}^N + H_{18,U}^N + H_{11}^N - H_{17,U}^N$$

$$- H_{20}^N - H_{14,UP}^N - H_{15,UP}^N \tag{5.22}$$

$$\frac{\partial N_P}{\partial t} + v_\varphi \frac{\partial N_P}{\partial \varphi} + v_\lambda \frac{\partial N_P}{\partial \lambda} = H_{18,P}^N + H_{4,P}^N + H_{14,UP}^N + H_{15,PL}^N$$

$$- H_{17,P}^N - H_{14,PL}^N - H_{15,UP}^N \tag{5.23}$$

$$\frac{\partial N_L}{\partial t} = Q_L + H_{12,L}^N + H_{14,PL}^N + H_{15,LF}^N - H_{14,LF}^N - H_{15,PL}^N \tag{5.24}$$

$$\frac{\partial N_F}{\partial t} = Q_F + H_{12,F}^N + H_{23}^N + H_{14,LF}^N - H_{13}^N - H_{15,LF}^N \tag{5.25}$$

where $V(V_\varphi, V_\lambda)$ is the wind speed, $v(v_\varphi, v_\lambda)$ is the current velocity in the ocean, Q_L and Q_F are functions describing the mixing of the deep waters of the ocean.

To simplify the calculation scheme presented in Figure 5.3, the advective processes in equations (5.19) through (5.25) can be described by a superposition of the fluxes H_{14}^N and H_{15}^N. The computer realization of the equations of the MGNC unit

Table 5.4. The characteristics of the reservoirs and fluxes of nitrogen in the biosphere (see Figure 5.3).

Reservoirs (Gt) and fluxes (10^6 t/yr)	Identifier	Estimate
Nitrogen supplies		
Atmosphere	N_A	$39 \cdot 10^5$
Soil	N_S	280
Photic and intermediate layer of the ocean	$N_U + N_P$	2,800
Deep and bottom layer of the ocean	$N_L + N_F$	36,400
Natural sources of the hydrosphere	H_1^N	0.392
Technogenic accumulation		
In fuel burning	H_2^N	22.8
In fertilizer production	H_9^N	41.8
Input due to dead organisms		
On land	H_3^N	42.2
In the upper layers of the World Ocean	H_{18}^N	5
In deep layers of the World Ocean	H_{12}^N	7.8
Input due to organisms functioning on land		
In the World Ocean	H_5^N	0.1
Biological fixation	H_4^N	0.3
On land	H_6^N	20.3
In the World Ocean	H_{17}^N	10
In the atmosphere	H_{10}^N	40
Denitrification		
On land	H_7^N	
In the World Ocean	H_{20}^N	49.8
Atmospheric fixation		
Over land	H_8^N	4
Over the World Ocean	H_{16}^N	3.6
Run-off from land into the World Ocean	H_{11}^N	38.6
Precipitation		
Vertical exchange processes in the oceans	H_{13}^N	0.5
Descending	H_{14}^N	0.2
Lifting	H_{15}^N	7.5
Anthropogenic emissions to the atmosphere	H_{19}^N	15
Removal of nitrogen from the cycle due to sedimentation	H_{21}^N	0.2
Input of nitrogen to the atmosphere in the process of rocks weathering	H_{22}^N	0.217
Input of nitrogen to the water medium with dissolving sediments	H_{23}^N	0.091

introduces into the equations (5.19)–(5.25) some corrections for the agreement between the dimensionalities of the variables in conformity with the spatial digitization of Ω. Therefore the estimates of the fluxes H_i^N given below for their consideration in the MGNC should be corrected following this criterion.

5.5.3 Atmospheric components of the nitrogen cycle

The atmospheric part of the nitrogen cycle is an example of a complicated mechanism of transformation of gas substances characterized by an intricate set of fluxes on the borders between the basic reservoirs of nitrogen. Nevertheless, the results obtained by many experts make it possible to formulate clear ideas about the flux diagram of nitrogen compounds in the atmosphere.

Nitrogen resides in the atmosphere both in a free state (N_2) and in the form of various compounds – ammonia (NH_3), nitrous oxide, nitrogen oxide, nitrogen dioxide, and other nitrogen oxides (NO_3, N_2O_3, N_2O_4, N_2O_5), which play an intermediate role in chemical reactions. From the available estimates, the active part of atmospheric nitrogen constitutes $3.92 \cdot 10^{12}$ t, that is, $N_A = 0.77 \cdot 10^4\,\text{t/km}^2$. Detailing the atmospheric reactions of nitrogen is incomplete because of insufficiently studied sources and behaviour of various forms of ammonia. The most important reactions in the atmosphere are the following:

$$\mathrm{NO_2 \rightarrow NO + O} \qquad \varphi K_a = 0\text{–}25\,\mathrm{h}^{-1}$$

$$\mathrm{O_3 + NO \rightarrow NO_2 + O_2} \qquad K_1 = 1320\,\mathrm{ppm}^{-1}\,\mathrm{h}^{-1}$$

The photochemical equilibrium is described by the relationship:

$$(\mathrm{NO})(\mathrm{O_3})/(\mathrm{NO_2}) = \varphi K_a / K_1$$

The time of relaxation in this case constitutes 16 s and therefore the equilibrium between NO, NO_2, and O_3 in the atmosphere can be considered stable. However, the equilibrium $N_2 + O_2 \leftrightarrow 2NO$ under anthropogenic conditions is connected with the transformation of NO into NO_2 over several hours. Therefore here, too, from the viewpoint of global modelling, a separate consideration of the components NO and NO_2 is inexpedient. In other words, we shall consider nitrogen of the atmosphere as a generalized component of the global model.

5.5.3.1 Atmospheric fixation

As a result of various physico-chemical processes taking place in the atmosphere, free nitrogen can move from the atmosphere to soil and water bodies. Fixation of atmospheric nitrogen due to electrical charges and photochemical processes constitutes annually not more than $0.035\,\text{t/km}^2$ (from more accurate estimates, it may constitute $0.027\,\text{t/km}^2$ for land and $0.01\,\text{t/km}^2$ for oceans). Since the nitrogen fluxes due to atmospheric fixation are mainly determined by meteorological conditions, it is quite natural to consider them independently for each region of land and each water body of the World Ocean in the form of the functions of temperature and precipitation.

The flux H_{16}^N of nitrogen fixed in the atmosphere over any ocean basin is described by the relationship:

$$H_{16}^N = [\lambda_1(\theta_1)^{\Delta T} + \lambda_2 R_W] N_A$$

where ΔT is the atmospheric temperature variation, θ_1 is the indicator of the temperature dependence of the rate of atmospheric fixation of nitrogen, R_W is precipitation, λ_1 and λ_2 are the coefficients.

The equation of atmospheric fixation over a land site Ω_{ij} is written by analogy to H_{16}^N:

$$H_{8,ij}^N = [\lambda_3(\theta_1)^{\Delta T} + \lambda_4 R_{W,ij}]N_A$$

where λ_3 and λ_4 are coefficients.

To estimate the coefficients λ_i $(i = 1, \ldots, 4)$, as a first approximation one can use average data on nitrogen fluxes and precipitation. If we assume $H_{16}^N = 9.96 \cdot 10^{-3}$ t/km^2/yr and $H_8^N = 0.027$ t/km^2/yr, estimate local precipitation over the ocean and land at 1.01 m/yr and 0.24 m/yr, respectively, and the convective precipitation over the ocean and land at 0.19 m/yr and 0.116 m/yr, respectively, we obtain $\lambda_1 = 0.00498$, $\lambda_2 = 0.00458$, $\lambda_3 = 0.0135$, and $\lambda_4 = 0.0285$. These estimates are easily specified with account of local data at a fixed moment in time for smaller regions and water bodies.

5.5.3.2 *Input of nitrogen into the atmosphere from geospheric sources*

The flux of nitrogen H_1^N is determined by the geothermal activity of the Earth. Its estimates testify to a necessity of consideration of this constituent in the global model. In particular, for instance, in the nitrogen fumaroles of Vesuvius the content of nitrogen by weight constitutes 98%, in gases of the lavas of the Hawaiian volcanoes there are only 5.7% of nitrogen, and over the globe an input of juvenile nitrogen averages $0.4 \cdot 10^6$ t/yr. Let H_1^N be a function of time approximating a statistical series of observations (Hernández *et al.*, 2001; Jones and Cox, 2001). A more strict account of this flux of nitrogen in the model can be realized by using the algorithms of parameterization of random processes, for instance, with the use the evolutionary modelling. However, within the global model, orientated towards describing the processes with a time step of decades, it is enough to use a consecutive approximation with an average annual step-by-step discretization.

The flux H_1^N can be, to some extent, interpreted as a compensation for the fluxes H_{13}^N and H_{21}^N.

5.5.4 The surface part of the nitrogen cycle in the biosphere

The nitrogen supplies on land consist of the assimilable nitrogen in soil $N_{S2} \approx 0.19 \cdot 10^4$ t/km^2, in plants ($12 \cdot 10^9$ t), and living organisms ($0.2 \cdot 10^9$ t). A diversity of nitrogen fluxes is formed here of the processes of nitrification, denitrification, ammonification, fixation, and river run-off. The intensities of these fluxes depend on climatic conditions, temperature regime, humidity, as well as chemical and physical properties of the soil. Many qualitative and quantitative characteristics of these dependences have been described in the literature. Let us consider some of them.

5.5.4.1 *Nitrification*

The return of nitrogen to the cycle due to living micro-organisms is one of the stabilizing natural processes. In order to simplify the whole process of transformation of ammonia salts into nitrates, we will consider the activity of heterotrophic micro-organisms and saprophagans as a standard process of organic matter decomposition. The rate of the organic matter decomposition and nitrification increases with an increasing temperature, reaching its optimum value at $T_a = 34.5°C$. Therefore for the flux H_3^N an approximation $H_3^N = \lambda_{N\kappa} M_\kappa$ can be assumed, where M_κ is the rate of the dying-off of the component κ and $\lambda_{N\kappa}$ is the content of nitrogen in the component κ.

5.5.4.2 *Denitrification*

The processes of denitrification of H_7^N on land represent an important channel for the input of nitrogen into the atmosphere. The intensity of these processes depends on temperature, humidity, pollution of soils with poisonous chemicals, and pH. The quantitative and functional characteristics of these dependences have been well studied. Within the global model it is only possible to take into account the factors of temperature and humidity:

$$H_7^N = \lambda_6 \theta_2^{\Delta T} W_S \frac{N_S}{k_1 + N_S}$$

where W_S is soil moisture, θ_2 is the temperature coefficient, and λ_6 and k_1 are empirical parameters. If we assume $H_7^N = 0.318\,\mathrm{t/km^2/yr}$, then $\lambda_6 = 0.496$ and $k_1 = 0.556$.

5.5.4.3 *Biological fixation*

In the biological cycle of nitrogen of importance are the processes of its fixation by micro-organisms and plants whose intensity is estimated at $148 \cdot 10^6$ t/yr. The rate of fixation, depending on the character of the medium, can vary reaching $3 \cdot 10^9$ t/yr in highly productive regions. The nitrogen flux H_{10}^N depends on the distribution of vegetation cover and can be described by the equation $H_{10}^N = \sigma_\kappa \lambda_\kappa \gamma_\kappa / \sigma_{ij}$, where σ_κ is the area under vegetation of type κ in the territory Ω_{ij} of the area σ_{ij}, R_κ is the productivity of plants of type κ, and λ_κ is the coefficient.

Fixation of nitrogen by plants directly from the soil via the root systems (flux H_6^N) occupies a principle place in the cycle of nitrogen, especially in the territories with cultural vegetation. For instance, an increase of the share of legumes in agriculture can raise H_6^N to 35 t/km^2/yr. Therefore consideration of this flux in the model is necessary and can be realized in the following form:

$$H_6^N = \sigma_\kappa R_\kappa \mu_\kappa / \sigma_{ij} \tag{5.26}$$

where μ_κ is the constant.

The rate of assimilation of nitrogen by the roots of plants is known to much depend on the soil temperature regime, decreasing a little with temperatures lowering to 8–10°C and dropping significantly at temperatures below 5–6°C. The motion of nitrogen from the roots to the upper parts of the plant also slow down with decreasing temperature. The formula (5.26) reflects this regularity through the respective reactions of plant productivity (time lag neglected).

On land, plants assimilate about $30 \cdot 10^6$ t N annually from the atmosphere and more than $5.3 \cdot 10^6$ t N directly from the soil. Approximate estimates of the productivity of various types of vegetation average $R_\kappa = 710$–$3243\,\mathrm{t/km^2/yr}$. Hence, from (5.26) we have $\mu_k = 0.134 \cdot 10^{-5}$–$0.506 \cdot 10^{-4}$.

5.5.4.4 *The loss of nitrogen through washing-out from soils*

On global scales, means of nitrogen migration include the transport of its compounds between land and oceans due to water run-off. The annual input of nitrogen from land into the World Ocean is estimated at $38.6 \cdot 10^6$ t. If the total sink to the ocean from land is described by the function W_{SO}, then the nitrogen flux H_{11}^N can be approximated by the expression:

$$H_{11}^N = \lambda_N N_{S2}[1 - \exp(-k_N W_{SO})]$$

where λ_N and k_N are the coefficients. The functional form foresees that the nitrogen flux from land to the ocean is equal to zero in the absence of the run-off and its stabilization at a level λ_N, with the run-off volume considerably increasing. To estimate the parameters λ_N and k_N, it is necessary to take into account the spatial heterogeneity of the types of soil–vegetation formations, relief, and other geophysical parameters. In particular, the content of nitrogen compounds in water differs as a function of the run-off territory. The river waters in forest regions in temperate climate contain 0.4 mg/l of nitrates, for an arid climate this value is 1.45 mg/l. The concentration of nitrates increases sharply in the drainage waters of irrigation systems (5.5 mg/l), in the river waters of densely populated regions (25 mg/l), and reaches a maximum in the soil solutions of the salted and irrigated soils (200 mg/l). Ground waters contain between 10 to 100 mg/l of nitrates. With the total run-off of water into the World Ocean reaching $50 \cdot 10^3\,\mathrm{km^3}$, 30% of which constitutes underground run-off, the total flux of nitrogen per unit area of the ocean is 0.107 t/yr. Assuming $W_{SO} = 0.337$ m and that a 95% level of the sink saturation is reached at a fivefold increase of W_{SO}, we obtain $k_N = 1.367$ and $\lambda_N = 0.708$.

The surface part of the nitrogen cycle contains a constant process of removal of nitrogen from the biosphere into deposits, in particular, as a result of accumulation of saltpetre (pottasium nitrate) on the Earth's surface through erosion and oxidation. From the available estimates, $H_{21}^N \approx 3.9 \cdot 10^{-4}\,\mathrm{t/km^2/yr}$, with $H_{21}^N < H_{22}^N$, but $H_{21}^N + H_{13}^N \cong H_1^N + H_{22}^N + H_{23}^N$. This relationship follows from the fact that during the geological period the loss of nitrogen had been balanced by its input. Of course, in the present biosphere, with the changing intensity of most of the fluxes enumerated in Table 5.4, this balance has started to alter due to increasing fluxes H_9^N and H_{19}^N.

5.5.5 The hydrospheric constituent of the global nitrogen cycle

In seawater, nitrogen is present in the form of dissolved gas, ions of ammonium NH_4^+, nitrite NO_2^-, nitrate NO_3^-, and in the form of various organic compounds. Inorganic nitrogen compounds are assimilated by algae and phytoplankton and thus transfer into organic forms that serve as food for other living organisms. An expenditure of inorganic nitrogen supplies is compensated by atmospheric precipitation, river run-off, and mineralization of organic remains in the process of living organisms life cycles. According to Ivanov (1978), the nitrogen fluxes in the seawater can be presented schematically in Figure 5.4. Of course, not all nitrogen fluxes existing in nature have been taken into account here. A diversity of the means by which nitrogen transformation occurs in water has been studied inadequately, although the available information is sufficient for the global model. The processes, such as replenishing the nitrogen supplies in water due to the transformation of detritus and functioning of living organisms, the nitrogen exchange between photic and deep layers of the ocean, as well as nitrogen fixation at photosynthesis and denitrification, have been thoroughly studied and described in the literature. Also, there are rough estimates of nitrogen supplies in the ocean, according to which one can assume, on average, that $N_U = N_P = 0.77 \cdot 10^4\ \mathrm{t/km^2}$ and $N_L = N_F = 10^5\ \mathrm{t/km^2}$. More detailed spatial distributions of nitrogen supplies in the hydrosphere can be calculated from the data on biomass, dissolved organic matter, and concentration of dissolved oxygen. The volume relationships of dissolved nitrogen are related to the volume of oxygen as $\mathrm{ml}\ N_2/\mathrm{l} = 1.06 + 1.63\ \mathrm{ml}\ O_2/\mathrm{l}$ (Ivanov, 1978).

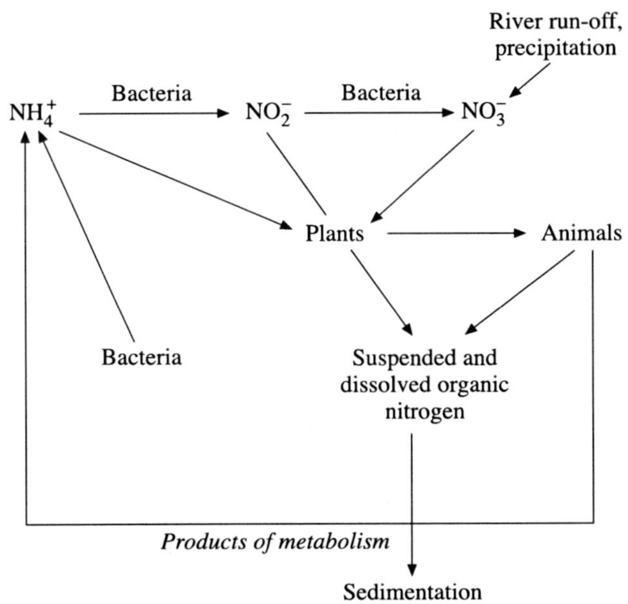

Figure 5.4. The scheme of nitrogen fluxes in the marine medium.

The nitrogen supplies in water bodies are replenished due to bacterial decomposition of organic sediments and dissolved organic matter. Consider the component D denoting the content of dead organic matter in the water medium. On its basis, one can write the following relationship: $H_{18}^N = \lambda_D D(\varphi, \lambda, z, t)$, where λ_D is the indicator of the nitrogen content and the rate of detritus transformation. The free nitrogen supplies in the water medium are also replenished in the process of functioning of various organisms. With account of phytoplankton Φ and nekton r, we have:

$$H_4^N = H_r^N + H_\Phi^N \qquad H_r^N = \lambda_r T_r \qquad H_\Phi^N = \lambda_\Phi T_\Phi$$

where T_r and T_Φ are the characteristics of metabolic processes, in nekton and phytoplankton, respectively, and λ_r and λ_Φ are the coefficients. To determine the averages of these coefficients, assume $T_r = 0.194\,\mathrm{t/km^2/yr}$, $T_\Phi = 0.125\,\mathrm{t/km^2/yr}$, and $H_r^N = H_\Phi^N = 0.83 \cdot 10^{-3}\,\mathrm{t/km^2/yr}$. Then $\lambda_r = 0.00428$ and $\lambda_\Phi = 0.00664$.

The process of denitrification in the water medium delivers considerable amounts of nitrogen into the atmosphere: $H_{20}^N = \lambda_5(\theta_2)^{\Delta T} N_U(\varphi, \lambda, z, t)$, where λ_5 and θ_2 are the constants.

The biological fixation of nitrogen in the water medium constitutes about $10 \cdot 10^6$ t/yr, reaching $20.7 \cdot 10^6$ t/yr in the photic layer of the ocean, and $(36\text{–}1800) \cdot 10^4\,\mathrm{t/km^3/yr}$ in small lakes. For the World Ocean, $\mathrm{H}_{17}^N = 0.0277\,\mathrm{t/km^2/yr}$, on average. Assuming $H_{17}^N = \lambda_R R_\Phi$, where R_Φ is the phytoplankton produce averaging $168.8\,\mathrm{t/km^2/yr}$, we obtain $\lambda_R = 0.164 \cdot 10^{-3}$.

The characteristic feature of the nitrogen cycle in the water medium is its transport due to gravitational sedimentation, vertical convection, turbulent diffusion, and convergence. The processes of nitrogen transport by migrating animals are almost negligible and can be neglected in the global model. The simplest form of description of the vertical fluxes of nitrogen is reduced to the model $H_{14}^N = \lambda_\kappa \Delta N_\kappa$ and $H_{15}^N = \lambda_\rho \Delta N_\rho$, where $\kappa = (U, P, L)$ and $\rho = (P, L, F)$.

5.5.6 Anthropogenic factors of nitrogen oxide in the biosphere

The present contribution of human activity to the general biospheric cycle of nitrogen has reached a level of which the consequences of the introduced changes become unpredictable and probably rather catastrophic. The epidemiological studies testify to the growth of respiration diseases in territories with high concentrations of nitrogen and sulphur oxides as well as photochemical oxidizers. The harmful effect of nitrogen oxides on living organisms starts manifesting itself when the level of $940\,\mathrm{mkg/m^3}$ is exceeded. In general, the consequences of nitrogen pollution of the biosphere are more complicated. For instance, on the one hand, the technogenic accumulation of nitrogen from the atmosphere due to fertilizer production plays a positive role by raising the productivity of land and water ecosystems, whilst on the other hand, it causes an undesirable eutrophication of water basins. Removal of nitrogen from the atmosphere for industrial and agricultural needs is compensated for by a technogenic input of nitrogen into the atmosphere via burning of solid and liquid fuel. A considerable share is contributed here by mechanical transport, emitting nitrogen oxides reaching in the USA, $11.7 \cdot 10^6$ t/yr. However, even observation data for this physical equilibrium cannot explain the chemico-biological balance. Therefore in this multifunctional hierarchical set of global fluxes of nitrogen, the most vulnerable elements and correlations should be revealed, which is only possible within a well-planned numerical experiment.

The quantitative estimate of the main stages of the nitrogen cycle taking account of anthropogenic factors enables us to see the integral effect of breaking the global balance of nitrogen. Tables 5.4 and 5.5 demonstrate the sizes of this imbalance. From the available data on the global distribution of the violations of the nitrogen cycle it is impossible to reliably estimate the contribution of the industrial synthesis of nitrogen compounds and their scattering over the globe into its biogeochemical cycle. During the last years of the 20th century industry increased the total amount of nitrogen circulating in the biosphere by 50%. As a result, the natural equilibrium between the processes of nitrification and denitrification has turned out to be violated with the excess of nitrates reaching $9 \cdot 10^6$ t.

Preliminary estimates of an increasing anthropogenic pressure on the nitrogen fluxes between the biospheric elements suggests a hypothesis about a strong correlation existing between the fertilizers production H_9^N and population density G, technogenic accumulation of nitrogen at fuel burning H_2^N and mineral resources expenditure R_{MG}, anthropogenic input of nitrogen into the atmosphere H_{19}^N, and intensity of emissions of general pollutants Z_{VG}. The quantitative characteristics of the factors of these dependences can be obtained from the known trends. From some estimates, the amount of nitrogen oxides emitted into the atmosphere is proportional to the weight of the used fuel, which is characterized by a 4% annual increasing trend. The scale of industrial fixation of nitrogen over the last 40 years has increased by a factor of 5, reaching a value that could have been fixed by every ecosystem on the Earth before the use of the present agricultural technology. In 1968 the global industry produced about $30 \cdot 10^6$ t of fixed nitrogen, and in 2000 this value reached 1 billion tonnes.

Let us formalize these correlations as the following models:

$$\left.\begin{aligned} H_9^N &= \min\{U(K)\bar{G}(K,t); N_A\sigma_K/\sigma\} \\ H_2^N &= \lambda_{AG}R_{MG} \qquad H_{19}^N = \lambda_{GA}Z_{VG} \end{aligned}\right\} \tag{5.27}$$

where K is the number of an economic region, $\bar{G}$ is the average population density of the K region, σ_K is the area of the K region. The coefficients U, λ_{AG}, and λ_{GA} are determined from analysis of available information about the processes in (5.27). If we assume that $H_2^N = 0.154\,\mathrm{t/km^2/yr}$, $H_{19}^N = 0.102\,\mathrm{t/km^2/yr}$, and $H_9^N = 0.283\,\mathrm{t/km^2/yr}$, then at $\bar{G} = 24.4$ persons/km^2, $R_{MG} = 30.5$ conditional units/km^2/yr, and $Z_{VG} = 3.39\,\mathrm{t/km^2/yr}$, we obtain $U = 0.283$, $\lambda_{AG} = 0.504 \cdot 10^{-2}$, and $\lambda_{GA} = 0.03$.

The anthropogenic interference into the nitrogen cycle can also have the medico-biological consequences expressed, for instance, through increased mortality with growing amounts of NO_2 at 190–320 μg/m^3, affecting a living organism during one hour more than once a month. From the data of the World Health Organization

Table 5.5. Estimates of some parameters of the global biogeochemical cycle of nitrogen in the biosphere.

Parameter	Estimate
Nitrogen content in the atmosphere (%)	
By volume	78.084
By weight	75
Nitrogen content in some biospheric components (%)	
Soil humus	6
River water	$1.5 \cdot 10^{-3}$
Faeces	2.5
Vegetation dying	1
Marine organisms	5
Content of dissolved nitrogen in sea water, ml/l	
Near the equator	14.1
High latitudes	8.2
Average content of nitrogen in sea water (g/m^3)	0.3
Content of nitrates in river waters (t/km^3)	
Arid regions	1.45
Densely populated regions	25
Forests of temperate climate	0.4
Concentration of nitrates (t/km^3)	
Drainage waters of the irrigation systems	5.5
Soil solutions of the salted irrigated soils	200
Ground waters	10–100
Fall-out of fixed nitrogen onto land (10^6 t/yr)	
with precipitation	25
aerosols	15
Assimilation of nitrogen by land plants (10^6 t/yr)	
from the atmosphere	30
from the soil	5.3
Input of nitrogen to the ocean due to detritus lysis (10^6 t/yr)	5
Intensity of the nitrogen motion from the photic layers of the oceans to deep layers with descending water, sedimenting algae, and animal carcasses (10^6 t/yr)	0.2
Global production of nitrogen fertilizers (10^6 t/yr)	31.6–90.1
Removal of nitrogen from the soil with yields (10^6 t/yr)	90–200
Emissions of nitrogen oxides in some countries (10^6 t/yr)	
USA	22.8
Japan	2.4
Great Britain	2.43
The Netherlands	0.32
Concentration of NO_2 in the stratosphere at the altitudes (ppb)	
10 km	0.2–0.5
30 km	4–12
Background value of N_2O concentration in the atmospheric layer up to 16 km (ppm)	325
Observed variations of the background N_2O concentration in the atmospheric layer up to 16 km (ppm)	0.08–0.35

Parameter	Estimate
NO concentration near the Earth's surface (ppb)	
Over the oceans	0.004
Over industrial regions	1
Background content of NO in the atmosphere (ppb)	
In the layer up to 7 km	0.03–0.06
At altitudes 35–45 km	5–20
NO_2 concentration in the surface air layer (ppb)	
Over the oceans	0.1–2.6
Over the continents	0.8–16

Table 5.6. Basic reactions of the global biogeochemical cycle of nitrogen and their power characteristics.

Reaction	Reaction formula	Energy output (kcal)
Ammonification	$CH_2NH_2COOH + 3/2\ O_2 \rightarrow 2CO_2 + H_2O + NH_3$	176
Nitrification	$NH_3 + 1/2O_2 \rightarrow HNO_2 + H_2O$	66
	$KNO_2 + 1/2\ O_2 \rightarrow KNO_3$	17.5
Fixation	$N_2 \rightarrow 2N$	−160
	$2N + 3\ H_2 \rightarrow 2\ NH_3$	12.8
Respiration	$C_6H_{12}O_6 + 6O_2 \rightarrow 6CO_2 + 6H_2O$	686
Denitrification	$C_6H_{12}O_6 + 6KNO_3 \rightarrow 6CO_2 + 3H_2O + 6KOH + 3N_2O$	545
	$5C_6H_{12}O_6 + 24KNO_3 \rightarrow 30CO_2 + 18H_2O + 24KOH + 12N_2$	570
	$5S + 6KNO_3 + 2CaCO_3 \rightarrow 3K_2SO_4 + 2CaSO_4 + 2CO_2 + 3N_2$	132

(WHO) the natural background concentration of NO_2 over the continents constitutes 0.4–9.4 mkg/m^3.

Finally, the human impact on the nitrogen cycle can affect the structure and intensity of the energy exchange in the biosphere. As seen from Table 5.6, a lot can be done in this direction depending on the intensification of either reaction.

5.6 BIOSPHERIC BUDGET OF OXYGEN AND OZONE

The oxygen cycle in nature is composed of characteristic biogeochemical transitions between the reservoirs of basic constituents circulating in the biosphere. Therefore the block-scheme of oxygen exchange is similar to that of sulphur, nitrogen, carbon, and phosphorus (Figure 5.5). However, oxygen refers to constituents spread over the globe most widely, which makes it one of the substantial components of the biogeochemical cycle. Its amount in the Earth's crust, including the hydrosphere, reaches

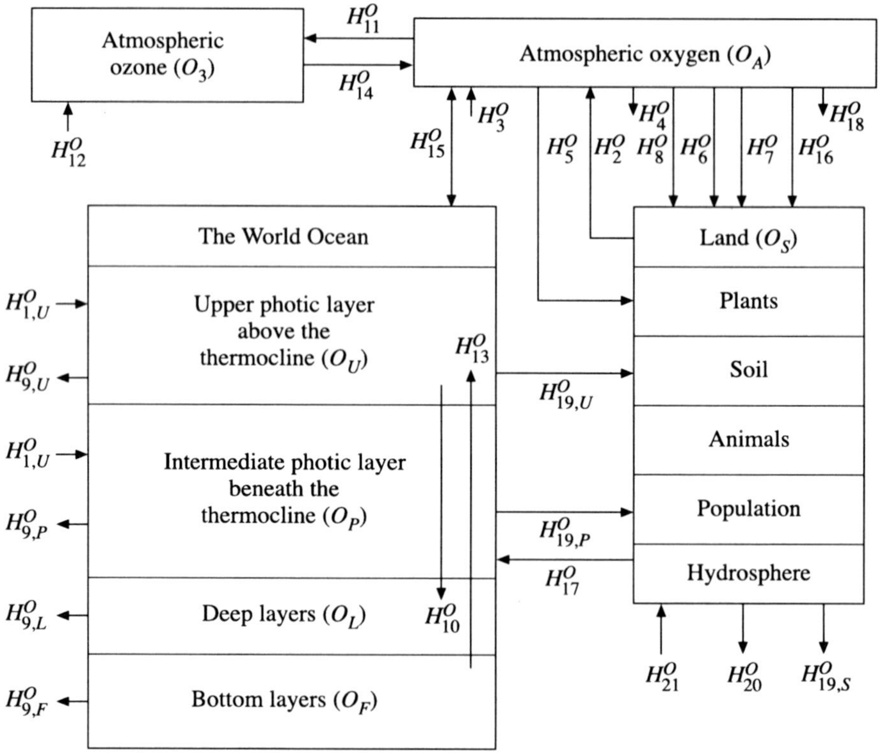

Figure 5.5. Oxygen fluxes in the biosphere.

49% by mass. The lithosphere (without the ocean and the atmosphere) contains 47.2% oxygen, water – 88.89%. In the ocean water, oxygen constitutes 85.82%, with living substance at 65% by mass. These estimates testify to the significance of oxygen for the biosphere. Presently about $39 \cdot 10^{14}$ t O_2 circulate in the biosphere.

Oxygen is present in the biosphere in the form of molecular oxygen (O_2), ozone (O_3), atomic oxygen (O), and as a constituent of various oxides. On the one hand, oxygen maintains life on the Earth due to the process of respiration and formation of the ozone layer, and on the other hand, it is the product of living organisms. This fact hinders a description of its cycle, since it requires a synthesis of the descriptions of various processes. Here an attempt has been made to describe the cycle and to derive a model of the global oxygen cycle (MGOC).

Many authors believe that in the near future nothing threatens the stability of the global biogeochemical cycle of oxygen. This statement is not valid for ozone, whose concentration and spatial distribution suffered serious changes during the past decades (Varotsos and Kondratyev, 1998a, b, c; Varotsos *et al.*, 1998; Varotsos, 2002b, 2003a, b). According to Kondratyev and Varotsos (2000), the available observations of the vertical profile of atmospheric ozone show a very complicated spatial and temporal variability that depends on many characteristics of the nature–society system. The MGOC unit at a parameterization of ozone fluxes follows a numerical model by Aloyan and Arutiunian (Kondratyev and Varotsos, 2000), with a required correlation being taken into account. This correction consists in the substitution of some functional dependences for scenarios reflecting the dynamics of the changes in concentrations of the chemicals which are not described in the global model of the carbon cycle.

5.6.1 Oxygen sources

Presently and in the geological past there have been two sources of oxygen – endogenic and photosynthetic. Without dwelling upon the respective scientific discussions and all existing concepts, we shall try to describe the sources of oxygen in the present biosphere, following numerous studies in this field.

The basic source of atomic oxygen is the photosynthesis of plants, whose equation has the form:

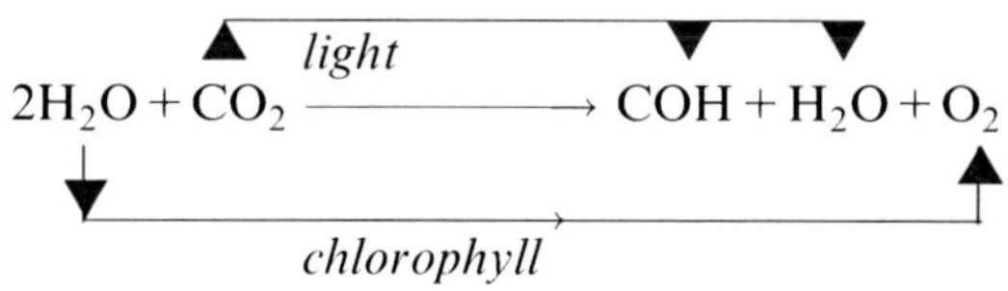

Photosynthesis produces $50 \cdot 10^9$ t of oxygen annually more than, that is, of an order of $3.3 \cdot 10^{14}$% of its supplies in the atmosphere. Hence, we see that only due to photosynthesis can the oxygen supplies in the atmosphere be totally renewed over a time period of 300,000 years. About 80% of the total amount of oxygen produced by photosynthesis results from phytoplankton functioning with land vegetation communities producing only 20%. If we denote the produce of phytoplankton Φ and land surface of the type k by $R_\Phi(\varphi, \lambda, t)$ and $R_k(\varphi, \lambda, t)$ at the Earth surface point (φ, λ) at a time moment t, then the oxygen fluxes to the hydrosphere and from land to the atmosphere can be described by relationships:

$$H_1^O = a_\Phi R_\Phi \qquad H_2^O = a_k R_k \qquad H_{21}^O = a_S R_\Phi$$

where the coefficients a_Φ, a_k, and a_S depend on phytoplankton species and the type of vegetation. For their averaging we use the data on the fluxes: $H_1^O = 140\,\mathrm{t\,O_2/km^2/yr}$, $H_2^O = 70\,\mathrm{t\,O_2/km^2/yr}$, $H_{21}^O = 600\,\mathrm{t\,O_2/km^2/yr}$, $R_\Phi = 401.3\,\mathrm{t/km^2/yr}$, $R_k = 102.4\,\mathrm{t/km^2/yr}$. Then $a_\Phi = 0.35$, $a_k = 0.68$, and $a_S = 1.49$. Of course, these estimates have a considerable spatial and temporal scatter. In particular, using the data on the productivity of some oceans, we obtain the values of the coefficient a_Φ: Atlantic Ocean – 0.53; Indian Ocean – 0.25; Arctic Ocean – 11.1; and Pacific Ocean – 0.64.

Apart from photosynthesis, photolysis can be a source of oxygen in the atmosphere, that is, the decomposition of water vapour under the influence of UV radiation in the upper layers of the atmosphere. However, the intensity of this source in present conditions is negligible. Nevertheless, let us denote this flux by $H_3^O = a_H W_A$, where W_A is the water vapour content in the atmosphere and a_H is the empirical coefficient. If we assume that in the upper layers of the atmosphere a constant share of W_A can reside, then at $H_3^O = 0.0039\,\mathrm{t\,O_2/km^2/yr}$ and $W_A = 0.025\,\mathrm{m}$, we have $a_H = 1.56 \cdot 10^{-7}/\mathrm{yr}$.

Vernadsky (1944) considered rock metamorphism, basaltic volcanism, and underground radioactive waters as possible sources of oxygen. However, there are no sufficiently reliable estimates of these fluxes and therefore it is impossible to parameterize them.

5.6.2 Processes of oxygen assimilation

The oxidation process both on land and in the water medium is the basic consumer of oxygen on the Earth. The ability of oxygen to react with many elements of the Earth's crust forms the fluxes of oxygen leaving the biospheric reservoirs. The balance between the income and expenditure fluxes of oxygen has been reached in the course of biospheric evolution.

Oxygen is spent on respiration of plants, animals, humans, and on the dead organic matter decomposition both in the hydrosphere and on land. To parameterize the income elements of the oxygen balance, we use the following models: $H_5^O = a_1 T_k$, $H_6^O = a_2 T_F$, $H_7^O = a_3 T_G$, $H_{19}^O = a_6 T_R$, $H_8^O = a_Q R_D$, and $H_{20}^O = a_5 R_S$, where T_m is the energy expenditure on respiration ($m = k, F, G, R$) and R_ζ is the rate of the dead organic matter decomposition ($\zeta = Q, D, S$).

5.6.3 Ozone

Atmospheric ozone constitutes $0.64 \cdot 10^{-6}$ of the atmospheric mass and belongs to the family of optically active gases. It absorbs UV solar radiation in the range 200–300 nm, strongly affecting thereby the thermal regime of the stratosphere. Ozone has a number of vibration–rotation bands of absorption in the IR spectral region (9.57 μm) and partially absorbs visible radiation in the Chappuis band (0.6 μm). The formation and destruction of ozone have been described in detail (Kondratyev and Varotsos, 2000; Kondratyev, 1999a).

Ozone forms in the upper stratosphere from molecular oxygen under the influence of UV solar radiation. In the lower stratosphere and troposphere, the source of ozone is the decomposition of nitrogen dioxide under the influence of UV and visible radiation. The formation of the vertical profile of ozone concentration is connected with its meridional and vertical transports. The general characteristic of this profile is the total amount of ozone measured by the thickness of its layer in the Dobson units (1 DU = 0.001 cm).

The ozone destruction is a complex set of photochemical reactions with participation of hydrogen, nitrogen, and chlorine. From the available estimates, 50–70% of ozone is destroyed by nitrogen compounds, 20–30% by oxygen, 10–20% by water-containing particles of HO_x, and less that 1% by chlorine compounds. The prevailing role of nitrogen compounds in ozone destruction has been confirmed (Wauben *et al.*, 1997) for all latitudes. The equation of photochemical equilibrium between concentrations of ozone and nitrogen oxides is $[NO] \cdot [O_3]/[NO_2] = \mu$, where the equilibrium constant μ depends on the solar radiation intensity and can range from 0 to 0.02.

There are various approaches to a parameterization of the process of formation and destruction of the ozone layer. The complexity of derivation of the dynamic models of the ozone cycle in the atmosphere is connected with its participation in more than 75 chemical reactions, a qualitative and quantitative description of which is impossible without deriving detailed models of numerous minor gas components of the atmosphere. Nevertheless, there are empirical models of the ozone layer, which make it possible, in the present climatic situation, to obtain adequate spatial distributions of ozone. For instance, Bekoriukov and Fedorov (1987) derived a simple empirical model of the total ozone content confirmed by the observation data for the southern hemisphere:

$$O_3(\varphi, \lambda) = \sum_n \sum_{n \leq m} P_n^m(\varphi)[a_{n,m} \cos(m\lambda) + a_{n,-m} \sin(m\lambda)] \tag{5.28}$$

where P_n^m are non-normalized spherical functions of the nth degree, of the mth order, $a_{n,m}$ and $a_{n,-m}$ are the empirical coefficients whose values are given in Bekoriukov and Fedorov (1987) and in Krapivin and Kondratyev (2002).

There are also static models to describe the vertical profile of the ozone density distribution. One such model is the Kruger formula:

$$O_3(h) = 51.4 \exp[-(h - 40)/4.2](\text{mkg/m}^3)$$

Other models of ozone have been reviewed by Kondratyev and Varotsos (2000).

The simplest dynamic model of the ozone layer can be written in the form of a balance equation that reflects its income–expenditure components. The ozone supplies are replenished due to the impact of UV radiation on oxygen $(H_{11}^O = e_3 O_A)$ and nitrogen dioxide $(H_{12}^O = e_2 N_A)$. The ozone layer is destroyed at a rate $H_{14}^O = O_3/T_3$, where T_3 is the lifetime of ozone molecules depending on atmospheric pollution: $T_3 = T_3^{O_3} - e_1 B$. The lifetime $T_3^{O_3}$ of the ozone molecules in a perfect atmosphere averages 50–60 days. Nitrogen oxides participating contrary to the H_{12}^O cycle of ozone destruction contribute much to the magnitude of B.

5.6.4 Anthropogenic impact on the oxygen cycle

Studies of the history of biospheric evolution reveal a close correlation between the oxygen production intensity and the development of life on Earth. And although the expected relative oscillations of the oxygen concentration in the near future do not exceed 10%, the considered impacts on the biosphere do not cover all potential anthropogenic trends, and therefore cannot be considered reliable. Therefore let us analyse the constituents of possible mechanisms for violation of the natural balance of oxygen. Naturally, our concern is not only for an increase but also a decrease of the oxygen content in the atmosphere (Table 5.7).

The oxygen cycle is complicated by its ability to take part in a lot of chemical reactions giving a multitude of epicycles. This fact makes the oxygen cycle sufficiently stable but hinders an assessment of its stability.

Table 5.7. Some factors of impacts on the natural cycle of oxygen.

Factor	Functional manifestation of the factor
Thermal pollution of water media	Decrease of the content of dissolved oxygen due to its emission and increased expenditure by aerobic bacteria and cold-blooded animals.
Discharge of municipal sewage, wastes of food industry, and agriculture.	Increased rate of consumption of dissolved oxygen in fermentation of additional organic substances.
Reduction of the size of aquatic animals.	Decreased expenditure on respiration.
Reconstruction of the structure of the surface vegetation cover.	A wide spectrum of oscillations of the fluxes H_2^O and H_5^O.
Liquidation of land vegetation	A decrease of the fluxes H_2^O and H_5^O. The burning of vegetation increases the flux H_{18}^O.
Burning of fuels and wastes.	The flux H_{18}^O increases.
Gas pollution of the atmosphere.	The flux H_4^O increases and the fluxes H_{11}^O, H_{12}^O, and H_{14}^O are sharply violated.

Anthropogenic forcing on numerous epicycles of oxygen manifests itself both directly through its involvement in other cycles of substances at fuel burning and production of various materials, and indirectly through environmental pollution and biospheric destruction. Therefore a parameterization of the anthropogenic impact on the oxygen balance is realized within other units of the global model. The flux H_{18}^O taken into account in scheme Figure 5.5 completely covers direct consumption of oxygen both in industry and in agriculture. Assume $H_{18}^O = y_1 R_{MG}$, where R_{MG} is the rate of the natural resources expenditure, and y_1 is the coefficient ($\approx$0.084).

The fluxes H_{15}^O and H_{16}^O are strongly affected by anthropogenic forcing. Their variations are caused by discharges of high-temperature industrial sewage containing considerable amounts of oxidizers as well as by oil-polluted water bodies. The quantitative characteristics of the change of oxygen dissolved in water as a function of temperature have been studied comprehensively (see Table 5.8). The empirical formula to calculate the concentration of the seawater dissolved oxygen has the form: $[O_2 \text{ dissolved}] = 80/(0.2T_O - 7.1)$, where $[O_2]$ is expressed in mg/l and T_O in °C. The estimates of oxygen solubility in water are well known (Krapivin and Kondratyev, 2002).

The fluxes H_9^O and H_{19}^O, balancing in natural conditions the oxygen fluxes into the water medium, at an anthropogenic forcing increase, as a rule, due to more active aerobic bacteria and increasing metabolic needs of animals. For instance, a 10°C increase of water temperature increases the oxygen expenditure on respiration of marine animals by a factor of 2.2.

One of the negative manifestations of the anthropogenic impact on the oxygen cycle is a depletion of the ozone layer, especially marked in the polar regions. There are various hypotheses on the causes of sharply changing concentrations of ozone, as well as discussions on the so-called "ozone hole" over the Antarctic. The main cause of all violations is connected with the progressing human activity accompanied by the growing volumes of long-lived components emitted to the atmosphere (e.g., freons). The consequences of these violations are very serious, and the real scales of danger threatening life on the Earth can be estimated only with the use of the global model of the nature–society system.

A diversity of the anthropogenic impacts on the global biogeochemical cycle of oxygen is determined by the direct and indirect causes of breaking the natural balance of oxygen. According to the equation of photosynthesis, the gramme-molecular amounts of assimilated CO_2 and emitted O_2 are equal. Also equal are the gramme-molecular amounts of assimilated O_2 and emitted CO_2 for the dead organic matter decomposition and fuel burning. Hence, for time periods of tens and hundreds of years, a change of the CO_2 amount in the atmosphere is accompanied by the same change of O_2, but in the opposite direction. For instance, a CO_2 doubling in the atmosphere leads to a decrease of the amount of O_2. But, since the volume concentration of CO_2 in the atmosphere is now estimated at 0.031%, and that of O_2 at 20.946%, in this case a decrease of O_2 will constitute only 0.15% of the total O_2 content in the atmosphere.

Imagine the following situation. Let the total biomass of the biosphere ($\sim 9.6 \cdot 10^{11}$ t C), all organic matter of soil ($\sim 14 \cdot 10^{11}$ t C), and all fossil chemical fuel, the explored supplies of which constitute $128 \cdot 10^{11}$ t of conditional fuel ($64 \cdot 10^{11}$ t C), be burnt. Then the amount of CO_2 in the atmosphere increases by a factor of 12.5, and that of O_2, respectively, decreases but only by 1.75%. Hence, the amount of oxygen during hundreds of years has to be practically constant.

However, it should be borne in mind that the region of excess anthropogenic emissions of CO_2 and, hence, O_2 assimilation is concentrated in a small area of cities and forest fires. Since the concentrations in the atmosphere do not equalize instantly, a gradient of O_2 concentrations can appear around these sites, when the oxygen provision will be insufficient for animals and humans. Therefore the model of the global oxygen cycle (MGOC unit) reflecting the spatial heterogeneities in the distributions of O_2 concentrations enables us to reveal such dangerous territories.

An interaction between the cycles of oxygen, nitrogen, sulphur, phosphorus, and carbon manifests itself through the processes of oxidation and decomposition. The level of detailing the global model units does not permit one to reflect all the diversity of these processes. Therefore in the simplest case, when only averaged characteristics of the oxygen cycle elements are taken into account, the scheme in Figure 5.5 of the global O_2 fluxes can be presented as the scheme in Figure 5.6. The indicated stability of the O_2 concentration in the atmosphere makes it possible to simplify a description of the MGOC unit, using a single balance equation:

$$\frac{\partial O}{\partial t} + V_\varphi \frac{\partial O}{\partial \varphi} + V_\lambda \frac{\partial O}{\partial \lambda} = k_0 R_F + k_L R_L - \nu_L T_L - b_G G - \nu_F T_F - \nu_G T_G - \mu_Q R_Q$$

where k_0 and k_L are the indicators of the rate of O_2 emission due to photosynthesis in the ocean and on land, respectively, ν_s ($s = F, G$) is the indicator of the role of

Table 5.8. Estimates of the characteristic parameters of the biogeochemical cycle of oxygen and ozone in nature (Krapivin, 2000a; Kondratyev and Varotsos, 2000).

Parameter	Estimate
Total mass in the atmosphere (10^9 t)	
Oxygen	$1.184 \cdot 10^6$
Ozone	3.29
Oxygen supply in the Earth's crust (t)	$13 \cdot 10^{18}$
Concentration of oxygen in the present atmosphere by volume (%)	$20.94 \div 0.02$
Oxygen supply in the ocean water (t)	
In bound state	$1.28 \cdot 10^{18}$
In dissolved compounds	$2.6 \cdot 10^{15}$
In free state	$7.48 \cdot 10^{12}$
Partial density of ozone at normal pressure and temperature (mkg/m^3)	2.14
Input processes in the oxygen balance (10^9 t O_2/yr)	
Photosynthesis in the oceans	56
Photosynthesis on land	174
Wild vegetation	152
Agricultural vegetation	21
Algae	1
Water photolysis	0.013
Melting of metals from oxides	0.3
Expenditure processes in the oxygen balance (10^9 t O_2/yr)	
Oxidation of organics in the World Ocean	59.2
Oxidation of organics on land	
By subaerial natural systems	151.2
In water reservoirs, lakes, and rivers	1.07
In burning of oil, coal and gas	16
In using the raw material, fodder, food, and biofuel	29.5
In anthropogenic decrease of humus	1.5
At dust-heaps and burials	2.6
Other processes	0.4
Fossilization in the Earth crust with enlargement of the continents	0.15
The ratio of the rate of deposition of suboxidized organic carbon to vegetation productivity	$2 \cdot 10^{-4}$
Photodecomposition of water vapor in the atmosphere (t/km^2/yr)	0.008
Distribution of the organic carbon deposits (%)	
Continents	69.6
Shelf and mainland slope	29.0
The World Ocean bed	1.4
Relative concentration of water vapour in the stratosphere	$5 \cdot 10^{-6}$
Ozone lifetime (days)	
Northern hemisphere	39
Southern hemisphere	85
Minimum permissible concentrations of oxygen in the atmosphere (% of the present level)	
Humans	31
Mammals	14–31
Birds	22–31
Reptiles	10–20
Amphibians	3.1–11
Fish	2.5–11
Bacteria	1

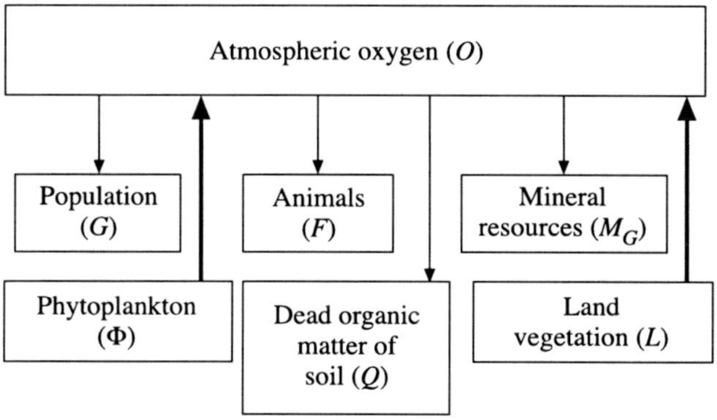

Figure 5.6. A simplified scheme of the biogeochemical cycle of oxygen in the biosphere.

respiration of land vegetation ($s = L$), animals ($s = F$), and humans ($s = G$) in the removal of oxygen from the atmosphere, and μ_Q is the rate of O_2 consumption at the decomposition of the soil dead organic matter.

5.6.5 The model block-scheme of the oxygen budget in the biosphere

Assuming the scheme of the oxygen fluxes in nature shown in Figure 5.5 to be balanced, write the equations of the model in the following traditional form of the balance relationships (Table 5.9):

$$\frac{\partial O_A}{\partial t} + V_\varphi \frac{\partial O_A}{\partial \varphi} + V_\lambda \frac{\partial O_A}{\partial \lambda} = H_2^O + H_3^O + H_{14}^O + H_{15}^O + H_{16}^O - \sum_{i=4}^{8} H_i^O - H_{11}^O - H_{18}^O$$

$$\frac{\partial O_3}{\partial t} + V_\varphi \frac{\partial O_3}{\partial \varphi} + V_\lambda \frac{\partial O_3}{\partial \lambda} = H_{11}^O + H_{12}^O - H_{14}^O$$

$$\frac{\partial O_S}{\partial t} = H_{21}^O - H_{16}^O - H_{17}^O - H_{19,S}^O - H_{20}^O$$

$$\frac{\partial O_U}{\partial t} + V_\varphi \frac{\partial O_U}{\partial \varphi} + V_\lambda \frac{\partial O_U}{\partial \lambda} = H_{1,U}^O + H_{13,PU}^O + H_{17}^O - H_{9,U}^O - H_{10,PU}^O - H_{15}^O - H_{19,U}^O$$

$$\frac{\partial O_P}{\partial t} + V_\varphi \frac{\partial O_P}{\partial \varphi} + V_\lambda \frac{\partial O_P}{\partial \lambda} = H_{1,P}^O + H_{10,PU}^O + H_{13,PL}^O - H_{9,P}^O - H_{10,PL}^O - H_{13,PU}^O - H_{19,P}^O$$

$$\frac{\partial O_L}{\partial t} = Q_L + H_{10,PL}^O + H_{13,LF}^O - H_{9,L}^O - H_{10,LF}^O - H_{13,PL}^O$$

$$\frac{\partial O_F}{\partial t} = Q_F + H_{10,LF}^O - H_{9,F}^O - H_{13,LF}^O$$

Here Q_L and Q_F denote the oxygen fluxes resulting from mixing of the deep and bottom layers of the ocean. The oxygen exchange between the hydrosphere and the atmosphere (fluxes H_{15}^O and H_{16}^O) depend on its partial pressures on the water–air

Table 5.9. Estimates of the reservoirs and fluxes of oxygen and ozone in the MGOC unit used in its adjustment (Kondratyev and Varotsos, 2000).

Reservoirs (t/km^2), fluxes ($t/km^2/yr$)	Identifier	Estimate
Oxygen in the upper photic layer of the World Ocean	O_U	$0.8 \cdot 10^8$
Oxygen in the transition layer of the ocean	O_P	$0.7 \cdot 10^9$
Oxygen in deep ocean	O_L	$3 \cdot 10^4$
Oxygen in the bottom layer of the ocean	O_F	$9 \cdot 10^3$
Oxygen in the atmosphere	O_A	$0.24 \cdot 10^7$
Oxygen in the surface part of the hydrosphere	O_S	$0.6 \cdot 10^8$
Ozone	O_3	0.23
Photosynthesis in the ocean	H_1^0	108–388
Photosynthesis on land	H_2^0	70–100
Photodecomposition of water in the atmosphere	H_3^0	0.008
Oxidation processes in the atmosphere	H_4^0	0.009
Respiration of plants	H_5^0	0.07–0.1
Respiration of animals	H_6^0	50–60
Respiration of humans	H_7^0	70–80
Oxidation–restoration processes in soil	H_8^0	1
Oxidation processes in the World Ocean	H_9^0	164
Descending of the oxygen-saturated waters	H_{10}^0	190
Decomposition and destruction of O_2 in the atmosphere	H_{11}^0	
Formation of O_3 from NO_2	H_{12}^0	0.23–22.2
Lifting of dissolved oxygen in upwelling zones	H_{13}^0	36
Decomposition and destruction of ozone in the atmosphere	H_{14}^0	1.48–1.66
Exchange on the atmosphere–ocean border	H_{15}^0	18–140
Exchange on the atmosphere–inland water bodies border	H_{16}^0	18–140
Transport of oxygen into the ocean by river run-off	H_{17}^0	50
Anthropogenic consumption of oxygen	H_{18}^0	60–90
Expenditures of O_2 on the metabolism of aquatic animals	H_{19}^0	0.2
Oxidation processes in the continental water bodies	H_{20}^0	90–200
Photosynthesis in the continental water bodies	H_{21}^0	100–400

border. The direction of the fluxes H_{15}^O and H_{16}^O depends on the relationship between temperatures T_a, T_U, and T_S. Due to high concentrations of oxygen in the atmosphere, the partial pressure varies negligibly, and therefore the fluxes H_{15}^O and H_{16}^O can be considered to depend only on oscillations in the concentrations of O_U and O_S: $H_{15}^O = k_{OU}(T_U - T_a)O_U$; $H_{16}^O = k_{OS}(T_S - T_a)O_S$. If we assume $O_U = 5.5$ ml/l and $O_S = 2.1$ ml/l, then at $T_U - T_a = T_S - T_a = 2°C$, $H_{15}^O = 18\, t/km^2/yr$, and $H_{16}^O = 140\, t/km^2/yr$ we obtain $k_{OU} = 0.5 \cdot 10^{-4}$ km/°C/yr and $k_{OS} = 0.1 \cdot 10^{-2}$ km/°C/yr.

The ocean layers exchange oxygen by circulation processes, and as a result, depending on latitude, longitude, and season, the intensity of the fluxes H_{10}^O and

H_{13}^O can sharply change. In any case, this intensity mainly depends on the velocities v_A of the vertical water uplifting and v_H of its lowering. In the zones of upwellings the flux H_{13}^O prevails and, on the contrary, the flux H_{10}^O prevails in the zones of convergence. The velocities v_A and v_H range from 0 to 0.1 m/s. The most characteristic values of these velocities range from 10^{-2} to 10^{-4} m/s. For instance, near California, $v_A \approx 0.77 \cdot 10^{-5}$ m/s, and in the region of the Bengal uplifting $v_A = 0.25$ m/s. Thus for the fluxes H_{10}^O and H_{13}^O we use the following approximations: $H_{10}^O = \lambda_{\kappa\gamma}(O_\kappa - O_\gamma)$ and $H_{13}^O = \mu_{\gamma\kappa}(O_\gamma - O_\gamma)$, where $\lambda_{\kappa\gamma}$ and $\mu_{\gamma\kappa}$ are the coefficients of local mixing ($\kappa = U, P, L; \gamma = P, L, F$).

In the shelf zone of the ocean the oxygen balance formation is affected significantly by the run-off from the continents. Despite the complexity of this process, the following simple parameterization can be accepted for this case:

$$H_{17}^O = n_{SU}(O_S - O_U)$$

5.7 AN ASSESSMENT OF THE ROLE OF AVIATION IN THE OZONE LAYER DYNAMICS

The problem of monitoring and predicting the ozone layer dynamics in the atmosphere is equally important as the problem of the atmospheric greenhouse effect (Varotsos and Kondratyev, 1998; Popovicheva *et al.*, 2000). In both cases there are contradictory estimates of the causes of the ecological danger at the observed levels of ozone concentration, ozone destruction or greenhouse gases. Despite such contradictory estimates, often not without a political nuance (Zuev, 2000), these problems attract the attention of experts in the fields of natural sciences, who try to create some information technologies which would ensure a high level of objectivity and reliability of the estimates of consequences of the anthropogenic interference into the global biochemical cycles of ozone, carbon dioxide, methane, water vapour, and other minor gas components (MGC). Here we consider the narrow but important problem of ozone layer changes over a limited territory caused by a transport aviation route over this territory. This problem has recently attracted growing attention. The impact of flights of subsonic (altitudes 9–13 km) and supersonic (16–20 km) aircraft on the ozonosphere has become substantial, at least, on a regional scale. Even more so since the volumes of global air transportation has increased by almost 5% annually, and the amount of emitted nitrogen oxides, sulphur compounds, and other MGC increases by about 4% annually. According to average global estimates, the NO_x emissions ($NO+NO_2$) constitute now about 500 kt N/yr with a predicted increase to 1100 kt N/yr by 2015.

A totality of substances ejected by aircraft engines to the atmosphere include the components, such as H_2SO_4, HNO_3, HNO_2, HNO, Cl, NO_3, ClO_2, CO, CO_2, CH_4, N_2O, H_2O, SO_2, SO_3, N_2O_5, CH_3Cl, Cl_2, CH_3NO_2, CH_3NO_3, $BrONO_2$, HNO_4, and $ClONO_2$, many of which cause the formation of polar stratospheric clouds, affect markedly the aerosol composition of the atmosphere, and intensify the greenhouse effect.

An analysis of the distribution of these components in the atmosphere requires an understanding of many factors in the field of photochemistry and atmospheric dynamics. Unfortunately, the available ideas about the rates of reactions with participation of the enumerated substances as well as about the coefficients of micro–macro-turbidity, and local synoptic characteristics are limited by data averaged in time and space. In this regard, many authors apply various simplifications, which make it possible to overcome information uncertainties.

Atmospheric ozone chemistry has been significantly studied (Kondratyev and Varotsos, 2000). Nevertheless, this knowledge is insufficient to derive a model of the biogeochemical cycle of ozone whose adequacy would not raise doubts. The main cause is the time-dependent nature of the environmental processes. Unfortunately, the use of both simple and complicated climate models that take into detailed account the compounds of atmospheric chemistry of ozone, do not give acceptable results. Therefore one of the possible approaches to raise the reliability of the estimates of the state of the ozone layer over a given territory is an application of technology which combines measurements and modelling and takes into account the expert estimates. In this case, to assess the vertical profile of ozone, all available information can be used on its formation and destruction, and the additional background information about anthropogenic and natural processes can be obtained from established correlations or scenarios.

One of the difficulties in the synthesis of the model of ozone dynamics and observational data is a necessity to adequately describe the location of the tropopause. An uncertainty of the aircraft flight route of subsonic aircraft relative to the tropopause appears because of inaccurate determination of the height of the tropopause. This is very important, since depending on whether the flight route is below the tropopause or in the stratosphere, the photochemical reactions with participation of ozone differ. With supersonic aviation, this problem is solved uniquely: all routes lie in the stratosphere. Therefore, to exclude from the model the elements of instability, a hypothesis is assumed on the seasonal change of the tropopause altitude, for instance, following the binary law: in spring and summer Z_1, and in autumn and winter Z_2.

This approach excludes the instability of the obtained estimates. Though, of course, there are many successful results of modelling the ozone photochemistry. Among them a number of models of the Lagrange type are rather efficient which take into account up to 75 chemical elements and compounds. Also efficient is a 3-D model MOZART (Model for Ozone and Related chemical Tracers) (Kondratyev and Varotsos, 2000).

In this section the problem of dynamic estimation of the vertical profile of the ozone concentration in the atmosphere over limited territory is considered in a narrow aspect, taking into account only one anthropogenic source of the effect on the ozonosphere – aviation. Since the set of substances ejected by aircraft engines is rather large, the consequences of the aircraft flight over this part of the Earth's surface include numerous changes of both gas and aerosol composition of the atmosphere. Of course, the scale of these changes is determined by the intensity of the aircraft flights and density of the flight corridors. Here a method of integration of these causes is proposed and an information system of the expert level is developed, which can be used in the systems of regional and national monitoring. Besides, the presentation of the input information, the spectrum of chemical reactions have been limited. An operator can specify a list of substances contained in exhaust gases.

Units of the simulation system to control the regional ozonosphere (SSCRO) are enumerated in Table 5.10. The controlled territory Ω is covered with a geographical grid with steps $\Delta\varphi$ and $\Delta\lambda$ by latitude and longitude, respectively. Over each cell $\Omega_{ij} = \{(\varphi, \lambda)\colon \varphi_i \leq \varphi \leq \varphi_{i+1};\ \lambda_j \leq \lambda \leq \lambda_{j+1};\ i = 1, \ldots, N;\ j = 1, \ldots, M\}$ a possible location (altitude and time) of the flight corridor is determined. For this purpose, in the database for SSCRO an indicator is formed of the flight load on Ω (type of engine, fuel, velocity, and flight altitude). The background concentration of ozone and the meteorological situation are assumed to be taken from the data of regional, national, and global systems of environmental monitoring.

The ozone concentration as a function of the spatial coordinates and time is calculated by the formula:

$$\frac{\partial O_3(\varphi,\lambda,z,t)}{\partial t} = Q + S + U - P - R \tag{5.29}$$

where z is the altitude, and the functionals in the right-hand part of equation (5.29) describe the following processes of ozone formation: Q is a change of the ozone concentration due to atmospheric motion and gravitational sedimentation, P and U are photochemical destruction and formation of ozone outside the passageway, respectively, R and S are photochemical destruction and formation of ozone within the flight corridor.

The equation (5.29), with initial data for a time moment $t = t_0$, is solved with account of the mosaic of the flight corridors. The functionals R and S are calculated for $\Omega_{i,j,k} = \{(\varphi, \lambda, z)\colon (\varphi, \lambda) \in \Omega_{ij}; z_k \leq z \leq z_{k+1}\}$ at a time moment t only in the presence of aircraft. Three zones are considered for interaction of the products of fuel combustion in the trail of the working aircraft engine: (1) immediately behind all engines (time duration Δt_1); (2) at a stage of mixing the exhaust gases with the atmosphere (Δt_2); and (3) on the ways of penetration of the mixture to large-scale reservoirs (Δt_3). Hence, after aircraft transit the passageway exists during a time period $\Delta t = \Delta t_1 + \Delta t_2 + \Delta t_3$, after which $R(\varphi_i, \lambda_j, z, k, t) = S(\varphi_i, \lambda_j, z, k, t) \equiv 0$ and in equation (5.29) the functional Q, U, and P start working.

During a time period the Δt a lot of various processes of transformation of substances ejected by the engines occur within the flight corridor. There is a term "index of transformations in a jet", which is an integral estimate of the concentrations of these substances as a function of time. Let an exhaust take place at a moment t_0 (the moment of aircraft transit over a given point on the Earth's surface). Then the index of transformation of chemicals in the trail after the flight can be presented by a 3-step function:

$$J_N(t) = \begin{cases} J_{N1} & \text{for } t_0 \leq t < t_0 + \Delta t_1 \\ J_{N2} & \text{for } t_0 + \Delta t_1 \leq t < t_0 + \Delta t_1 + \Delta t_2 \\ J_{N3} & \text{for } t_0 + \Delta t_1 + \Delta t_2 \leq t < t_0 + \Delta t_1 + \Delta t_2 + \Delta t_3 \end{cases} \tag{5.30}$$

Table 5.10. The characteristic of the SSCRO units.

Unit	Functions of the unit
AADB	Algorithmic adaptation of the database to the structure of the controlled territory. A matrix structure is developed with elements reliably attaching the environmental elements to geographical coordinates, configuration of the territory of the region, location on it of the objects of the types of airports as well as to division of the territory on land and water.
IRDB	Information renewal of the database. A possibility is provided of an operational change of configuration and appearance of the territory of the region described in AADB.
MIF	Management of information fluxes between the SSCRO units. Dimensions of the model parameters are coordinated; dimensions of input data are coordinated with the scales assumed in SSCRO. For instance, the formula $1\,\text{ppmv} = 10^{-3}[M(M_i\rho)]\,\text{mkg}\cdot\text{m}^{-3}$, where M_i is the molecular weight of the ith chemical element. The formulae of the type of $1\,\text{mkg}\,O_3/m^2 \Rightarrow 0,467\cdot 10^{-7}\,\text{atm/cm}$ are also re-calculated.
CFSU	Control of the functions of the system units. Depending on the availability of the needed information in the database about correlations between various processes, a version is selected from the alternative versions that does not contradict the database.
PADB	Parametric agreement of the models and database. Signals of the user's interface are analysed for an efficient removal from the database of the coefficients of models or in the case of disagreements the model is substituted for the scenario.
MPR	Modelling the photochemical reactions in the flight corridor with a selection of three stages: (1) in the nearest zone after an ejection of the products of fuel combustion from the nozzle of the aircraft engine; (2) scattering from the jet of combustion products; and (3) complete mixing with the surrounding atmosphere.
MPTO	Modelling the processing of propagation and transformation of ozone in the interaction of the flight passageway (vortex trail) with the surrounding atmosphere.
MFDO	Model of the formation and destruction of ozone with an account of all flight corridors over the controlled territory.
CCAB	Calculation of corrections for the atmospheric balance of ozone due to account of the effects of land cover and sea surface.
FBLO	Formation of the background level of ozone either from the data of the regional and global monitoring system or with the use of the respective model.
FS	Formation of scenarios of location of the flight corridors and of the timetable of their load.
MUII	Management of the user's information interface. Provision of computer experiments.

Table 5.11. A sample fragment of the SSCRO. Averaged estimates of J_{N2} and J_{N3} (%) as functions of database season and temperature T(K) for the northern hemisphere (Karol, 2000).

Atmospheric layer (km)	January				July							
	$T = 217$ K		$T = 223$ K		$T = 223$ K		$T = 233$ K		$T = 223$ K		$T = 233$ K	
	J_{N2}	J_{N3}	J_{N2}	J_{N3}	J_{N2}				J_{N3}			
					D	N	D	N	D	N	D	N
9–12	87	59	81	55	82	88	58	81	64	69	38	56
16–20	69	54	62	48	74	77	53	68	59	64	34	50

The values J_{Ni} $(i = 1, 2, 3)$ depend on the time of day, season, and on many other parameters (temperature, altitude, and geographical coordinates). The empirical estimates of J_{Ni} of the type indicated in Table 5.11 are introduced to the database of the system and used in calculations of $J_N(t)$. With further improvement of SSCRO it will be expedient to include a unit, which would give theoretical estimates of J_{Ni}.

Since the aircraft flies at a velocity V_a along the route $x(\varphi, \lambda, z)$, at a time moment t_0 it is at some point x_0, and all its engines eject $V_M(t_0)$ of the substance of the M type. With account of (5.30) we obtain:

$$V_M(t) = \begin{cases} L_1 & \text{for } t_0 \leq t < t_0 + \Delta t_1 \\ L_2 & \text{for } t_0 + \Delta t_1 \leq t < t_0 + \Delta t_1 + \Delta t_2 \\ L_3 & \text{for } t_0 + \Delta t_1 + \Delta t_2 \leq t < t_0 + \Delta t \end{cases}$$

where

$$L_1 = V_M(t_0) - \frac{t - t_0}{\Delta t_1}[V_M(t_0) - J_{N1} V_M(t_0)]$$

$$L_2 = J_2 V_M(t_0) + \frac{t_0 + \Delta t_1 + \Delta t_2 - t}{\Delta t_2} V_M(t_0) \cdot (J_{N1} - J_{N_2})$$

$$L_3 = J_3 V_M(t_0) + \frac{t_0 + \Delta t_1 + \Delta t_2 - t}{\Delta t_3} V_M(t_0) \cdot (J_{N_2} - J_{N3})$$

After some time Δt an apparent effect of the aircraft flight is considered to cease. All processes of transformation and destruction of ozone within the flight corridor after the flight again become natural. The zone of the vortex trail behind the flying aircraft has a circular section of diameter r, and during the time period $\tau = \Delta t_1 + \Delta t_2$ its interaction with the surrounding atmosphere can be considered negligibly small. At the third stage, this interaction begins with slight contacts of two media. At any rate, the scenario of interaction of the flight corridor with the surrounding atmosphere needs to be specified and developed, in particular, by forming a set of scenarios.

NO_x is the most important component of exhaust gases. During the lifetime of the aircraft trail NO_x gets oxidized by hydroxyl OH, present in the vortex trail, giving HNO_3 and HO_2NO_2. As laboratory studies have shown, the processes of formation and destruction of ozone are also markedly affected by heterogenic mechanisms of the impact on atmospheric chemistry. This impact manifests itself both within the flight corridor and in a free atmosphere. In particular, the reaction $N_2O_5 + H_2O \rightarrow 2HNO_3$ with sulphate aerosols mainly resulting from the aircraft flight, reduces the rate of ozone destruction due to the NO_x cycle, but raises the negative role of Cl_x and HO_x in O_3 destruction.

The second important component of the exhaust gases is SO_2, whose ejection by the engines doubles the area occupied by sulphate particles in the atmosphere of the flight corridor, leading to an increase of O_3 losses. In a number of field experiments (Kraabol and Stordal, 2000) on the F-16 fighter and in laboratory experiments with the engine F-100 using several types of aviation fuel with a high (~1150 ppm S), moderate (~170–300 ppm S), and low (~10 ppm S) content of sulphur, the SO_2 emission changed from 2.49 g SO_2/kg for the fuel with a high sulphur content to 0.01 g SO_2/kg for the fuel with a low sulphur content. For these experiments the following relationship was derived:

$$0.02 \leq \frac{[SO_3]}{[SO_2] + [SO_3]} \leq 0.14$$

The results of studies carried out in Weisenstein *et al.* (1998) show that the mechanisms of evolution of the composition of the engines exhaust gases at their interaction with the atmosphere have been poorly studied, and therefore it is important to further develop the kinetic models describing the role of aircraft flights in changing the atmosphere.

The functional Q in (5.10) is written following the traditional scheme:

$$Q = -v_\varphi \frac{\partial O_3}{\partial \varphi} - v_\lambda \frac{\partial O_3}{\partial \lambda} - v_z \frac{\partial O_3}{\partial z} + \frac{\partial}{\partial \varphi}\left(D_\varphi \frac{\partial O_3}{\partial \varphi}\right) + \frac{\partial}{\partial \lambda}\left(D_\lambda \frac{\partial O_3}{\partial \lambda}\right) + \frac{\partial}{\partial z}\left(D_z \frac{\partial O_3}{\partial z}\right)$$

where $V(v_\varphi, v_\lambda, v_z)$ is the wind speed and $D(D_\varphi, D_\lambda, D_z)$ is the coefficient of eddy diffusion.

The units CCAB, MFDO and MPTO divide the functional Q with account of the output information from the unit AADB (see Table 5.10). As a result, the air mass mixing is realized at two stages: (1) mixing of atmospheric compartment of the zone of the aircraft flight with the environment; and (2) mixing of the cells $\{\Omega_{i,j,k}\}$ selected by the unit AADB. At the first stage the volumes and location of the zone of the effect of aircraft are calculated:

$$\omega = \{(\varphi, \lambda, z) \colon \varphi_0 \leq \varphi \leq \varphi_1; \lambda_0 \leq \lambda \leq \lambda_1; z_0 \leq z \leq z_1\}$$

where $\varphi_1 = \varphi_0 + v_{a\varphi}\Delta t/k_\varphi$; $\lambda_1 = \lambda_0 + v_{a\lambda}\Delta t/k_\lambda$; $\Delta z = z_1 - z_0$ is the diameter of the zone of the impact of aircraft ($\Delta z = r$), φ_0 and λ_0 are latitude and longitude of the aircraft location at a time moment t_0, k_φ and k_λ is the number of kilometers within 1°

of latitude and longitude, respectively, and z_1 and z_0 are the lower and upper boundaries of the flight corridor.

The obtained space ω agrees with the adjacent multitude of atmospheric units $\{\Omega_{ijk}\}$. Then the ozone content is averaged over ω and the adjacent compartments $\{\Omega_{ijk}\}$ with their volumes taken into account.

The second stage realizes a 2-step procedure of re-calculation of the ozone concentration over the whole space $\Xi = \{(\varphi, \lambda, z)\colon (\varphi, \lambda) \in \Omega; 0 \leq z \leq z_H\}$, where z_H is the altitude of the atmospheric boundary layer ($z_H \sim 70$ km), whose consideration is important in estimating the state of the regional ozonosphere. These two steps correspond to the vertical and horizontal constituents of the atmospheric motion. This division is made for convenience, in order that the user of the expert system could choose a synoptic scenario. According to the available estimates (Karol, 2000; Kraabol *et al.*, 2000; Meijer and Velthoven, 1997), processes of the vertical mixing prevail in the dynamics of the ozone concentration. It is here that due to uncertain estimates of D_z there are serious errors in model calculations. Therefore the units CCAB, MFDO and MPTO (see Table 5.10) provide the user with the principal possibility to chose various approximations of the vertical profile of the eddy diffusion coefficient (D_z). The database of SSCRO contains the versions of estimates of D_z (m^2/s) used already by many authors in the models of the ozone dynamics ($\approx 3\,m^2/s$; $\approx 0{,}5(V_\varphi + V_\lambda)$; $\approx 10\,m^2/s$ for $0 \leq z \leq 10$ km $\approx 10 + 2{,}57(10\text{-}z)$ for $10 < z \leq 13$ km; $\approx 70\text{–}1{,}22(70\text{-}z)$ for $13 < z \leq 70$ km).

Since the model retrieval of the background situation over a given territory has numerous versions that require information on regional and global processes, the SSCRO foresees a possibility to apply different models of ozone dynamics. The scenario is considered basic, and is formed by the operator from the data of the regional or global ozonometric network. The operator can prescribe a discrete function $O_{3,i,j,k,s} = Q_3(\varphi_i, \lambda_j, z_k, t_s)$, a priori considering the value of $O_{3,i,j,k,s}$ an average estimate of the function $Q_3(\varphi_i, \lambda_j, z_k, t_s)$ in space Ω_{ijk} for a time period $t_{s-1} \leq t \leq t_s$.

Possibilities are foreseen in choosing various versions of approximation of the O_3 function over the whole territory by a number of latitudinal and meridional distributions.

In this case only the ozone–air mixing ratio can be prescribed. From this ratio the O_3 function is reconstructed, provided the SSCRO database contains information on the vertical profile of air density at any point $(\varphi, \lambda) \in \Omega$. The respective computer dialogue procedure makes this choice automatic.

One of the important elements of the SSCRO functioning is the replenishing of the database in the period of its adaptation to conditions of a given region. The database includes information about the flight timetable over the region and other characteristics of the engines, cruising altitude, airspeed, location of airports, air route, etc. An example is given in Table 5.12.

All background information is concentrated in the form of matrix structures, such as $F = \|f_{ij}\|$, $D = \|d_{ij}\|$, $C = \|c_{ij}\|$, and $B = \|b_{ij}\|$, where f_{ij} is a vector whose components contain all information needed for the ith arrival flight (f_{i1} – time of landing, f_{i2} – arrival direction, f_{i3} – type of engine, f_{i4} – airspeed, f_{i5} – cruising

Table 5.12. Estimates of some parameters in the SSCRO database.

Parameter	Estimate
Tropopause altitude (km)	
Summer (Z_1)	10
Winter (Z_2)	13
Cruise altitude (km)	
Subsonic aviation	9–12
Supersonic aviation	16–20
Diameter of the vortex trail zone after the aircraft (m)	100–150
Indices of the supersonic transport aircraft emission	
$EL(NO_x)$	15
$El(CO)$	1.5
$El(CH_4)$	0.2
$El(H_2O)$	1230

altitude, f_{i6} and f_{i7} – latitude and longitude of the airport of entry and other possible characteristics. The d_{ij} vector contains similar information about departing aircraft, the b_{ij} vector describes information about air routes and transit aircraft, and finally, the b_{ij} vector decodes the f_{i3} component, giving the volume of the burnt fuel, its type and composition of exhaust gases. The transit flights, with the intermediate flight stop at the airports of the region, are taken into account by the identificators F and D separately, before their landing and after take-off.

Model estimates of the impact of the transport aircraft on the atmosphere and climate, which can be obtained with the use of SSCRO will make it possible, with available synoptic information, to solve the problem of optimization of the transport flight corridors and flight timetables. Considering various scenarios of the aircraft load on the regional ozonosphere using the SSCRO, it is possible to determine the location of the flight corridors, which, in other similar conditions, will reduce the consequences of this load. A possibility appears to specify the compounds of other biogeochemical processes with participation of greenhouse gases.

5.8 THE ROLE OF PRECIPITATION IN THE GLOBAL CYCLE OF CARBON DIOXIDE

A specification of the global model of the CO_2 cycle in the biosphere includes a consideration of the role of the process of its washing out from the atmosphere due to precipitation. This process is unidirectional, although studied inadequately. Nevertheless, there are direct observational verifications of CO_2 absorption by rain droplets (Egan *et al.*, 1991). In particular, one of the confirmations of the fact that precipitation washes out CO_2 from the atmosphere is the presence in rain of a considerable amount (up to 15 mg/l) of hydrocarbonate ion HCO_3^-. A combined analysis of precipitation amount and variations of the atmospheric CO_2 concentra-

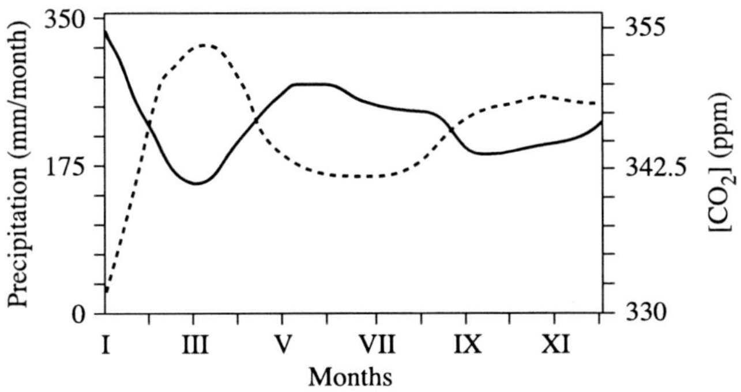

Figure 5.7. Variations of precipitation amount (solid curve) and CO_2 concentration in the atmosphere (dashed curve) estimates from integrated data of the observatory observations within the framework of the Global Carbon Project (GCP).

tion over the same territory performed using the data of the global observational network, has made it possible to reveal a persistent correlation between these processes. Figure 5.7 shows the curves of changes of average monthly precipitation and concentration of atmospheric CO_2. We see that the dependence between changes of atmospheric CO_2 and precipitation is sufficiently stable. A detailed analysis of this dependence for various latitudinal belts or for other configurations of limited territories reveals similar patterns independent of geophysical coordinates. Here one should point to a high sensitivity of correlation to the duration and type of precipitation. For instance, during a shower the HCO_3^- concentration in precipitation can either double or halve depending on the presence or lack of thunderstorms. Moreover, this ratio depends strongly on the duration of the precipitation period. Observations showed that with an increasing duration of rain the concentration of HCO_3^- ions decreases. In other words, the interaction between CO_2 concentration and moisture content in the atmosphere is an important component of the global carbon cycle.

A formalization of the role of rain in the global CO_2 cycle requires a model of CO_2 absorption by water droplets falling at some velocity u (Wang *et al.*, 1997; Wacker, 1995; Chate and Kamra, 1997). The most widely used version of such a model is an equation of gas balance on the surface of rain droplets:

$$\frac{dC}{dz} = \frac{3\Delta}{ur^2}(1 + 0,3\sqrt{Re}\sqrt[3]{Sc})(C_A - C^*)$$

where Δ is the coefficient of CO_2 diffusion in the air, r is the droplet's radius, C^* is the balanced concentration of CO_2 in a droplet, C_A is the CO_2 concentration in the atmosphere, z is the altitude, $Sc = \nu/\Delta$ is the Schmidt number, ν is the kinematic viscosity, and $Re = 2ru/\nu$ is the Reynolds number.

A diversity of the forms of precipitation over the globe complicates a consideration of their role in the global CO_2 cycle. This problem can be solved in two ways. The first way is the formal numerical description of the totality of the processes of precipitation formation. The second way is connected with the use of the present means of the global observation of precipitation. In both cases the forms of rain should be clearly classified as functions of meteorological situations. The rain rate can range widely from 1 mm/hr to 8 mm/hr and even more. Besides, there is a certain correlation between the precipitation rate and the size of rain droplets. With the low-intensity rain $r \in [0.1; 0.5]$. A shower can be characterized by the formation of droplets up to $r \sim 6$ mm.

Thus the problem of assessment of the role of precipitation in washing out of CO_2 from the atmosphere is urgent, and to solve it, the global model should take into account a change of hydrological cycles separately, over the World Ocean and over land, since these regions of the planet differ in their interaction with the atmosphere.

5.8.1 Modelling the global water cycle

Water belongs to substances most widespread in nature. It is present in various forms practically all over the globe, and plays an important role in energy and mass-exchange between the continents, oceans, and atmosphere. The problem of assessing the role of water in the global carbon cycle is only a small part of the general global problem of the nature–society interaction.

Oceans, polar ice caps, glaciers, lakes, rivers, soils and the atmosphere contain $1.4\text{–}1.5 \cdot 10^9\ \text{km}^3$ of water. This mass is in constant dynamic interaction with other biospheric components and determines thereby the factors of environmental variability. The developed methods of numerical experiments should be used to assess the role of these factors in present conditions and to show the significance of water balance in stabilization of numerous climatic and biogeocoenotic processes. Here an attempt has been made, through a systematization of the information about the water balance of the planet, to create a version of the model of the biospheric water balance (MBWB) able, within the general approach to modelling the carbon balance, to take into account the role of water fluxes.

An important moment of the MBWB functioning is its conjunction with the methods of determination of various parameters of the water cycle. Such methods are based on the use of surface, satellite, and airborne measurements. The unit MBWB as a global model makes it easier to understand the role of the oceans and land in the hydrological cycle, to choose the main factors that control it, as well as to trace the dynamics of its interaction with plants, soil, and topographic characteristics of the Earth's surface. It is based on an account of interaction between the elements of the water cycle, as well as having natural and anthropogenic factors taken into account through information interfaces with other units of the global model (Krapivin and Kondratyev, 2002).

Consider the block-scheme of the global water exchange and write the respective equations. The basic regularity of the global water exchange is the invariability of water supplies on the Earth on the time period of hundreds of year, that is, we can reliably write the balance equation $W_E = W_S + W_O$, where W_E, W_S, and W_O are water supplies on the Earth, on land and in the oceans, respectively. A compartment of the atmosphere is related to the respective region of water basin. The relationship is valid:

$$\frac{dW_E}{dt} = \frac{dW_S}{dt} + \frac{dW_O}{dt} = 0$$

or $dW_S/dt = -dW_O/dt$. Hence, the trend in changes of water supplies on land is contrary to the similar trend in the oceans.

With the water supply in the atmosphere $W_A = W_{AO} + W_{AS}$, we obtain $W_E = W_A + W_{S1} + W_{O1}$: where W_{AO} and W_{AS} are water supplies in the atmosphere over the oceans and land, respectively; $W_{S1} = W_S - W_{AS}$; $W_{O1} = W_O - W_{AO}$. The balance equation will be:

$$\frac{dW_E}{dt} = \frac{dW_A}{dt} + \frac{dW_{S1}}{dt} + \frac{dW_{O1}}{dt} = 0$$

As is seen, the structure of the trends in the ratios of water supplies is complicated and to analyse it, additional considerations are needed. This complication becomes considerable with further subdivision of the biosphere.

Within the MBWB, small corrections for the water exchange between the Earth and space are not taken into account. The model of the global water cycle can be based on the method of describing the hydrology of comparatively large territories. In this case the basic unit of such a territory is a compartment Ω_{ij} of the Earth surface of size $\Delta\varphi_i$ by latitude and $\Delta\lambda_j$ by longitude.

The state of the water component of the compartment Ω_{ij} with the coordinates (φ_i, λ_j) can be characterized by the magnitude of an equivalent liquid water column over unit area. Possible water fluxes across the border of Ω_{ij} are shown in Figures 5.8 and 5.9. The intensities of these fluxes depend on the phase state of water,

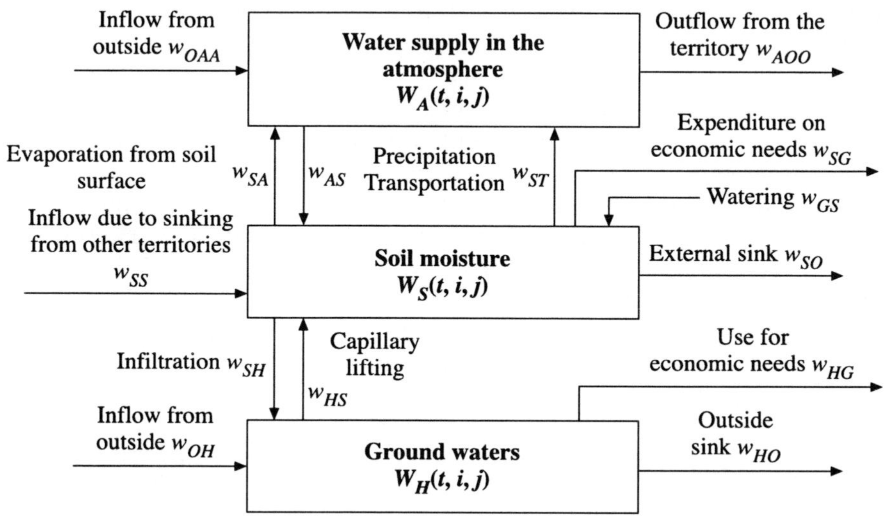

Figure 5.8. Water fluxes across the border of a limited land territory.

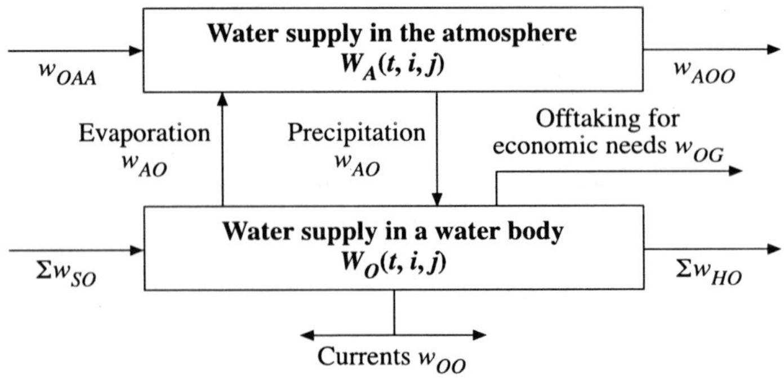

Figure 5.9. Water fluxes across the border of a limited territory with a water object.

temperature, wind speed, and other geophysical and ecological factors. It is difficult to take into account all the small details of these fluxes within the global model because their interactions have been studied inadequately. Therefore the degree of detailing chosen here has been orientated towards an account of the most important components of their states. Water is considered in liquid, solid, and gas phases. Within the compartment Ω_{ij} there is only one state although in the future with the available required information, a vector parameter can be introduced, which determines the share of precipitation over Ω_{ij} in the form of snow, pellets of snow, granulated snow, pellets of ice, ice rain, hail, rain, drizzle, wet snow, and other forms.

The water balance consists of a mosaic structure of local balances at the level of Ω_{ij}. The proposed description of water fluxes enables us to trace their balance at any level of spatial digitization – region, water basin, continent, ocean, hemisphere, or biosphere. Clearly, the general balance of evaporation and precipitation at the level of the biosphere is maintained. In other cases, on average, with decreasing spatial sizes of the selected unit of the biosphere, one should expect an increasing difference between the precipitation amount and evaporation. In this case the water transport through the atmosphere, with river run-off and sea currents will serve as an equalizer. Though the quantitative estimates of all these parameters have been well studied, the water cycle dynamics can be described only with the use of the model. As a first approximation, to assess the role of precipitation in the global CO_2 cycle, one can use only components W_{AU} and W_{AS}. However, with account of the spatially heterogeneous distribution of CO_2, the biosphere should be digitized.

With the notations assumed in Figures 5.8 and 5.9, the balance equations of the water cycle at the level of Ω_{ij} are written as follows:

$$\frac{dW_S(t,i,j)}{dt} = w_{AS} + w_{GS} + w_{SS} + w_{HS} - w_{SO} - w_{SG} - w_{ST} - w_{SA} - w_{SH}$$

$$\frac{dW_H(t,i,j)}{dt} = w_{SH} + w_{OH} - w_{HO} - w_{HG} - w_{HS}$$

$$\frac{dW_O(t,i,j)}{dt} = \sum_{(k,n)\in I_{kn}} [w_{IO}(t,k,n) + w_{HO}(t,k,n)]$$
$$+ w_{AO} + w_O - w_{OA} - w_{OG} - w_{OR} - w_T$$

$$\frac{dW_A(t,i,j)}{dt} = w_{OAA} - w_{AOO} + w_{SA} + \begin{cases} w_V & \text{for water surface} \\ w_{ST} & \text{for land} \end{cases}$$

Detailing of the right-hand parts of these equations with the functional presentations of fluxes in conditions of changing parameters of the environment, will determine the level of the qualitative and quantitative reliability of the model. In particular, the model can be simplified by approximating the average value of W_O:

$$\bar{W}_O = \begin{cases} 2500 + 350\sqrt{t} & \text{for } 0 \le t \le 70 \\ 6400 - 3200\exp(-t/62) & \text{for } t > 70 \end{cases}$$

where the average depth of the World Ocean is measured in metres, the age of the ocean t is calculated in millions of years. Variations of the ocean volume can be also approximated by the formula: $\Delta V = \Delta W_O A_O + 59.5(\Delta W_O)^2$, where $A_O = 361.06 \cdot 10^6\ \text{km}^2$.

5.8.2 Modelling the water cycle on the atmosphere–land border

The land–atmosphere exchange processes include evaporation of the soil moisture, evaporation from the leaf surface, stems and trunks of plants, transpiration, precipitation, and evaporation of unstable water accumulations on the surface. The water flow from the soil through the plant is the least studied link in this chain. The importance of the process of transpiration in the global water cycle may be illustrated by available estimates, which show that the process of transpiration takes more water than photosynthesis. For instance, from average estimates, to grow a 20 t yield (wet mass), the plants extract from the soil about 2,000 t of water, with only 3 t of the used water being atomic hydrogen bound with atomic carbon in photosynthesis.

The model description of the process of transportation requires an understanding of the role of physical and physiological factors in this process. A simplified idea about this role is reduced to the following. If the plant roots are in sufficiently wet soil, then the rate of transpiration is a function of temperature, humidity, wind speed, and insolation. Beyond some threshold of soil moisture, when the water supply in soil ceases, the role of the indicated physical factors sharply diminishes, being inferior to the physiological factors – the type of plants, the construction of roots, phase of development, type of soil, and soil layer thickness. This threshold can vary from 5 to 50 cm of the precipitable water. At any rate, if for a given type of plant the water is not a limiting factor (i.e., water is not limited), then, as a first approximation, the total growth of plants can be considered proportional to total potential transpiration for the whole period of growth. The latter is proportional to the amount of the incoming solar radiation.

At present about 12% of the total evaporation from the Earth's surface is used by plants in the process of photosynthesis. In this process about 2250 km^3 of water is used annually with the return coefficient being 0.75. Therefore the simplest description of transpiration will be: $w_{ST}(t,i,j) = \beta_{ij} W_S(t,i,j)$, where b_{ij} depends on the vegetation productivity. The values of β_{ij} are 0.67 for forests, 0.44 for a meadow steppe, and 0.25 for agricultural crops. However in real situations W_S is a limiting factor in a more complicated dependence of the impact on transpiration rate through the rate of photosynthesis R_p. In other words, $w_{ST} = k_p R_p$, where k_p is the transpiration coefficient for the plants of type p. As a first approximation, one can use $R_p = \varepsilon_p r_p$, where ε_p is the share of solar energy, assimilated by the pth type of plants in the process of photosynthesis. The value ε_p depends on the presence of water accessibility for plants:

$$\varepsilon_p = \varepsilon_{p,0}[1 - \exp(-\varepsilon_{p,1} W_S(t,i,j)]$$

where $\varepsilon_{p,0}$ is the value of ε_p at a sufficient amount of water, $\varepsilon_{p,1}$ is the coefficient reflecting a reduction of solar energy assimilated by plants with decreasing water accessibility. On average, $\varepsilon_{p,0}$ is reached at $W_S = 10$ mm. Assuming $\varepsilon_p/\varepsilon_{p,0} = 0.9$, we have $\varepsilon_{p,1} = 0.23$. In this case $r_p = 9.6$ kg/km^2/day of dry substance (or 37 kg of wet phytomass). The coefficient k_p is estimated for each type of plant. The coefficient k_p is equal to 368 for maize, 397 for sugarbeet, 435 for wheat, 636 for potatoes, 462 for cotton (kg H_2O/kg of pure substance).

As one of the models of the process of transpiration we can write $w_{ST} = Y_S (24a + b)$, where Y_S is the specific water return, a is the rate of ground water lifting (cm/hr), b is the average daily change of ground water level (cm).

Within the MBWB unit, to describe the process of the atmosphere–land interaction, the fluxes w_{SA}, w_{ST}, and w_{AS} are used, whose parametric descriptions serve as the basis for this unit. Information on precipitation w_{AS} is usually included in the information bulletins of hydrometeorological services. The pre-history of the distribution of precipitation in the form of a set of matrices $W_{AS}(\theta) = \|w_{AS}(\theta,i,j)\|$, where $(i,j) \in \Psi$ an θ are discrete time moments of the precipitation record by the hydrometeorological service, are used to derive the functional $w_{AS}(t,i,j) = F(W_{AS}(\theta_1),\ldots,W_{AS}(\theta_N),t)$. This is performed with the use of extrapolation, a grouped account of arguments, and evolutionary modelling. Such an approach requires data on precipitation over the discrete geographical grid $\Delta\varphi \cdot \Delta\lambda$. However, this database can be also modelled by simulating the global cloud field ensuring precipitation. The simplest parameterization of clouds consists in prescription of the threshold $W_{A,\max}$, beyond which an excess of atmospheric moisture transforms into water and precipitates. To reduce inevitable errors (mainly caused by overestimation of precipitation), it is expedient to introduce the matrix of threshold $W_{\max} = \|W_{A,\max}(i,j)\|$ where $(i,j) \in \Psi$. Then:

$$w_{AS}(t,i,j) = \begin{cases} 0 & \text{for } W_A(t,i,j) \le W_{A,\max} \\ W_{A,\max} - W_A(t,i,j) & \text{for } W_A(t,i,j) > W_{A,\max} \end{cases}$$

If the value $W_{A,\max}(i,j)$ corresponds to a real critical value of the moisture content in the atmosphere over Ω_{ij}, then w_{AS} will be overestimated. It is assumed here that at $W_A(t,i,j) > W_{A,\max}$ the cloud fills the whole cell Ω_{ij}, which does not always correspond to reality. Moreover, the fact that a considerable share of moisture, even with an exceeded critical level, can remain in the cloud and evaporate, is left out of the account. Therefore to take these special features into account, an adaptive coefficient $\alpha_W < 1$ should be introduced.

Divide precipitation into two basic types – solid and liquid. This division can be made by the thermal principle and by seasons. The thermal principle is more preferable due to flexibility at sharp climate changes and possible shifting of the scale of seasonality. The synoptic division of the periods with different precipitation is justified in various regions of the globe. The average daily temperature in the early period of solid precipitation is below zero, ranging from $-4°$C to $-7°$C. On the border of this division the precipitation of the mixed type is observed. The relationship between the types of precipitation is described by the formula: $x_T = a - bT$, where x_T is the share of solid precipitation, T is temperature, and a and b are empirical coefficients. For the Atlantic climatic zone $a = 50$ and $b = 5$.

To parameterize the process of evaporation from land, numerous formulae are used. Here a simple dependence is assumed:

$$w_{SA}(t,i,j) = \rho_{ij}[1 - \exp(-\delta_{ij}/\rho_{ij})]/\sigma_{ij}$$

where δ_{ij} is a maximum possible rate of evaporation in the region and ρ_{ij} is the total amount of moisture getting to the soil per unit time.

The choice of the concrete model of the evaporation process is determined by the character of the applied database. Evaporation from the soil surface substantially depends on the type of vegetation cover. The evaporation in the forest and in the field differs by 30–40%. This is connected with the heterogeneous impacts on the water regime, within various vegetation covers, of such factors as soil freezing, snow melting intensity, soil structure, radiation budget, and others. To take into account the dependence of the rate of evaporation from land on temperature T_{ij}, the character of vegetation cover, and soil properties, the following formula is used:

$$w_{SA}(t,i,j) = \delta^*(T_{ij}/\bar{T}_{ij})[1 - \lambda\exp(-\lambda_1 A_{ij}/\bar{A}_{ij})][1 - \lambda_2\exp(-\lambda_3 X_{ij}/\bar{X}_{ij})]$$

where $\bar{T}_{ij}$ is the surface air temperature in the region Ω_{ij}, averaged over the period considered, $\bar{A}_{ij}$ and $\bar{X}_{ij}$ are the average power of the soil layer and the density of vegetation cover, respectively, and $\delta^*, \lambda, \lambda_1, \lambda_2$ and λ_3 are the empirical coefficients.

A detailed analysis of possible models of evaporation from the land surface with account of various types of vegetation cover and changes of climatic parameters is given in the works of Bras (1990), Chock and Winkler (2000), and Karley *et al.* (1993). In particular, there are formulae to calculate evaporation as a function of the height and density of vegetation cover, wind speed, and temperature. For instance, the following dependence is proposed for the rate of complete evaporation:

$$E_T = \begin{cases} (\Delta Q_N + m\gamma L_j)/[\Delta + (1+n)\gamma + I(1-C)] & \text{for } T < 0°\text{C} \\ (\Delta Q_N + \gamma L_j)/[\Delta + \gamma] & \text{for } T \geq 0°\text{C} \end{cases}$$

where Δ is the rate of change of saturated vapour pressure as a function of temperature, Q_N is the amount of energy reaching the evaporating surface, γ is the psychometric coefficient ($\approx$0.66 mb/K), I is the share of complete evaporation due to precipitation caught by foliage, and C is the compensation coefficient due to transpiration. The coefficients m, n, and C are functions of height h and type r_s of vegetation:

$$m = 53\ln^2(20/h + 2.5) \qquad C = (\Delta + \gamma)/[\Delta + (1+n)\gamma] \qquad n = r_s[m(1 + U/100]/250$$

The indicator of the type of vegetation r_s (m/s) for some types is estimated at: 40 (sun-flower and alfalfa), 70 (barley and potatoes), 250 (citrus plants), 130 (cotton), 80 (maize and rice), 50 (sugar beet), 60 (wheat), 400 (tundra), 200 (subtropical meadows), 100 (temperate-zone meadows), 100–300 (tropical forests), 200–300 (coniferous forests), and 100–150 (deciduous forests in middle latitudes). Typical values of the parameters n, m, and I are: for grass ecosystems, $n \approx 2.5$, $m \approx 3.5$, $I \approx 0.2r$; for woodlands, $n \approx 30$, $m \approx 5$, $I \approx 0.3r$ (temperate latitudes) and $I \approx 0.15r$ (tropics); where r = precipitation.

Albedo is an important parameter to calculate the solar radiation energy participating in the process of evaporation. The relationship between the height of plants and albedo, as a first approximation, is described by linear dependence. With the height of plants reaching ~20 m, albedo decreases from 0.25 to 0.1. The albedo values for some types of the Earth cover are known: heather – 0.14; fern – 0.24; natural pastures – 0.25; shrubs – 0.21; savannah – 0.17; deciduous forests in mid-latitudes – 0.1; coniferous forests and orange-groves – 0.16; eucalyptus forests – 0.19; wet tropical forests – 0.13; and waterlogged forests – 0.12. The albedo of agricultural fields varies between 0.15 (sugar cane and fruit trees) and 0.26 (sugar beet, barley, and cucumber).

The surface part of the land–atmosphere water exchange is connected with the subdivision of the phase space, at least, into two levels – soil and ground waters. The soil level plays the role of the buffer reservoir between precipitation and ground waters. The simplest parameterization of fluxes between these levels is reduced to linear dependences: $w_{SH}(t,i,j) = \lambda_{ij} W_S(t,i,j), w_{HS}(t,i,j) = \mu^{ij} W_H(t,i,j)$. However, a more strict description of the soil level is dictated by natural heterogeneity of the structure of Ω_{ij}, where small water bodies and land sites with a certain relief can be located. According to the landscape–hydrological principle, to simulate Ω_{ij}, it is necessary to choose facies and sites of water surfaces, which are connected with typifying the floristic background, and whose concrete condition is determined by micro-relief, type and properties of soil, surface moisture, depth of ground waters, and other factors. It is possible to choose m_{ij} of facies and n_{ij} of water bodies. In this case the soil moisture forms not only due to the fluxes shown in Figures 5.8 and 5.9, but also due to leakage and filtration of water from the water bodies and aqueducts located in Ω_{ij}.

An important factor of the surface part of the water balance is infiltration of precipitation into the soil both during rain and in run-off. The rate of taking up water into the soil w_{SH} is described by the formula: $w_{SH} = k_S l$, where k_S is the coefficient of filtration and l is the hydraulic slope. If we denote the volume mass of the soil as κ, which, on average, varies within 1.4–1.5 g/cm^3, then for k_S it is convenient to use the Azizov formula: $k_S = 256.32\kappa^{-7.28} - 1.27\kappa^{1.14}$ (cm/day). The parameter l is calculated using the formula $l = (z_0 + z_1 + z_2)/z_0$, where z_0 is the depth of the column of washing-out, z_1 is the capillary pressure, and z_2 is the height of the water layer on the soil surface. At $z_0/z_1 \leq 2$ an approximation $w_{SH} = k_S + t^{-1/2}(0.5k_S z_1 D)^{1/2}$ is valid, where D is the soil moisture deficit and t is time. Other approximations of the function w_{SH} are known, such as the Horton empirical formula: $w_{SH} = [w_{SH}(t_0, i, j) - kS] \exp(-\beta t) + k_S$; the Popov formula, $w_{SH} = r \exp(-rt/D) + k_S$; and the Kostiakov formula, $w_{SH} = k_S + \alpha t^{-n}$, where α, n, and β are calibration parameters an r is the rain intensity.

The interaction between the regions of the assumed grid of the biosphere surface division $\{\Omega_{ij}\}$ is realized through fluxes w_{OAA}, w_{AOO}, w_{SS}, w_{SO}, w_{OH}, and w_{HO}.

5.8.3 Modelling the water cycle on the atmosphere–water surface border

The processes of transport on the atmosphere–water border surface have been well studied. The transport of moisture from the surface of the water body into the atmosphere is one of the fragments of complicated physical processes of mass-and energy-exchange across the water–air interface. These processes are functions of many climatic parameters and, to a large extent, are regulated by eddy motions in the surface layer of the atmosphere determined by the wind field.

The possibility of estimating the water transport from the water surface into the atmosphere consists of an assessment of the water content of the lower part of the surface layer of the atmosphere, which is in the form of spray and water vapour. The eddy flux of water through a unit surface can be described by the relationship:

$$W_V = -\rho K_W(\partial q/\partial z) = -(\rho w)'q' \approx -\rho\langle w'q'\rangle$$

where W_V is the vertical eddy flux of water vapour (g/cm^2/s), K_W is the coefficient of the eddy transport of water vapour (cm^2/s) (Table 5.13), q is the specific air humidity (g/g), ρ is air density (g/cm^3), z is the vertical coordinate, w is the vertical constituent of wind speed (cm/s), and w' and q' are pulsations of w and q values, respectively. Let p be the atmospheric pressure, then we express q through average water vapour elasticity e: $q = 0.621e/p$.

Evaporation from the water body surface depends on air temperature and can be described by the function $w_{SA} = w^* T^\omega$, where w^* and w are the empirical parameters. If measurements are made of wind speed θ (m/s), saturated water vapour pressure at a temperature of the evaporating surface E_1, and atmospheric pressure p (mm Hg),

Table 5.13. The coefficient K_W (cm^2/s) of water vapour diffusion in the atmosphere at the pressure 1000 mb as a function of temperature T.

T °C	−20	−10	0	10	20	30	40
KW	0.197	0.211	0.226	0.241	0.257	0.273	0.289

then to estimate the rate of evaporation, both the Dalton law: $w_{SA} = A(E_1 - e)/p$ and the Shuleikin formula: $w_{SA} = C\theta(E_1 - e)$ can be used where A and C are the parameters related as $A = C\theta/p(C = 0.45 \cdot 10^{-6}\ \text{g/cm}^3/\text{mb})$. The models by Horton and Power described above are rather efficient.

5.8.4 Simulation of water fluxes in the atmosphere

Atmospheric processes of the moisture transport directly connected with the temporal variations of the meteorological elements, play an important role in the global water cycle. The global atmospheric circulation can be described by the Monin model (Monin and Krasitsky, 1985):

$$\frac{\partial \nu_\delta}{\partial t} + V_z \frac{\partial \nu_\delta}{\partial z} + V_\delta R^{-1} \frac{\partial \nu_\delta}{\partial \delta} + V_\lambda R^{-1} \sin^{-1}\delta \frac{\partial \nu_\delta}{\partial \lambda} = R^{-1}(V_\lambda)^2 \operatorname{ctg}\delta + 2\Omega V_\lambda \cos\delta + (R\rho)^{-1}\frac{\partial p}{\partial \delta} + f_\delta$$

$$\frac{\partial \nu_\lambda}{\partial t} + V_z \frac{\partial \nu_\lambda}{\partial z} + V_\lambda R^{-1} \frac{\partial \nu_\lambda}{\partial \delta} + V_\delta R^{-1} \sin^{-1}\delta \frac{\partial \nu_\lambda}{\partial \lambda} = -R^{-1} V_\delta V_\lambda \operatorname{ctg}\delta - 2\Omega V_\delta \cos\delta - (R\rho \sin\delta)^{-1}\frac{\partial p}{\partial \lambda} + f_\lambda$$

where Ω is the angular rate of the Earth's rotation, $\delta = \pi/2 - \varphi$ is an addition to latitude, λ is longitude, V_z, V_δ, and V_λ are the components of the velocity of the atmospheric motion, R is the Earth radius, and f_δ and f_λ are the components of acceleration due to friction expressed through the friction stress tensor u_{sl}:

$$\rho f_\delta = \frac{\partial u_{\delta z}}{\partial z} + (R\sin\delta)^{-1}\left(\frac{\partial u_{\delta\delta}}{\partial \delta} + \frac{\partial u_{d\lambda}}{\partial \lambda}\right) - R^{-1}u_{\lambda\lambda}\operatorname{ctg}\delta$$

$$\rho f_\lambda = \frac{\partial u_{\lambda z}}{\partial z} + (R\sin\delta)^{-1}\left(\frac{\partial u_{\lambda\delta}}{\partial \delta} + \frac{\partial u_{\lambda\lambda}}{\partial \lambda}\right) - R^{-1}u_{\lambda\delta}\operatorname{ctg}\delta$$

The velocity field equations are closed by prescribed zero boundary conditions on the Earth's surface, determined by the equation of relief in the form of the prescribed function $z = h(\delta, \lambda)$ and by an added equation of the state of humid air $p = \rho T[r_d + q(r_V - r_d)]$, where $r_d = 0.287$ Joule/g/K and $r_V = 0.46$ Joule/g/K are the gas constants of dry air and saturated water vapour and $q \approx 3$–4% is the specific humidity. The distribution of temperature $T(\delta, \lambda)$ and function $q(\delta, \lambda)$ can be described by the respective equations of evolution, and the data of the global archive may be used for relevant parameters.

A sufficiently complete description of the models of the general atmospheric circulation has been given in Nicolis and Nicolis (1995). A simplified description of the atmospheric part of the hydrological cycle is made possible by the equation $\partial W_A/\partial t + \nabla Q = E - P$, where W_A is the vertically integrated specific air humidity in this column and E and P are evapotranspiration and precipitation at the soil level, respectively.

A further simplification of the model of the hydrological cycle is connected with selection of the following three types of prevailing directions: western, eastern, and meridional. For such an approximation, data are used on the amplitude of the wind speed oscillations and the number of days of the seasonal directions of the moisture transport in the atmosphere. If the mass of water vapour in air column over the area σ_{ij} is $a = W_A\sigma_{ij}$, then, for instance, for the eastern orientation of the atmospheric circulation, the water flux between the adjacent cells of the Earth's surface digitizing grid will be $w_{AO} = 2a\theta/d_{ij}$, where θ is the wind speed and d_{ij} is the diameter of Ω_{ij}. Following this scheme, it is easy to recalculate the moisture supplies at each step of time digitization, since it is unnecessary to solve the problems of numerical integration of partial differential equations. The background information about W_A, σ_{ij}, θ, and d_{ij} is accumulated in the database from different sources. The function W_A is calculated by the balance equation or can be prescribed based on other data. In particular, with changing temperature T and partial pressure of water vapour e, then W_A can be estimated from the relationship: $W_A = meh(1 + \alpha T)^{-1}$, where h is the height of the effective atmospheric layer and m and α are the proportion coefficients ($m = 0.8$; $\alpha = 1/273$ when measuring W_A in g/m^2; and T in °C).

5.8.5 Simulation of water fluxes in the World Ocean

First place among all the water reservoirs on the Earth is occupied by the World Ocean whose present volume exceeds 50 times the volume of water in glaciers (which are second in volume to the World Ocean). This comparison is important for understanding the correlation between the steps of hierarchy of water basins and determining their structure in the model. Within a priori scenarios of anthropogenic activity and possible changes in the biosphere, the correlation between these steps is important. For instance, 1.6% of the global supplies of water are accumulated in the Antarctic. A comparison of these supplies with the volume of the Arctic Ocean where the water content is some 20% less than in the Antarctic glacier cover, allows us to highlight the inadequacy of the global model of the hydrological cycle without inconclusion in it of the role of the Antarctic (Keeling and Visbeck, 2001).

The hydrology and sea currents of the Southern Ocean with account of the effect of the glacier cover have been described in numerous monographs, and the circulation models of different complexity and degree of detailing have been derived to simulate them. Such models for the World Ocean, on the whole, are based on account of configurations of the non-penetrating boundaries and topology of the straits. Numerous numerical experiments with such models have made it possible to reveal the principal structure of the global oceanic circulation consisting of a hierarchy of the closed ring circulations with the centres of lifting and descending waters and water nuclea, and including the geometry of water basins with the straits between them. To describe the water circulation in the Southern basin, it is necessary that Drakes Passage be taken into account. A scheme of the hydrological fields circulation in the World Ocean, which is acceptable for simulation, has been proposed by Seidov (1987) and Chahine (1992). The model is a system of equations and boundary conditions taking into account the outline of the shores,

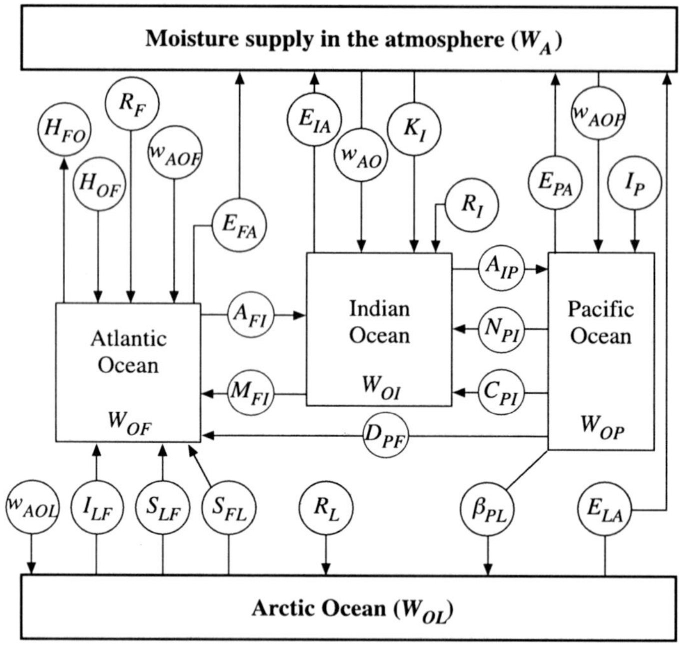

Figure 5.10. Elements of the global water balance with the role of ocean taken into account. Notations: w_{AOL}, w_{AOF}, w_{AOI}, w_{AOP} – precipitation; H_{FO}, H_{OF} – Strait of Gibraltar; R_F, R_I, R_L – rivers; E_{FA}, E_{IA}, E_{PA}, E_{LA} – evaporation; A_{FI}, A_{IP} – the Antarctic Current; M_{IF} – the Cape Igolny Current; C_{PI} – the East-Australian Current; β_{PL} – Bering Strait; I_{LF} – Arctic ice; D_{PF} – Drake Passage; I_P – Antarctic ice; N_{PI} – the Indonesian seas; S_{LF}, S_{FL} – the straits.

the bottom relief, as well as ice formation and melting. However, on a global scale, to simulate the oceanic circulation, a simplified scheme is necessary reflecting mainly the role of the straits. Such a scheme is shown in Figure 5.10. The quantitative characteristics of the constituents of this scheme are given in Table 5.14. The final unit responsible for the modelling of the World Ocean circulation has the following form:

$$\sigma_{OF} H_{16}^{O} = H_{OF} + R_F + I_{LF} + S_{LF} - S_{FL} - H_{FO} - A_{FI} + M_{IF} + D_{PF} + (w_{AOF} - E_{FA})\sigma_{OF} + A_F$$

$$\sigma_{OI} \frac{dW_{OI}}{dt} = A_{FI} + C_{PI} + N_{PI} + K_I + R_I + (w_{AOI} - E_{IA})\sigma_{OI} - A_{IP} - M_{IF}$$

$$\sigma_{OP} \frac{dW_{OP}}{dt} = A_{IP} + R_P + (w_{AOP} - E_{PA})\sigma_{OP} + I_P - B_{PL} - D_{PF} - C_{PI} - N_{PI}$$

$$\sigma_{OL} \frac{dW_{OL}}{dt} = R_L + B_{PL} + (w_{AOL} - E_{LA})\sigma_{OL} + S_{FL} - I_{LF} - S_{LF}$$

Table 5.14. Quantitative estimates of water fluxes for the scheme in Figure 5.10 (10^3 km^3/yr).

Flux	Estimate	Flux	Estimate
w_{AOL}	3.6	I_{LF}	0.57
S_{LF}	436	S_{FL}	400
R_L	5.14	A_F	0.3
H_{OF}	23.97	R_F	19.33
w_{AOF}	72.5	E_{FA}	96.6
A_{FI}	6,780.24	M_{IF}	952
D_{PF}	5,771.09	C_{PI}	437
N_{PI}	66.86	I_{OI}	0.765
A_{IP}	6,338.74	E_{IA}	115.4
w_{AOI}	84	R_I	5.386
K_I	0.005	R_P	13.12
E_{PA}	200.4	w_{AOP}	206.7
I_P	0.975	E_{LA}	1.7
β	80.5		

$$\sigma \frac{dW_A}{dt} = (E_{PA} - w_{AOP})\sigma_{OP} + (E_{FA} - w_{AOF})\sigma_{OF} + (E_{IA} - w_{AOI})\sigma_{OI} + (E_{LA} - w_{AOL})\sigma_{OL}$$

Within this large-scale approach to the formation of the MBWB ocean unit, the dependences of the fluxes of water in its different phases on the environmental parameters remain uncertain. Apparently, the mass exchange between the reservoirs s and l can be described by the simplest linear scheme: $w_{sl} = |W_{OS}\sigma_{OS} - W_{OL}\sigma_{OL}|/T_{sl}$, where T_{sl} is the time for equalizing the levels W_{OS} and W_{OL} and σ_{OS} and σ_{OL} are the areas of water basins s and l. For the scheme in Figure 5.10 we have:

$$A_{FI} = \max\{(V_{OF} - V_{OI})/T_{FI}, 0\}$$

$$M_{FI} = \max\{(V_{OI} - V_{OF})/T_{IF}, 0\}$$

$$A_{IP} = \max\{(V_{OI} - V_{OP})/T_{IP}, 0\}$$

$$N_{PI} = \max\{(V_{OP} - V_{OI})/T_{PI}, 0\}$$

$$C_{PI} = \max\{(V_{OP} - V_{OI})/T^*_{PI}, 0\}$$

$$D_{PF} = \max\{(V_{OP} - V_{OF})/T_{PF}, 0\}$$

$$S_{LF} = \max\{0, (V_{OL} - V_{OF})/T_{LF}\}$$

$$S_{FL} = \max\{0, (V_{OF} - V_{OL})/T_{FL}\}$$

$$\beta_{PL} = \max\{0, (V_{OP} - V_{OL})/T_{PL}\}$$

where $V_{OS} = W_{OS}\sigma_{OS} (S = F, I, P, L)$.

To estimate the flux K_l, take into account the information on the moisture balance in the region of the Red Sea. According to available estimates, the input of water to the Red Sea via the Suez Canal and by precipitation can be neglected. Not a single river flows into the Red Sea. The main component of the flux K_l through Bāb al Mandab is rather persistent. Hence, we can assume $K_l = \max\{0, w_{AK}\sigma_{KMP} - E_{KMA}\sigma_{KM}\}$, where w_{AK} and σ_{KMP} are the level and the area of the mainland run-off to the Red Sea, respectively and E_{KMA} is the evaporation from the area σ_{KM} of the Red Sea. The water expenditure through the Strait of Gibraltar H_{FO} (H_{OF}) is determined by the relationship of the levels of W_{OF} and the Mediterranean Sea. In order not to complicate the structure of the model, the level of water in the Mediterranean Sea is determined by its watershed and the difference between precipitation and evaporation. Since the intra-annual distribution of the water inflow into the Atlantic Ocean varies within 10–20%, we can reliably assume $W_{FO} = -W_{OF} = \text{const}$.

5.8.6 The model of the hydrological regime of a territory

Consider the scheme of Figure 5.11 as the basis for modelling the hydrological regime of a limited territory Ω_L, occupied by the aqua-ecosystem under study. Each territory has a river network, water bodies, and land. According to the landscape–hydrological principle, to derive a simulation model in the zone of the hydrological system functioning it is necessary to select the facies, which are connected with typifying the floristic background, whose concrete appearance is determined by the micro-relief, type and properties of the soil, surface moistening, depth of ground waters, and other factors. In general, the territory Ω_L is characterized by the presence of m facies, and the water network has n heterogeneous sites. Bearing this in mind, according to the scheme in Figure 5.11, the closed system of balance equations has the form:

$$\sigma_{ij}\frac{dW_{A,ij}}{dt} = E_{ij} - R_{ij} + \sum_{k=1}^{n}(V_k - B_kS_k) + D_{ij} + \sum_{l=1}^{m}(L_l + T_l - W_l\sigma_l) \tag{5.30}$$

$$S_k\frac{dG_k}{dt} = Y_k - V_k + B_kS_k - H_k + J_k + \sum_{l=1}^{m}(K_{lk} - F_{kl} - V_{kl} - M_{kl})$$
$$- \Gamma_k + S_k(C_{k-1}V_{k-1}/\Delta_{k-1} - C_kV_k/\Delta_k) \tag{5.31}$$

$$\sigma_l\frac{d\Phi_l}{dt} = \sum_{k=1}^{n}(F_{kl} + V_{kl} + M_{kl}) + \sum_{k=1}^{m}\psi_l^k\theta_l - L_l - T_l - P_l - \theta_l + N_l + W_l\sigma_l \tag{5.32}$$

$$\sigma_{ij}\frac{dG_{ij}}{dt} = I_{ij} - Z_{ij} - D_{ij} + \sum_{k=1}^{n}(H_k - J_k) + \sum_{l=1}^{m}(P_l - N_l) \tag{5.33}$$

In the formulae (5.30)–(5.33) the following notations are assumed: σ_{ij}, σ_l, and S_k are the areas of the territory Ω_{ij}, of the lth facies and the kth compartment of the river

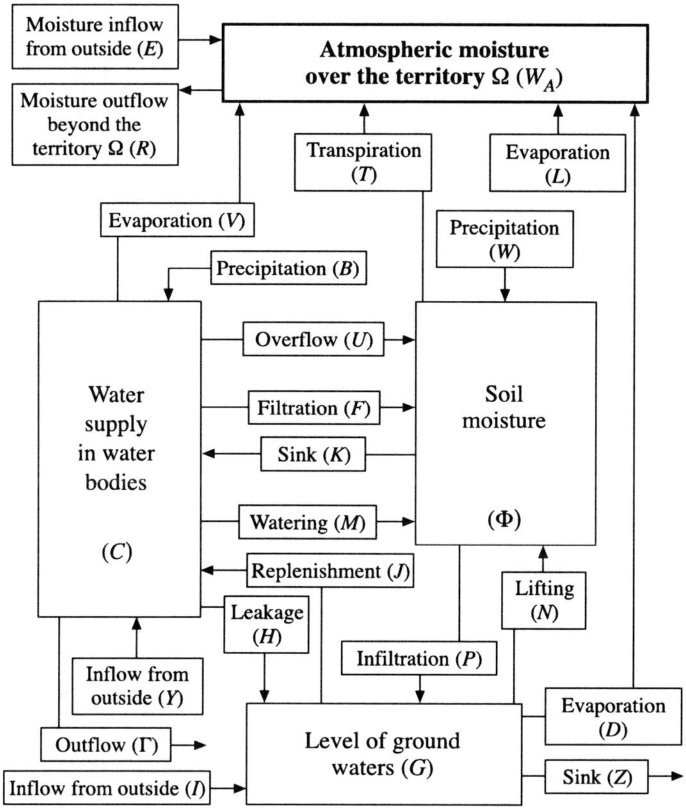

Figure 5.11. The block-scheme of the sample model of water balance in the limited territory.

network (km^2), respectively, Δ_k is the linear size of the kth compartment of the river network (km), $W_{A,ij}$, G_k, and Φ_l are, respectively, the levels of water in the atmosphere, in the kth compartment of the river network, and the lth facies on the territory Ω_{ij}, θ_{ij} is the level of ground water (m), ψ_l^k is the share of the run-off of the kth facies, getting to the territory of the lth facies, and the rest of the notations are given in the scheme in Figure 5.11.

An application of the model to other regions is realized via the variables E, R, Y_i, Γ_i, I, and Z. Besides, in analysis of the concrete situation, the configuration of the waterway and the level of the water body are taken into account. The required equations are written like those above, proceeding from the condition of the water volume balance. Functionally, all fluxes in the scheme in Figure 5.11 can be described based on the laws of hydrodynamics and with account of the available observational information. The inflow E_{ij} and outflow R_{ij} of moisture can be determined from the data of remote sounding. Between measurements, the information used about wind speed V_{ij} and the functions E_{ij} and R_{ij} are calculated by the formulae: $E_{ij} = E_{H,_{ij}}$ and $R_{ij} = W_{A,_{ij}l^*}/(l^* + k_1 V)$, where $l^* = 2\sqrt{\sigma/H}$, E_H is the atmospheric moisture on the windward border of the region and k_1 is the constant coefficient reflecting the contribution of wind into the circulation of precipitation.

Information on precipitation and run-off is introduced into the information catalogues of the hydrometeorological services. Based on these data, the respective model units can be derived. Assuming that the distribution of precipitation is proportional to relevant areas, we obtain:

$$B_k = W_{A,ij}\sigma_k/\sigma_{ij} \qquad W_l = W_{A,ij}\sigma_l/\sigma_{ij}$$

The model of the river run-off formation should take into account the watershed topography and the spatial distribution of its soil characteristics, as well as special features of vegetation cover. Let $\theta_l = (g_l + K_l \exp[-a_l X_l - C_l A_l])\sigma_l$, where X_l and A_l are, respectively, the vegetation density (m/km^2) and the soil layer thickness (m) over the area σ_l, g_l is the coefficient of the relief run-off in the lth facies, k_l is the coefficient of water penetration through vegetation and soil cover over the area σ_l, and a_l and C_l are the coefficients of precipitation retained by plants or soil in the lth facies, respectively. The parameters of this dependence can be determined from field measurements that establish, for a given type of soil and plants, a connection between precipitation intensity, the rate of uptake of water by the soil, and water resistance of its structure. So, for takyrs (saline lands), for instance, the run-off is equal to precipitation. This rather rough approximation can be specified, since the radiometric methods make it possible to classify the soil moisture, at least, into three types: firmly-bound, loosely-bound, and free water. The bound water is a film moisture adsorbed by the surface of ground particles with the film thickness of 6–8 molecular layers. The content of bound water is estimated at 2–3% in sands, and 30–40% in clay and loess. The bound water cannot be assimilated by plants, it does not dissolve salts. In the models considered these specific features are taken into account when determining the respective coefficients of evaporation and transpiration.

The run-off θ_l is distributed between facies and in the form of return water K_{lk} flows into the river. In a general form this is reflected in the formula (5.32) through the coefficients of the run-off distribution $\psi_s^l(\sum_{s=1}^{m+2}\psi_s^l = 1)$, where ψ_{m+1}^l is the share of the run-off from the lth facies beyond the region and ψ_{m+2}^l is the share of the run-off from the lth facies into the river. The coefficients $\omega_{lk}(\sum_{k=1}^{n}\omega_{lk} = 1)$ characterize the run-off distribution from the lth facies by the river compartments and are determined by the landscape relief and the spatial location of the facies and the waterway compartments. Thus $K_{lk} = \omega_{lk}\psi_{m+2}^l\theta l$.

Evaporation from the soil surface can be described by the formulae of Hitchcock, Horton, Weissman, and others (Bras, 1990). For instance, the formula by Priestly and Taylor for the latent heat of evaporation q_E is: $q_E = \alpha_S(q^* - q_i)/(S+\gamma)$, where q_i is the soil heat flux (W/m^2), q^* is the net radiation flux (W/m^2), $\gamma = 0.066 \cdot 10^3$ Pa/K is the psychometric constant, and S is the slope of the curve of the temperature dependence of the saturated moisture pressure (Pa/K).

$$\alpha = \begin{cases} 1.06 & \text{for wet soil} \\ 1.04 & \text{for dry soil} \\ >1.26 & \text{at a warm air advection over wet surface} \end{cases}$$

The Horton formula gives:

$$V = 0.36[(2 - \exp\{-0.44\theta\})l_V - l_a](\text{mm/day})$$

where θ is the wind speed (m/s), l_V is the vapour pressure near the water surface, and l_a is the water vapour elasticity.

The Power formula is written as:

$$V = 0.771(1.465 - 0.007\rho)(0.44 + 0.26\theta)(l_V - l_a)(\text{mm/day})$$

where ρ is atmospheric pressure (mm Hg).

A diversity of the forms for parameterization of the dependence of the rate of evaporation from the soil surface on the environmental parameters enables one to easily adapt the model of water balance to the information base.

The flux T in Figure 5.11 reflects the impact of vegetation cover on the hydrological regime of the territory. One of the simple models of transpiration is the following dependence:

$$T = y(24\alpha^* + \beta^*)(\text{cm/day})$$

where y is the specific water return of the soil, α^* is the rate of the ground water lifting (cm/hr), and β^* is the daily change of the level of ground water (cm).

Determine the constituents of the block-scheme in Figure 5.11 characterizing the processes of leakage and filtration of water from the river. Both leakage and filtration depend on the quality of the river bed and water level. Let:

$$H_i = \begin{cases} \mu_i C_i S_i & \text{for } 0 \le C_i \le C_{i,\min} \\ \mu_i C_{i,\min} & \text{for } C_i > C_{i,\min} \end{cases}$$

where μ_i is the coefficient of penetration of water through the river bed. Filtration F_i increases with increasing C_i between two critical values of $C_{i,\min}$, when there is no filtration, and $C_{i,\max}$, when it is at a maximum:

$$F_{i,\max} = \begin{cases} 0 & \text{for } 0 \le C_i \le C_{i,\min} \\ \mu_i(C_i - C_{i,\min})S_i & \text{for } C_{i,\min} < C_i < C_{i,\max} \\ \mu_i(C_{i,\max} - C_{i,\min})S_i & \text{for } C_i \ge C_{i,\max} \end{cases}$$

The distribution of water filtering from the river between facies depends on the distance r_{ij} between the ith compartment and the jth facies, as well as on the structure of soil and landscape relief. In particular, this dependence can be described by the function $F_{ij} = F_{i,\max}\chi(r_{ij})$, where $\chi(r_{ij})$ is the decreasing function meeting the condition:

$$\sum_{j=1}^{m} \chi(r_{ij}) = 1$$

Evaporation from the river surface depends on the environmental temperature and can be described by the formula $V_i = V_i^* T^\omega$ or by the relationship $V_i = \mu(\theta)(\rho_V - \rho_2)$, where $\mu(\theta)$ is the function reflecting the impact of the wind, ρ_V is the water vapour pressure at the temperature of the evaporating surface (mb), and ρ_2 is the absolute air humidity at a height of 2 m (hPa).

The volume of overflow is determined by the binary regime of the waterway functioning within a maximum possible water level $C_{i,\max}$, so that:

$$V_i^* = \begin{cases} 0 & \text{for } 0 \le C_i \le C_{i,\max} \\ C_i - C_{i,\max} & \text{for } C_i > C_{i,\max} \end{cases}$$

The distribution of U_i^* between facies depends on the landscape relief, characterized by the matrix of the relief run-off $\Psi = \|Y_{ij}\|$, and is written as:

$$\sum_{ijj=1}^{m} \Psi_{ij} = 1 \qquad \Psi_{ij} \ge 0$$

As a result, $U_{ij} = \Psi_{ij} U_i^*$.

Taking water for watering from the ith compartment of a waterway is an anthropogenic factor, and should be considered as free parameters $M_i^* = \sum_{j=1}^{m} M_{ij}$. To take into account a possible heterogeneity of the distribution of M_i^* between facies, take the matrix of the coefficients of the distribution of watering $v = \|v_{ij}\|(v_{ij} \ge 0, \sum_{j=1}^{m} v_{ij} = 1, i = 1, \ldots, n; j = 1, \ldots, m)$, so that $M_{ij} = v_{ij} M_i^*$.

The relationship between the surface water fluxes and ground waters strongly depends on the flux of water infiltrating down through the soil layer. This flux, called infiltration, with account of only the vertical heterogeneity of the soil can be described in a general form by the equation:

$$\frac{\partial P}{\partial t} = \frac{\partial}{\partial z}\left[p(P)\frac{\partial P}{\partial z} + K_z(P)\right]$$

Bras (1990) gave various versions of solutions for this equation. For practical use the following solution can be recommended:

$$f = f_c + (f_0 - f_c)\exp(-Pl^2 t) \tag{5.34}$$

where $f = (P_i - P_0)P/(\pi t)^{-1}$, f_c is the asymptotic value of the rate of filtration, and f_0 is the initial value of the rate of filtration.

The processes of infiltration and evaporation of ground waters depend strongly on the vertical profile of the soil layer. The following soil layers can be selected: saturated and unsaturated. The saturated layer usually covers the depths >1 m. The upper unsaturated layer includes the soil moisture in the zone of the plants roots, the intermediate level, and the level of capillary water. The water motion through these layers can be described by the Darsy law, and the gravitation term $K_z(P)$ in the equation (5.34) is calculated from the equation:

$$K_z(P) = 256.32\,\delta_s^{-7.28} - 1.27\,\delta_s^{1.14}(\text{cm/day})$$

where δ_s is the volume mass of soil (g/cm^3).

Thus the system of equations (5.30) through (5.33) with the indicated functional descriptions of water fluxes in the region under study at initial values of $W(t_0)$, $G(t_0)$, $C_i(t_0)$, and $\Psi_j(t_0)$ prescribed for a time moment t_0 enables us to calculate the characteristics of the water regime of the whole region for $t \ge t_0$. The initial values are provided by the monitoring system (Table 5.15). The regularity of surveys depends on the required accuracy of prognosis and can be realized by planning the monitoring regime. Based on the synthesis of the model and the remote sounding system the monitoring can be organized practically for any irrigated agroecosystem. In this case the problems appear in identifying the airborne measurements with the values of geophysical, ecological, and hydrological parameters. An example of the successful solution of such problems (Vinogradov, 1983) is a determination of the dependence between the coefficient of spectral brightness $\tau_J = \tau_z + (\tau_0 - \tau_z)\exp(-\alpha W^c) + dW^n$, where τ_0 is the coefficient of the dry soil brightness, τ_z is the coefficient of the brightness of soil with the moisture content close to a minimum of the field moisture capacity (when there is no free water in the soil), and α, c, d, and n determine the type of the soil ($\alpha, d, n < 1; c > 1$; for achromatic loamy soils we have: $\tau_z = 0.09$, $\tau_0 = 0.28$, $\alpha = 0.01$, $c = 2.3$, $n = 0.9$, and $d = 0.0001$). Obtaining these estimates is an important problem in the remote sounding of the environment.

Finally, note that the deterministic approach to modelling the water cycle in the zone Ω_L described here cannot be considered as the only possible one. Such an approach gives only average trends in the changes of the water cycle components. Their distribution and probabilistic prognosis can be obtained only on the basis of the dynamic–stochastic models of the water balance. When we model the global carbon cycle this approach enables us to take into account the sink of the atmospheric CO_2 over the region due to washing-out (taking into account Table 5.15 data).

5.9 METHANE AND THE CARBON CYCLE

Methane like carbon dioxide belongs to the family of greenhouse gases. The spectrum of its natural and anthropogenic sources are wide, and its greenhouse effect is substantial (Dementyeva, 2000). Before human interference, the natural cycle of methane had been balanced with respect to climate. By extracting natural combustible gases consisting 90–95% methane, humankind has contributed to this cycles factor of instability and uncertainty. Most of the authors estimated the level of the global emission of methane into the atmosphere at $535 \cdot 10^6\,t\,CH_4$/yr, of which $375 \cdot 10^6\,t\,CH_4$/yr is of anthropogenic origin ($50 \cdot 10^6\,t\,CH_4$/yr being from the rice fields). The anthropogenic input of methane is expected to increase within the next 20–30 years, though in some developed regions, measures are being taken to reduce the anthropogenic emissions of methane into the atmosphere. Nevertheless, the concentration of methane in the present atmosphere increases 7 times faster than the growth of the CO_2 concentration, so that its amount increases annually by 2%, that is, by 2020 the amount of methane in the atmosphere may double compared to 2000, which, from numerous estimates, will lead to a global warming of 0.2–0.4°C. As is the case of CO_2, these estimates will remain rather doubtful and contradictory until the global model mentioned above is synthesized. However, using our present knowledge, only the first steps are possible in modelling all these features of the global cycle of CH_4.

Table 5.15. Estimates of temperature in the beginning and in the end of the period with solid precipitation for some global regions.

Region	T_{AF} (°C)
Taimyr Peninsula	−2.4
South-east of the West-Siberian plain	−7.6
Western Siberia	−6.0
South-eastern coastline of Russia	−4.4
Kolyma	−4.0
North of 70°N	−4.0
The Caucasus	
500–1000 m above sea level	−6.2
1500–2000 m above sea level	−4.8
Finland	−3.0
Japan	−7.0

Table 5.16. Sources of the input of CH_4 into the terrestrial atmosphere.

Source of CH_4	Area of the source (10^6 km^3)	Rate of CO_4 formation (g/m^2/yr)	Average rate of CH_4 formation (10^6 t/yr)
Rice fields	1.35	206	280
Marshes	2.6	50–100	130–260
Freshwater lakes	2.5	50–100	1.25–25
Arid soils	30	0.44	10
Woodlands	44	0.01–0.09	0.4
Tundra	8	10	0.8–8
Oceans	361	0.012	4–6.7
Shelf zones	1.4	5–10	0.07–1.4
Animals			101–220
Termites			150
Fossil fuels			100
Dust heaps			20–40
Sewage			30–40

The sources of methane are oil, sedimentary and ejected rocks, bottom sediments of lakes, seas, oceans, and other objects of the hydrosphere, as well as soil, peatbogs, and rice plantations. As shown in the scheme in Figure 5.12 and in Table 5.16, numerous processes of methane transformation (70–80%) are of biogenous origin, mostly affected by humans. Of course, the significance of these processes varies depending on many natural and anthropogenic parameters. The relationships of individual elements of correlation between the cycles of CO_2, CH_4, and other chemicals vary, too. At any rate, it is clear that depending on the strategy of the nature–society system, in due course the composition of the terrestrial atmosphere can change substantially. Note that burning 1 m^3 of methane extracts

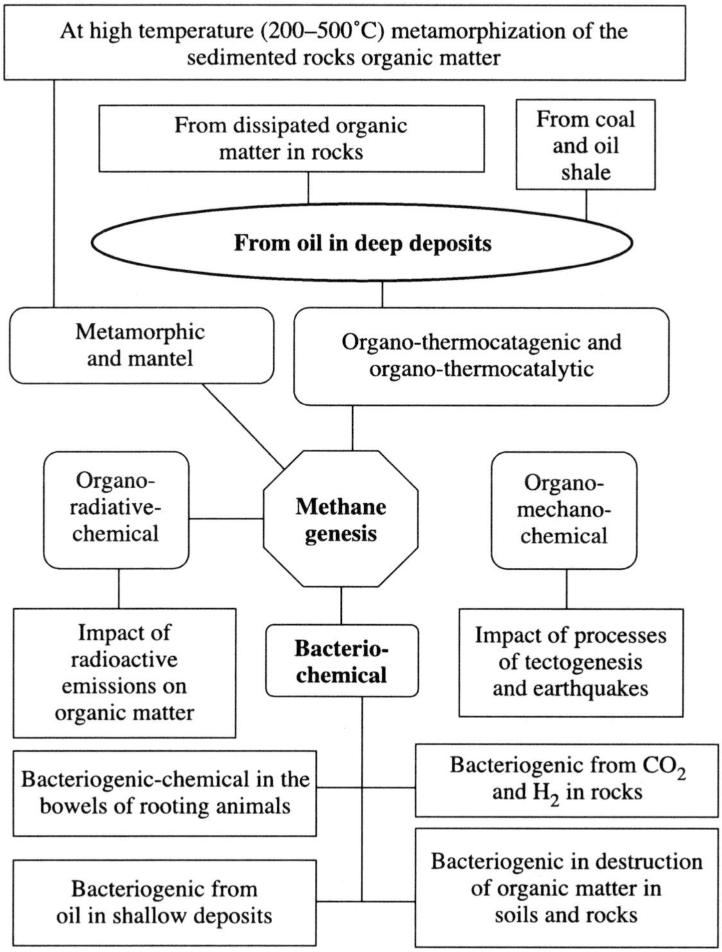

Figure 5.12. Methods of methane formation in nature (Zor′kin *et al.*, 1986).

$2\,m^3$ O_2 from the atmosphere. From the open dust heaps and municipal and industrial sewage, the atmosphere annually obtains about 2% anthropogenic methane ($\sim$270–460 · 10^6 t C). These integral estimates do not permit one to calculate an actual distribution of the CH_4 flux in the atmosphere. A certain contribution to detailing of the spatial distribution of the sources of methane has been made by the Second International Conference on the problems of methane held in Novosibirsk in 2000. The Proceedings of this Conference contain concrete data on the sources of methane in many regions of the globe. For instance, according to Byakola (2000), within the framework of the international project United Nations Environmental Programme/Gross Domestic Product (UNEP/GDP), an inventory of the sources and sinks of CO_2 and CH_4 has been made for the territory of Uganda (236 · 10^3 km^2). In Uganda, the main anthropogenic sources of methane are agriculture, municipal sewage, and biomass burning. In 1990, the stockbreeding and rice fields in Uganda contributed $205.45 \cdot 10^3$ t CH_4 and $23.45 \cdot 10^3$ t CH_4, to the atmosphere respectively. Agricultural waste burning added $3.55 \cdot 10^3$ t CH_4.

Naturally, Uganda is recommended to reduce the greenhouse gasses emissions, but it is not clear what the threshold of these emissions are. Of course, in the future stockbreeding and rice production in Uganda will develop, increasing thereby the volumes of CH_4 emitted to the atmosphere. Hence, a balanced correlation should be sought of the economy in the country and the state of the environment. This problem can be solved using new technologies of nature (Krapivin and Kondratyev, 2002). In particular, one of the ways to reduce CH_4 emissions is a secondary use of organic wastes, for instance, in paper production. In Uganda, up to 16% of urban waste is used in paper production.

The gas transport systems are one of the powerful anthropogenic sources of CH_4. The study by Coconea *et al.* (2000) contains information about methane emissions from pipelines in the territory of Romania, a country that signed the Lisbon Protocol in 1994 and now supports the Kyoto Protocol. Romania is the first country in Europe where in 1917 a 50-km pipeline was laid to transport natural gas. At present, natural gas constitutes 37% of the energy resources of the country, the share of oil and coal constituting 32.6% and 15.2%, respectively. Therefore the problem of anthropogenic input of CH_4 from the territory of Romania into the atmosphere is rather urgent. Here, like in Uganda, technology plays an important role, reducing the leakage of methane from the pipelines by 38.9% – 1994 values being 55.35% of those in 1987. On the whole, both extraction and distribution of coal, oil, and gas on the territory of Romania gives 56% of the entire amount of CH_4 emitted from this territory. Agriculture sits in second place at 29%.

One of the significant sources of CH_4 is the territory of Russia, which contributes about $47 \cdot 10^6$ t CH_4/yr to the atmosphere. This flux is expected to reach $78 \cdot 10^6$ t CH_4/yr by 2025. This increase will be caused by the developing infrastructure of the gas, oil, and coal industry. On global scales, these trends will be practically observed in all countries. In Table 5.17 the contribution of the coal industry into CH_4 production is estimated for various global regions. These estimates are determined by technologies used in the coal industry. On average, the contributions of various sources into the coal industry itself constitute: 70% – underground ventilation in the coal mines, 20% – underground drainage, 5% – surface loading and unloading operations, 4% – opencast mining of deposits, and 1% – derelict mines.

The global cycle of methane has been studied inadequately, and therefore its modelling faces a lot of unsolved problems. The CH_4 fluxes from the waterlogged territories have been studied the best. These fluxes constitute about 20% of the total input of methane to the atmosphere from all sources (Table 5.16). Note that almost 80% of sources of methane are of the biological nature, so that the anthropogenic interference into its natural cycle is also possible through violation of various biospheric processes. In particular, in the waterlogged territories, methane forms only due to biological processes. The hydrospheric sources of methane can be presented

Table 5.17. Emissions of methane by the coal industry in various countries (Gale and Freund, 2000).

Country	Coal production (10^6 t/yr)	CH_4 emissions (10^6 t/yr)	Specific rate of CH_4 emission (kg CH_4/t of coal)
Australia	229	0.8	3.5
England	68	0.5	7.4
Germany	280	1.0	3.6
India	263	0.4	1.5
China	1141	7.7	6.7
Poland	199	0.6	3.0
Russia	539	4.5	8.3
USA	859	4.3	5.0
Czechoslovakia	88	0.3	3.4
South Africa	182	1.0	0.5
Total	*4399*	*21.7*	*4.9*

by a multilayer model (Figure 5.13). This scheme describes the vertical structures of most of the water bodies. Methane forms in the layer of bottom deposits due to bacteria functioning, and in the zone with oxygen, methane is partially oxidized giving carbon dioxide $CH_4 + 2O_2 \rightarrow CO_2 + 2H_2 + E$. Bacteria taking part in the methane oxidation uses the released energy E for organic matter synthesis. The remaining methane gets to the atmosphere and, in contrast to CO_2, practically does not return to the water medium. This is somehow connected with the fact that the CH_4 solubility in water is almost 40 times lower than that of CO_2 (Eilrich *et al.*, 2000). The lifetime τ_H of methane in the atmosphere is estimated at about 5 years. Its extraction from the atmosphere takes place due to the participation of methane in photochemical reactions, resulting in methane oxidation first to CO, and then to CO_2. The cycle CO–OH–CP_4 plays an important role in the cycle of methane:

$$OH + CH_4 \rightarrow CH_3 + H_2O$$

$$OH + CO \rightarrow CO_2 + H$$

The participating OH-radicals form in the atmosphere during the water vapor photolysis. As a result, the simplest diagram of methane oxidation in the atmosphere is the following:

$$\overset{\quad OH}{\swarrow}$$
$$CH_4 \xrightarrow{\qquad\qquad} CO \rightarrow CO_2$$

The human interference into the processes described by this diagram breaks the natural stability of the balance $CH_4/CO/CO_2$. In particular, the reclamation of marshes is one such destabilizing factor. For instance, the drainage of 20% of the marshes leads to a natural reduction of CH_4 emissions from the marshes by 20%, and on the whole, the amount of methane is reduced by 4%, which does not practically influence climate, but causes changes in the biogeochemical cycles of ozone and carbon dioxide with unpredictable consequences. These estimates are important for the final conclusion about the level of the integral greenhouse effect. However, the solution of this problem is connected with many factors, neglection of which leads to serious errors. For instance, the CH_4 flux on the atmosphere–marsh border depends on the vertical profile of the temperature in the marsh body. In the simplest case, if we denote $T_W(z, t)$ as the temperature at a time moment t at a depth z and write the equation of heat conductivity:

$$\frac{\partial T_W(z,t)}{\partial t} = a^2 \frac{\partial^2 T_W(z,t)}{\partial z^2} \tag{5.35}$$

where $a^2 = Kc^{-1}\rho^{-1}$, K is the coefficient of heat conductivity, c is specific heat capacity, and ρ is the medium density, then an estimation of the flux F^1_{CH4} as a time function becomes dependent of the multitude of the poorly assessed characteristics of the environment.

Let the marsh surface temperature vary cyclically with frequency ω and amplitude A, decreasing with depth: $T_W(0, t) = A(z)\cos(\omega t)$, where:

$$A(z) = A(0)\exp\left(-\sqrt{\frac{\omega}{2a}}z\right)$$

The solution of equation (5.35) enables us to trace the temperature variations $T_W(z, t)$ and suggests the conclusion that in this case these variations weakly depend on $T_W(0, t)$. Even if $T_W(0, t)$ increases by 2°C, then according to (5.35), the amplitude of temperature changes with depth will rapidly decrease to 0.97°C, 0.33°C, and 0.01°C at depths 40 cm, 2 m, and 3 m, respectively. Hence, with a 2°C increase of the average global atmospheric temperature the flux F^1_{CH4} will increase by not more than 1.4%.

Comparing the global significance of the CO_2 and CH_4 cycles in the atmosphere–marshes system, note that the CO_2 cycle promotes a climatic stabilization, whereas the CH_4 cycle intensifies the climate changes. With a climate warming the marshes assimilate part of the CO_2 from the atmosphere and reduce thereby the greenhouse effect. On the contrary, when the climate warms due to increasing F^1_{CH4}, the greenhouse effect intensifies.

The western Siberian region of Russia is characterized by numerous intensive natural and anthropogenic sources of methane formation. These are marshes, tundra, permafrost, and oil and gas deposits. In this region the flux F^1_{CH4} varies widely both during a year and shorter time periods. From measurements carried out by Jagovkina *et al.* (2000) at the coastline of Yamal in June 1996, the CH_4 concentration in the atmosphere at a height of 2 m varied from 1.83 ppmv on 18 June to 1.98 ppmv on 23 June, with an average daily value of 0.032 ppmv.

The peatbogs of Siberia are quite special in the global cycle of methane. They play a unique role in the biogeochemical cycles of methane and carbon dioxide. On the one hand, they are a non-anthropogenic source of CH_4 and CO_2, but on the other hand, they are intensive assimilators of carbon from the atmosphere and extract it from the natural cycle over a long time. The marshes of western Siberia, for instance, contain 20–30% of the global carbon supplies. The intensity of CH_4 emissions from the marshes are, on average, almost 2000 times weaker than that of CO_2. 35–50% of all methane emitted from the territory of Russia falls within the marshes. The western Siberian marshes emit to the atmosphere no more than $1.7 \cdot 10^6$ t CH_4/yr, which does not exceed 1% of the global CH_4 flux. The spatial heterogeneity of the flux F^1_{CH4} is high, which is determined by diverse characteristics of the marsh ecosystems. In particular, the upper oligotrophic coniferous–shrubby sphagnum swamps emit 0.9–10 mg C/m^2/hr (Dementjeva, 2000). This estimate is rather approximate, since the scattering of such estimated values between various authors constitutes hundreds of percent. For instance, a drained sphagnum swamp of the transitional type can emit 142–204 g C/m^2/hr, and the rush-sphagnum bogs 83.5–309 mg C/m^2/hr.

The main mechanism for the formation of methane in a marsh is connected with the functioning of special groups of micro-organisms. An amount of methane produced by diffusion is emitted to the atmosphere, but most methane remains in the peat layer and is gradually emitted to the atmosphere.

Humankind is interfering with the natural biogeochemical balance of greenhouse gases practically everywhere in the world. One of the aspects of this is a reduction in areas of marshland and their transformation into agricultural fields. Diverse human agricultural activity adds to the atmosphere 20% of all the anthropogenic flux of greenhouse gases. In the USA it is 30%. Stockbreeding contributes considerably to this flux. In California and Wisconsin each hectare of pasture emits annually 502 kg CH_4 (or 10,511 kg CO_2) and 134 kg CH_4 (or 2814 kg CO_2), respectively. In New Zealand such emissions of CH_4 are estimated at 291 kg CH_4 (or 6110 kg CO_2) (Johnson and Ulyatt, 2000).

Among the Kyoto Protocol signatories, England takes ninth place by volume of reduced emissions of greenhouse gases. The decreasing trend of methane emissions is part of the general reduction of emissions of six greenhouse gases (CO_2, CH_4, N_2O, hydrofluorocarbons, perfluorocarbons, sulphur hexafluoride) from 1990. In 2000 the greenhouse emissions decreased by 15% compared to 1990. By 2010, the CH_4 emissions will constitute 20,134 t/yr. This reduction will be reached mainly due to new technologies in processing wastes and in the coal industry. On the whole, in England, according to the developed scenario, emissions of methane will decrease by 14% in agriculture, by 82% in the coal industry, by 29% in the oil and gas industry, and by 73% in waste processing by 2010. The possibility of realization of this scenario is confirmed by the decreasing trend in CH_4 between 1998 and 1990. For instance, during this period, emissions of methane in the coal industry have decreased by 64%, and in waste processing by 29%. In 1990, the share of waste processing in England constituted 32% of all CH_4 emissions, only 3% of these emissions being connected with sewage processing.

In agriculture, emissions of CH_4 in England constituted $1037 \cdot 10^3$ t in 1990 and $998 \cdot 10^3$ t in 1998. The scenario of reduction of the F^1_{CH4} flux from the territory of

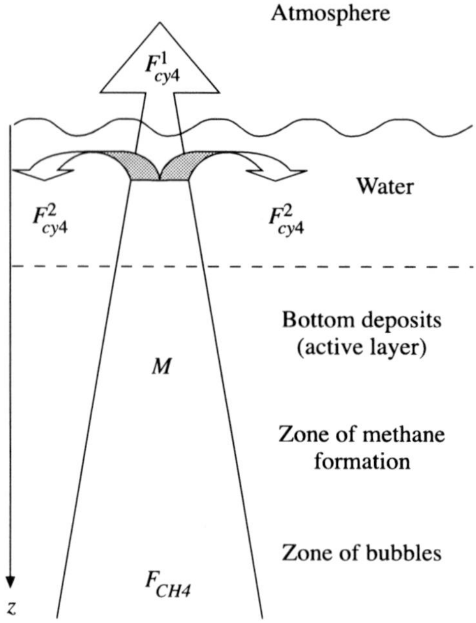

Figure 5.13. The block-scheme of the formation and transport of methane in the waterlogged area. Notations: F^1_{cy4} – methane flux on the border atmosphere–water object; F^2_{cy4} – oxidation of methane in the aerobic zone; F_{CH4} – energy of the source of methane; M – methane concentration.

England due to improved technologies in agriculture foresees emissions of 902–983 · 103 t CH_4 in 2010. The CH_4 emission from burning agricultural waste is completely excluded, and in stockbreeding, emissions of methane are reduced by 8% compared to 1990.

In the coal industry in England, emissions of methane in 1990 constituted $819 \cdot 10^3$ t with the main contribution to this flux made by underground operations. This constituted 24% of the whole flux of methane from the territory of England. In 1998 the F^1_{CH4} flux decreased to $264 \cdot 10^3$ t and by 2010 it should decrease to $218 \cdot 10^3$ t. A similar decreasing trend of methane emissions from the territory of England also appears in the oil and gas industry. According to the scenario, the contribution of these sectors of energy production into the F^1_{CH4} flux will decrease from $540 \cdot 10^3$ t in 1990 to 349–$464 \cdot 10^3$ t in 2010 (Meadows, 2000).

According to Bazhin (2000), the F^1_{CH4} flux in every water basin with a vertical stratiform structure forms in an active layer beneath the water layer. Practically, all aquatic systems have such a structure. The layer, where methane forms, has two areas. In the bottom area located at a depth h, methane takes the form of bubbles. Above this layer, due to diffusion, the concentration of methane decreases, and the bubbles disappear. Denote $D_{CH4}(z)$ as the coefficient of methane diffusion at a depth z, then the stationary behaviour of the whole system shown in Figure 5.13 is described by the equation:

$$\frac{d}{dz}\left[D_{CH4}(z)\frac{d}{dz}M(z)\right] - F_{CH4}(z) + F^1_{CH4} + F^2_{CH4} = 0$$

Model calculations and field measurements performed by Bazhin (2000), show, for instance, that on the rice fields $hb = 1.3\,\text{m}$, $F_{CH4} = (1.3\text{–}1.7)\cdot 10^{-12}\,\text{mol/cm}^{-3}/\text{s}$. According to Khalil *et al.* (2000), the rice fields play a significant role in the gas balance of the atmosphere due to emissions of CH_4, CO, N_2O, H_2, and $CHCl_3$. For instance, in China the rice fields deliver these gases to the atmosphere at the following rates ($\text{mg/m}^2/\text{hr}$): CH_4 – 900–50000, CO – 80–100, H_2 – 5–30, N_2O – 50–1000, and CHCl – 1–8. A wide scatter of these estimates is explained by highly unstable fluxes of these gases due to rice growing technology. For instance, the use of sulphates on the rice fields increases emissions of methane by 12.0–58.9% depending on other characteristics of these fields (Liping *et al.*, 2000).

Thus an estimation of the F^1_{CH4} flux as a function of a given territory with account of natural and anthropogenic processes taking place there, requires first of all a detailed inventory of methane sources as well as natural and technogenic systems functioning on this territory. Examples of such an inventory given above serve the basis for development of studies in this direction.

The dynamics of the CH_4 content H_A in the atmosphere can be parameterized by a simple balance relationship:

$$\frac{\partial H_A}{\partial t} + V_\varphi\frac{\partial H_A}{\partial \varphi} + V_\lambda\frac{\partial H_A}{\partial \lambda} = F^1_{CH4}(t,\varphi,\lambda,\Xi) - \frac{H_A(t,\varphi,\lambda)}{\tau_H}$$

where Ξ is the identifier of the type of natural or technogenic system.

6

Modelling the global carbon cycle

6.1 INTRODUCTION

Many scientists have tried to model the global carbon cycle (Nefedova, 1994; Nefedova and Tarko, 1993; Alexandrov and Oikawa, 2002; Kondratyev *et al.*, 2002a, b; Krapivin and Vilkova, 1990; Nitu *et al.*, 2000a, b; Svirezhev, 2002). Most of the models are based on the data for the pre-industrial period and predict the CO_2 concentration for this century. The number of factors taken into account in the models is constantly growing, and the models become more and more adequate. One of the first sufficiently complete models of the global CO_2 cycle has been proposed by Bjorkstrom (1979), which takes into account the dynamic interaction between the carbon reservoirs in the biosphere and its fluxes within them. For the first time, the unit of the World Ocean has been realistically represented, where the ocean is considered as a multilayer composition of the uniformly mixed reservoirs with the CO_2 exchange between them being described by linear laws. The models of the unit type have been analysed by many experts, who have promoted a successful division of the Earth environment into heterogeneous compartments and derivation of the respective box models (Pervaniuk, 2001). Using the ability of carbon to reside in the atmosphere for a long time, which makes it possible to describe the atmosphere by a point model, Bacastow (1981) derived a global 4-reservoir model, which approximates the dynamics of the CO_2 exchange between the northern and southern hemispheres.

Each of the available models of the global carbon cycle differs by a set of assumptions and therefore is orientated towards selection of some effect. Models allowing a mutual parametric correlation can be used in a more complicated model. For instance, a simple numerical model of gas exchange on the ocean–atmosphere border in the case of wind-driven roughness at wind speeds of up to 7 m/s enables us to include in the global model a unit to calculate an established CO_2 flux between the water surface and the atmosphere. The models of the ocean carbonate system described in literature exemplify this. There are also other models of the CO_2 cycle in natural systems (Riedo *et al.*, 2000; Zonneveld, 1998).

All known models of the CO_2 cycle should be modified by detailing the spatial distribution of the soil–plant formations and by specifying the exchange processes both in the ocean and on the atmosphere–ocean border. Also, most accurate parameterizations of all known elements of the biogeochemical cycles of carbon should be combined into a single system. Such an attempt has been made in the block-scheme of the model described in the work by Krapivin and Kondratyev (2002). The basic sources of CO_2 are the functioning of land and marine animals, photochemical reactions, decomposition of dead organic matter, as well as anthropogenic activity. The period of CO_2 migration into the upper atmospheric layers is substantially less than the time step of the model. It is here that the problems appear of an adequacy of the obtained estimates of the role of the land and water ecosystems in the assimilation and burying of excess amounts of carbon dioxide in the atmosphere (Alekseev *et al.*, 1990; Alverson, 2000; Bezborodov *et al.*, 2003; Hoerling and Kumar, 2003; Huber and Caballero, 2003; McNeil *et al.*, 2003; Mirtskhoulava, 2003; Soon and Baliunas, 2003).

A thorough analysis of numerous publications on the role of increasing CO_2 concentrations in the atmosphere which leads to an enhanced greenhouse effect and to possible climate warming in the near future, shows that predicted respective cataclysms in the environment are only hypotheses. These conclusions are made on the basis of rough models of the global biogeochemical CO_2 cycle that do not take into account the spatial mosaic of most of the effects and feedbacks considered above. Solving such problems is impossible, however, because despite numerous studies of the greenhouse effect, at present there are no global databases on the greenhouse gases complete enough to reflect the seasonal and spatial structure of their fluxes. Besides, even the fluxes enumerated in Table 2.9 have not been modelled and, hence, the global models of CO_2 are only of "educational" character. For example, the results obtained by Alekseev *et al.* (1992) from studies of the CO_2 exchange in the coastal waters of the Caspian Sea, found that even for such a small space, the mosaic of CO_2 fluxes is characterized by variations of the Δp parameter within -10 to 234 ppm and of the fluxes H_2 and H_3 from 1.2 to 32.3 mmol/m^2/day. This confirms the necessity for a thorough analysis of local fields p_m and p_a to derive accurate models of gas exchange in the ocean–atmosphere system. Similar conclusions are suggested for the soil–plant–atmosphere system. Here it is necessary to consider the effect of a maximum number of the types of plants and soils. The available models consider no more than 30 types, whereas there is an atlas of 172 types of plants which could be considered (Eliasson *et al.*, 1999).

Inaccuracies of the global models of the greenhouse effect due to CO_2 increase are bound to occur when we consider the lack of regular reliable data on the biospheric cycle of other greenhouse gases (GHGs), including ozone and water vapour. This is confirmed by a large scatter of the predicted estimates of climate warming by 2075 compared to the pre-industrial period: $1.1°\mathrm{C} \leq \Delta T_\Sigma \leq 4.2°\mathrm{C}$.

The recent detailed analysis of the causes and consequences of the greenhouse effect was made at the 4th International Conference on the greenhouse gases control held from 30 August to 2 September, 1998 in Sweden (Eliasson, 1999). The papers presented at the conference questioned the reliability of numerous predictions of the possible greenhouse effect dynamics. The authors discussed the biological, energetic, chemical, geological, and social aspects of the greenhouse effect. The problem of a complex consideration of all these aspects, with account of additional feedbacks reflecting the correlations between them, remains neglected. A solution for this problem has been suggested by Krapivin and Kondratyev (2002). They proposed to combine in the global model the functioning of several spatially heterogeneous natural–anthropogenic systems and processes. This global model will make it possible, in an adaptive regime, to search for an efficient mechanism for monitoring the greenhouse effect and to reliably assess the role of the land and ocean biocoenoses.

The prediction of emissions of anthropogenic CO_2 and other GHGs into the atmosphere is only possible with thorough and regular monitoring of the spatial structure of a huge number of parameters of the nature–society system, with an adaptive mechanism of the use of the developed global model of this system (Kondratyev *et al.*, 2000). Only an adaptive and evolutionary approach to the solution of this problem can reduce the uncertainty in the estimates of the nature–society system parameters and guarantee a harmonic development of human civilization. A thorough analysis of the publications of numerous experts raises hopes that within the Global Carbon Project (GCP) a constructive step forward will be made in the synthesis of a global model which will adequately reflect the whole spectrum of direct correlations and feedbacks in the environment and will also take into account all developed models of various processes (Yang *et al.*, 2002; Porte and Bartelink, 2002; Yemshanov and Perera, 2002; Sternberg and De Angells, 2002; Peng *et al.*, 2002b; Qi *et al.*, 2002; Huesemann *et al.*, 2002).

6.2 THE GLOBAL MODEL OF THE NATURE–SOCIETY SYSTEM DYNAMICS

The present stage of scientific and technical progress, with the attendant realization of the global ecological situation that is characterized by insufficiency of energy, geological, biogeocoenotic, and other resources, leads us on to the problem of information about global ecology: knowledge of the conditions under which humankind and the natural world have co-evolved. This information resource for thousands of years had been weakly correlated to total human activity, which till the beginning of the industrial era had been relatively low. Then, with a historically rapid approach to the situation when the commercial attitude to the biosphere had become the governing strategy of humankind and when an ecological deadlock had been realized, the significance of the information resource reached its maximum.

This resource includes not only "academic" knowledge. Every ecological problem is "open", it is part of the system of the global problems of today, the main problem being a necessity to preserve the homeostasis of humankind

(Kondratyev, 1998b). It means that the "danger for the biosphere" realized at the end of the 20th century has brought forth the problem of survival of *Homo Sapiens*, and hence emphasized the problem of the responsible attitude towards nature. Both ecological and moral problems have clashed. The rhetorical question – why do we live and work? – has acquired a deep social aspect and is connected with the harmonious relationship between society and nature. The fate of the biosphere will depend on how rapidly humankind solves these problems and finds an optimal balance between a "wise" and an "unwise" attitude to the environment. As the model estimates have shown, 90% of humankind should take part in the solution of these problems. At the present historical stage, it is unlikely that so many people can deliberately change their position as the master of nature and develop new harmonious relationships between society and nature. To reach global harmony, it is necessary to concentrate on negative ecological and socio-economic changes, so that the ecological knowledge is put into practice in the form of concrete technologies ensuring a high quality of decisions are made in the field of nature protection activities.

At the present stage of scientific and technical progress, intensive developments have been realized in this field, analysis of which reveals the characteristic features of the ecological knowledge as well as the problems of the applied methods in order to formulate the basic requirements for an efficient information technology. One of the necessary prerequisites for creating the systems to monitor the environmental conditions was an availability of data of different quality and, hence, a multitude of numerical models of various types (balance, optimizing, evolutionary, statistical, etc.). Derived on the basis of parameterization and, as a rule, linearization of the laws of natural phenomena, these models include a wide spectrum of determinate and probabilistic descriptions of the geological, ecological, oceanographic, biogeochemical, and biogeocoenotic processes of global, regional, and local nature. Most of them are directed towards a theoretical understanding of the specific features of the higher living systems on the basis of available knowledge, and only a small part is aimed at the first steps to an objective assessment of the present global ecological situation. Differing in the goals and mathematical apparatus of description, many models have to be rough due to limitations, inadequacy, and uncertainty of the information base, as well as because of a lack in present-day instrumental systems in the field of numerical experiments. An increase of the number of the biospheric parameters in the models in order to raise their adequacy makes them multiparametric, that is, the problem of the "multivariance curse" appears.

As a basic instrument to solving these problems, the method of simulation modelling has been considered, which makes it possible to adjust different data that refer to various mathematical formalisms, and to remove the problem of an excess number of parameters. The required model is constructed on the basis of empirical information not restricted a priori by any mathematical apparatus, which ensures a flexible formalization, unavoidable when the important laws of phenomena are unknown.

The development of simulation modelling by broadening the information base, combining the formal and informal methods in the stage-by-stage synthesis of the required model, and finally, an active man–computer dialogue will provide an efficient approach to the system–ecological modelling. However, it appears that it is not that easy. If we compare the information queries in the field of ecological problems and the existing information provision for their solution (different numerical and simulation models, principles of ecological information processing), we see that not all levels of natural and anthropogenic complexes have a well-developed apparatus for their description as well as for construction of efficient information approaches to obtain the required estimates of problematic situations. Difficulties appearing in this regard are not only of technical character consisting in an accumulation of models of different types. These features manifest themselves especially in global modelling, which reveals a substantial and principally ineradicable inadequacy of knowledge about natural processes, which is shown both by the fragmentary character of empirical data and by the lack of knowledge about the laws that dictate the natural processes of evolution. Now, it is clear that a mechanical set of hierarchies of models and a desire for accumulation of empirical databases are not adequate. This is just an attempt to reanimate the old school reasoning that searched for an all-inclusive pattern of development of biospheric processes without hope of success, without the possibility of explaining the permanent self-organizing ability of living systems, and without a substantially improved understanding of the mechanism of the nature–society system functioning. The situation is such that it is necessary to use computer technologies that combine the methods of both the evolutionary and simulation modelling, as it has been described in Kondratyev *et al.* (2002a). This makes it possible to take into account the internal dynamics (evolution) of the modeled processes and to synthesize models in conditions of inadequate and partially reliable data.

Traditional approaches to the global model construction face difficulties in an algorithmic description of many socio-economic and natural processes, so that we deal with an information uncertainty. The developed approaches to global modelling neglect this uncertainty, and as a result, the models do not adequately consider real processes. A combined use of the evolutionary and simulation modelling makes it possible to remove this shortcoming by synthesizing a combined model, the structure of which is adapted on the basis of the pre-history of the biospheric and climatic components. The model can be realized in various classes of models, with the use of traditional computer programs and special processors of the evolutionary type. The form of such a combination can vary and it depends on the spatial and temporal adequacy of the global databases.

An accumulated experience in global modelling contains a lot of examples of insurmountable difficulties when attempting to find a means of description of the scientific and technical progress and of the human activity in its various manifestations. Also, difficulties appear in modelling the climate characterized by a superposition of processes with different temporal scales. As for an adequate description in the global model, it is impossible here to clearly define the limits of the information provision and the limits of the required spatial and structural resolution. Therefore without dwelling upon philosophical analysis of global problems and without trying to recommend a comprehensive approach to global modelling, we shall discuss only one of the possible ways, showing how the evolutionary modelling in special processor realization enables us to overcome the difficulties mentioned above.

An adjustment of the evolutionary model by the pre-history of natural rhythms makes it possible to obtain a model, which implicitly follows various regularities of the nature–society system dynamics in the past and makes it possible to make forecasts in the same temporal rhythm. The special processor version of the model overcomes all difficulties of the algorithmic and digital character caused by multi-dimensionality of the global model and a multitude of parametric uncertainties.

The synthesis of the global model of the nature–society system is based on its consideration as a self-organizing and self-structuring system, in which the elements are coordinated in time and space by the process of natural evolution. The anthropogenic constituent in this process breaks this integrity. Attempts to parameterize, on a formal level, the process of co-evolution of nature and humans, as elements of the biosphere, are connected with the search of a single description of all processes in the nature–society system, which would combine all spheres of knowledge in perceiving the laws of the environment. Such a synergetic approach forms the basis of numerous studies in the field of global modelling (Krapivin *et al.*, 1982; Krapivin and Kondratyev, 2002; Kondratyev and Krapivin, 2002).

Cover the Earth surface Ω with the geographical grid $\{\varphi_i, \lambda_j\}$ with mesh widths $\Delta\varphi_i$ and $\Delta\lambda_j$ by latitude and longitude, respectively, so that within a cell of land surface $\Omega_{ij} = \{(\varphi, \lambda): \varphi_i \leq \varphi \leq \varphi_i + \Delta\varphi_i; \lambda_j \leq \lambda \leq \lambda_j + \Delta\lambda_j\}$ all the processes and elements are considered homogeneous and parameterized by point models. In the case of a water surface on the territory of the cell Ω_{ij} the water mass is divided into layers Δz_k thick, that is, 3-D volumes $\Omega_{ijk} = \{(\varphi, \lambda, z): (\varphi, \lambda) \in \Omega_{ij}, z_k \leq z \leq z_k + \Delta z_k\}$ are selected, inside of which all elements of the ecosystem are distributed uniformly. Finally, the atmosphere over the site Ω_{ij} by height h is digitized either by levels of atmospheric pressure or by the layers Δh_s thick.

Interactions in the nature–society system are considered as interactions between the nature–society elements within these spatial structures as well as between these structures. The complex model of the nature–society system takes into account the spatial hierarchy of the hydrodynamic, atmospheric, ecological, and socio-economic processes with account of division of the whole volume of the environment into the structures Ω_{ij} and Ω_{ijk}. The cells of this division are the support grid in the digital schemes in solutions based on dynamic equations or in the synthesis of data series in the learning procedures of the evolutionary types.

The cells Ω_{ij} and Ω_{ijk} are heterogeneous in parameters and functional characteristics. It is through this heterogeneity that the global model is attached to databases. Moreover, to avoid an excessiveness in the structure of the global model, it is supposed a priori that all elements and processes of the nature–society system considered in the model have the spatial digitization inherent in them. An ambiguity of spatial digitizations in various units of the global model is removed on a logarithmic level of agreement of data fluxes from the monitoring system. As a result, the structure of the model is independent of the structure of database and hence, does not change with a change of the latter. A similar independence is

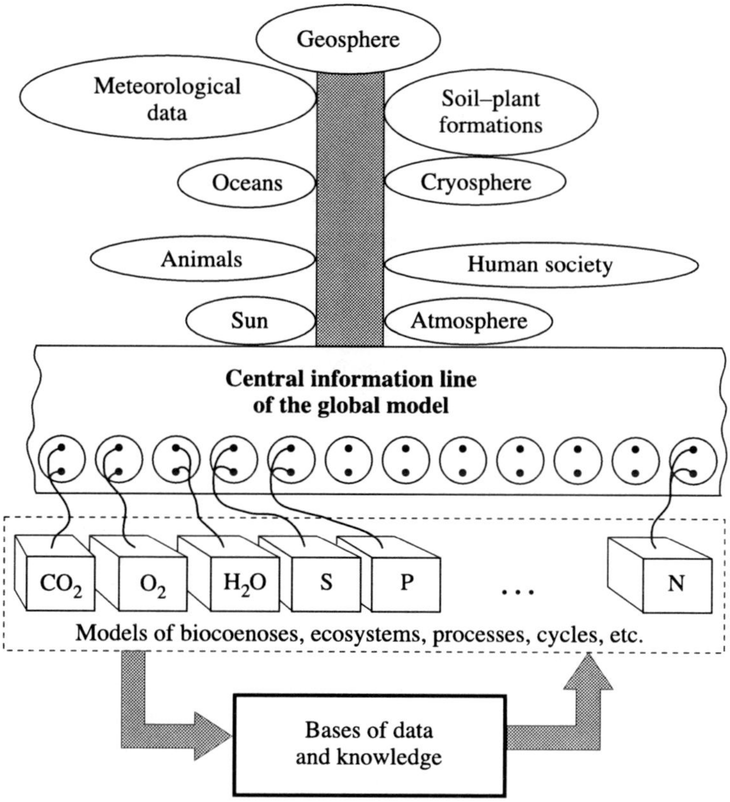

Figure 6.1. Organization of the global model of the nature–society system functioning.

provided between the units of the model. This is accomplished by a data exchange between them only via inputs and outputs under control of the basic data line as shown in the scheme in Figure 6.1. In case of removal of one or several units their inputs are identified with the respective inputs to databases or knowledge bases. Then the use of the model in the regime of numerical experiment can be schematically presented in the form of the operator–model dialogue. Here by the user's choice the spatial image of the modelled medium and the regime of the numerical experiment management is formed. Of course, the user should possess a certain knowledge and understand the method of its structuring. For instance, he/she can use the list of key problems of global ecology or the list of the nature–society system elements recommended to be studied (by information presented in many global databases).

The specific features of the global model spatial structure are determined by a database. A simplest version of the point model is realized with the input information averaged over land surface and over the whole World Ocean basin. The spatial heterogeneity is taken into account through various forms of digitization of space. The basic form of the spatial division of land and oceans is a uniform grid $\Delta\varphi \times \Delta\lambda$. In the real version of the model the cells Ω_{ij} can be integrated, so that various forms of the spatial structure of the considered elements and processes in the biosphere can be present in each cell. This flexible prescription of the spatial structure of the biosphere makes it possible to easily adapt the model to heterogeneities of databases and to carry out numerical experiments with an actualization of separate regions.

Depending on special features of the natural process under consideration the structure of regional division can be identified with climatic zones, continents, latitudinal belts, socio-administrative structures, and natural zones. For instance, for climatic processes, many investigators aim at regions with sizes $\Delta\varphi = 4^\circ$ and $\Delta\lambda = 5^\circ$ (Krapivin and Vilkova, 1990), the biogeocoenotic processes are studied at $\Delta\varphi = \Delta\lambda = 0.5^\circ$ (Sellers *et al.*, 1995), the socio-economic structure is presented by nine regions (Krapivin, 1993), the atmospheric processes in the biogeochemical cycles of long-lived elements are approximated by the point models (Nitu *et al.*, 2000b), and the functioning of the ocean ecosystems is described by a heterogeneous structure of the spatial digitization of the shelf zone into the cells Ω_{ij} with selection of four parts of the World Ocean.

The structure of the division of the Earth surface into regions Ω_{ij} covers all the versions enumerated. It means that the general scheme of digitization of the processes in the nature–society system foresees the hierarchy of the levels, including the global, continental, regional, landscape, local levels, etc. The scheme of independent inclusions of the units on all these levels, combining them through parametric interfaces, does not prevent an increase of the number of the model units due to new components, which specify the models of the considered processes. The upper level model can serve as the information base for the lower level model and vice versa, the results of modelling at a lower level can be used to form the information base for the higher level models. The mechanism of the information exchange between the models of various levels reduces the requirements of the global database and broadens possibilities of the nature–society system model.

The structure of the global model includes the auxiliary units ensuring user–model interaction and operating with the database. In particular, these are units realizing the algorithms of the spatial and temporal interpolation or correlating the user's operations with the bank of scenarios. Note that some scenarios, at the user's will, can be transformed into model units. This duality (excessiveness) is characteristic of the scenarios of climate, demography, anthropogenic activity, scientific and technical progress, and agricultural production. The user interface provides a possibility of choosing the structure (Ω_{ij}) in the default regime. Otherwise, the required spatial structure can be formed from the base elements by averaging and interpolation.

Thus the synthesis of the global model version requires a preliminary analysis of the present situation with the use of global databases and knowledge bases. Here the scientists face principal difficulties – the absence of an adequate base of knowledge about the climatic and biospheric processes, as well as incomplete data on the global processes on land, in the atmosphere, and in the oceans.

Another principal difficulty is connected with the inability of present scientific knowledge to formulate the requirements of global databases required for reliable assessments of the environmental conditions and reliable predictions of its development for a sufficiently long time period. Moreover, there is no approach to formation of databases aimed at deriving a global model. Also, there are no approved criteria for assessing the state of environmental elements. For instance, there are a large number of separate studies of global or regional carbon balance without a uniform base of knowledge (Field, 2001; Florkowski *et al.*, 1998; Forest *et al.*, 2002; Kolber *et al.*, 2001; Lafreur, 1999; Lal, 2001; Lovett, 2002; Morgan, 2000; McCarl and Schneider, 2001; Vay *et al.*, 1999; Lenaz *et al.*, 2000; Loehle, 1993; Lowe *et al.*, 1998; Marland *et al.*, 2001; Metzger and Benford, 2001; Milford *et al.*, 2001; Monnin *et al.*, 2001).

Many scientists have tried to find solutions to the problems mentioned (Ivanov-Rostovtsev *et al.*, 2001; Gorshkov *et al.*, 2000). An efficient way of solving these problems is to create a single, worldwide, adaptive eco-information system, making use of a hierarchical structure for data accumulation and forming a multilevel global database (Kondratyev *et al.*, 2000a, b). An adaptive character of this system is provided by the corrected regime of data collection and by the changed parameters and structure of the global model.

The global model can be constructed with account of the existing structure of databases, whose formation is continued within the framework of the International Geosphere–Biosphere Programme (IGBP) and numerous national ecological and nature-protection programmes. The available system of the World Data Centers favours a rapid use of accumulated information about the global processes and simplifies the synthesis of the geo-information monitoring system.

According to Kondratyev (1999b), to control the global geospheric–biospheric system of the Earth, a regular monitoring of a large collection of variables is needed. With an increasing probability of strong global changes, the spectrum of these variables will vary, and the global prediction system should be constantly modified. The methodological grounds for the completeness of information of the complex of variables for the monitoring system can be objective only in the case of the functioning of the geo-information monitoring system. Many variables can be calculated on the basis of the respective models and they need not be measured. However, the planning of measurements and the development of models are realized in parallel, and so far, there are no hopeful results in optimal planning of observation systems. As follows from Marchuk and Kondratyev (1992), the databases and bases of knowledge enable one to synthesize and develop a series of global geo-information monitoring systems. The spatial and temporal variability of biospheric and geospheric processes are well described by the version presented, for instance, by Krapivin (1993).

Inclusion of the global model into the structure of the global monitoring system enables one to transfer it onto the level of an expert system. It means that there is a possibility of a complex analysis of a multitude of elements of the nature–society system for hypothetical situations, which can be both natural and anthropogenic.

6.3 THE ROLE OF LAND BIOTA IN CO_2 ASSIMILATION FROM THE ATMOSPHERE

The land biota is one of the sinks for atmospheric CO_2 (Mannion, 1997; Metting *et al.*, 2001; Potter *et al.*, 2001; Tian *et al.*, 1998; Gorshkov *et al.*, 2000; van Wesemael and Lambin, 2001; Peng *et al.*, 2002a, b). Unfortunately, the available data and knowledge of the processes of plant respiration suggest only rough integrated estimates of CO_2 flux in the vegetation cover. Actually, the role of plants in assimilation of atmospheric CO_2 changes sharply over a 24-hour period and is a complex function of such factors of the environment as temperature, illumination, and air humidity. Nevertheless, attempts to parameterize the functions of vegetation made in Chapter 3 make it possible to assess the role of all types of the soil–plant formation in CO_2 assimilation (mentioned in Table 2.5).

Figure 6.2 shows the role of forest vegetation in CO_2 dynamics. In addition to these results, note that experiments with global models make it possible to trace the dependence of the atmospheric gas composition on the structure of the planetary forest cover. From the available estimates, the total area of forests for $t_0 = 1970$ can be estimated at $\sigma_{L0} = \sigma_L(t_0) \approx 40.3\text{–}41.84 \cdot 10^6\ \text{km}^2$ (Watson *et al.*, 2000), with 1% constituting national parks and forest reserves. With the formulated scenario, assume $t_L = 2050$ and $\sigma_{X0} = \sigma_X(t_0) = 19.5 \cdot 10^6\ \text{km}^2$. As seen from Figure 6.2, the increasing rate of deforestation raises considerably the concentration of CO_2 in the atmosphere. Even with a 10% reduction of forest areas by 2050 compared to 1970, that is, $\sigma_{L1}/\sigma_{L0} = 0.9$, the atmospheric CO_2 can increase by 44% by the end of the 21st century. On the contrary, a 10% increase of the forested areas decreases the

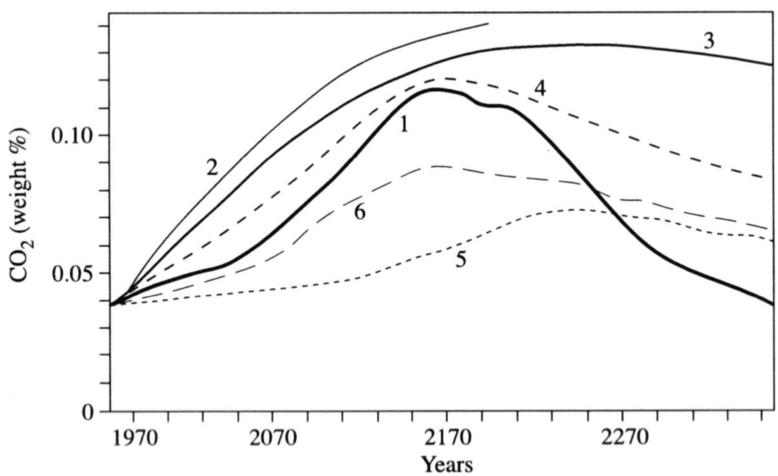

Figure 6.2. The dynamics of CO_2 concentration for different scenarios of changing forest areas within the framework of the scenario after Demirchian *et al.* (2002): 1 – rate of change in forest areas remains the same as that in 1970; 2 – by 2050 forests have totally disappeared, 3 – by 2050 the area of forest is reduced by 50%; 4 – by 10%; 5 – by 2050 forest areas increase by 50%; 6 – by 10%.

Table 6.1. The dynamics of CO_2 assimilation by plants in the territory of Russia. Emission of carbon on this territory in 1990 is assumed to be 1.6 Gt C/yr with an annual change after Keeling's scenario (Krapivin and Vilkova, 1990).

Vegetation formation (see Table 2.5)	Rate of CO_2 assimilation, 10^6 t C/yr				
	Years				
	1990	2000	2050	2100	2150
A	2.6	2.8	6.7	7.1	6.9
C	3.7	4.6	10.9	12.0	12.1
M	4.0	5.1	12.4	14.5	13.8
L	3.2	3.9	9.2	10.3	10.4
F	11.2	14.8	43.6	47.2	44.2
D	31.6	39.9	110.6	121.9	109.3
G	23.3	29.2	72.2	73.4	70.5
R	5.2	6.2	13.1	13.8	10.7
W	4.7	5.1	8.2	8.8	7.9
V	0.7	0.7	0.9	1.1	0.8
@	2.4	2.6	3.7	3.9	2.9
S	0.6	0.7	1.2	1.4	1.0
Q	1.5	1.6	2.2	2.3	1.8
Total	*94.7*	*206.1*	*294.9*	*317.7*	*292.3*

concentration of atmospheric CO_2 by 15%. With a 50% increase of forested areas by 2050, a decrease of atmospheric CO_2 will constitute 60% by 2100 relative to its possible value, with the scale of impacts on the forest ecosystems, observed by the end of the 20th century, preserved. Hence, variations of the forested areas in the biosphere even within ±10% can substantially change the dynamics of numerous components of the global ecosystem.

Table 6.1 exemplifies the calculation of CO_2 sinks into the vegetation cover in the territory of Russia. Such calculations, with the use of the global model of the nature–society system, demonstrate the dynamics of the CO_2 fluxes mosaic in the atmosphere–plant–soil system. Knowledge of this mosaic makes it possible to assess the role of concrete types of soil–plant formations in the regional balance of carbon, and on this basis, to calculate the global fluxes of carbon dioxide on the atmosphere–land border. Similar calculations are also possible for the atmosphere–ocean system.

Table 6.2 demonstrates the consequences of changing the global structure of the soil–plant formations for the dynamics of CO_2 assimilation by vegetation. As is seen, the anthropogenic change of the vegetation cover substantially changes the balance components in the global carbon cycle. Clearly, such experiments require a thorough analysis of data on the possibility of transformation of vegetation cover with account of climatic zones and biocenological consistency. Nevertheless, such hypothetical experiments are useful for general assessments of possible ranges of anthropogenic impacts on the global carbon cycle. For instance, natural and anthropogenic

Table 6.2. The dynamics of the ratio of integral rates of (H_6^C) CO_2 assimilation by vegetation cover from the atmosphere with natural distribution of soil–plant formations (Figure 3.3) and with its transformation according to the scenario in the second column.

Scenario		H_6^C (changed Figure 3.3)/H_6^C (Figure 3.3) Years		
Figure 3.3	Changed Figure 3.3	2003	2020	2050
A	L	2.79	2.14	1.97
C	L	0.96	0.94	0.95
M	L	1.67	1.15	1.01
F	D	1.68	1.57	1.11
G	D	2.11	1.67	1.45
R	D	3.98	3.70	2.87
P	Z	3.17	2.68	2.43
U	Z	22.52	20.73	17.95
W	Z	23.12	19.44	16.32
E	Z	100.14	77.75	68.54
H	Z	194.56	155.50	138.39
Q	Z	799.14	777.50	751.26
Y	Z	1.43	1.39	1.23
N	Z	69.98	62.20	56.59
J	Z	5.89	5.09	4.67
T	Z	26.54	25.92	23.58
I	Z	17.88	16.37	14.91
#	Z	0.91	1.12	0.97

cataclysms connected with forest fires annually introduce considerable changes into this cycle, because they change numerous fluxes and supplies of carbon over large territories. Tables 6.3– 6.5 give estimates of deviations in the content of carbon in the basic biospheric reservoirs with forests fires in different zones. Large-scale impacts on land biota are damped during 60–100 years. The biosphere turns out to be more resistant to impacts on the forests of southern latitudes and more sensitive to violations of forest areas in temperate latitudes. Hence, the northern hemisphere forests up to 42° N play an important role in stabilizing the carbon cycle in the biosphere.

The scenario of deforestation, as seen from the studies of numerous authors, evokes great interest in studies of the global carbon cycle and associated climate change. A diversity of possible real situations of the transformation of land cover is so large that it is impossible to assess all the consequences. Note only that, for instance, a liquidation of all northern taiga and mid-latitude taiga forests (types *F*, *D*) in 50 years will lead to a 53% increase of atmospheric CO_2 with subsequent negative consequences for the flux H_6^C. Similar consequences follow after removal of all wet evergreen and deciduous tropical forests (types *Z*, *Y*). In this case the indicated increase of atmospheric CO_2 will be reached 20 years later.

Table 6.3. Model estimates of the deviation of the carbon content provided all coniferous forests of the northern hemisphere (up to 42° N) are burnt.

Years after impact	Deviations in carbon content (Gt)			
	Atmosphere	Soil	Upper ocean	Deep ocean
0	140.9	−5.4	15.5	0.1
10	104.8	−33.1	29.8	3.1
20	83.1	−44.1	21.5	7.4
30	63.4	−43.5	19.0	8.5
40	47.2	−39.7	14.6	10.4
50	34.2	−34.6	11.6	11.7
60	24.1	−29.4	8.3	12.7
70	16.3	−24.7	6.2	13.5
80	10.2	−20.5	4.5	14.0
90	5.6	−17.1	3.3	14.3
100	2.1	−14.2	2.3	14.5
200	−9.0	−3.4	−0.8	13.5

Table 6.4. Model estimates of the deviation of the carbon content provided all forests of the northern hemisphere (up to 42° N) are burnt.

Years after impact	Deviations in carbon content (Gt)			
	Atmosphere	Soil	Upper ocean	Deep ocean
0	230.8	−7.9	24.9	0.1
10	173.2	−30.9	47.9	4.9
20	138.9	−67.6	39.2	10.0
30	107.9	−89.3	31.1	13.8
40	82.0	−64.3	24.1	16.8
50	60.9	−56.9	18.4	19.1
60	44.2	−49.0	13.9	20.7
70	31.1	−41.6	10.3	21.9
80	20.9	−35.0	7.5	22.8
90	12.9	−29.4	5.4	23.3
100	6.9	−24.7	3.7	23.6
200	−12.7	−5.9	−1.7	21.7

Land cover structure changes not only due to human activity. In some regions of the globe, hurricanes introduce considerable changes in the carbon balance of the forest ecosystems. So, over US territory every 3 years two hurricanes occur, which accelerate the transition of living biomass of trees into dead organic matter. If in the USA territory, 20 Tg C are removed from the forests every year, then 10–15% of this is the result of one hurricane (McNulty, 2002). Hence, the hurricanes accelerate the

Table 6.5. Model estimates of the deviation of the carbon content provided all tropical forests are burnt.

Years after impact	Deviations in carbon content (Gt) Atmosphere	Soil	Upper ocean	Deep ocean
0	396.0	−20.0	42.2	0.2
10	261.6	−93.7	73.9	8.0
20	162.2	−84.8	48.0	14.9
30	90.6	−38.6	28.2	19.1
40	45.3	−36.5	15.0	21.3
50	18.3	−21.6	7.5	22.4
60	2.9	−12.4	3.0	22.8
70	−5.8	−7.0	0.5	22.8
80	−10.6	−4.2	−0.9	22.6
90	−13.2	−2.6	−1.7	22.5
100	−14.5	−1.9	−2.1	21.8
200	−13.2	−2.3	−1.9	17.7

return of carbon to the atmosphere, and their global inventory is needed to estimate numerous components of the global carbon cycle more accurately.

6.4 THE ROLE OF THE WORLD OCEAN IN THE ASSIMILATION OF CARBON DIOXIDE FROM THE ATMOSPHERE

An estimation of the total contribution of the World Ocean to CO_2 assimilation from the atmosphere, as in the case of land, is only possible by spatial integration of the difference between fluxes H_3^C and H_2^C (McGillis *et al.*, 2001; Monahan and Dam, 2001; van Oijen, 1998; Sarma *et al.*, 2001; Seibel and Walsh, 2001). Table 6.6 gives average estimates of this difference. Even with these rough estimates, we see a mosaic character of the role of various basins of the World Ocean in atmospheric CO_2 assimilation. Quite significant are the water basins in the northern latitudes and in the zones of upwelling. The cold-water basins in the southern latitudes remain practically unknown, where, like in the Arctic Ocean, large territories are covered with ice. From available estimates, $\Delta H_{32} = 0.006$ Gt C/yr for the ice-covered water basins and 0.022 Gt C/yr for ice-free water bodies.

Some ideas about the role of the World Ocean in the CO_2 assimilation from the atmosphere can be obtained from the data in Figures 6.3 and 6.4. Figure 6.5 gives a spatial distribution of CO_2 assimilated from the atmosphere with account of the combined role of land and ocean ecosystems. We can see that the general role of the environment in stabilizing the CO_2 content in the atmosphere with the considerably increasing share of carbon varies negligibly. It means that the natural medium preserves persistent feedbacks and the level of variability of reactions of the land and

Table 6.6. The spatial distribution of $\Delta H_{32} = H_3^C - H_2^C$ (Gt C/km^2/solar year) estimated from averaged values of assimilation and emission of CO_2 on the atmosphere–ocean border during the whole industrial period (σ – the area of the World Ocean basin (10^6 km^2).

Water basin	σ	ΔH_{32}
South Atlantic Ocean	28.80	0.00138
Equatorial Atlantic Ocean	32.38	−0.00285
North Atlantic Ocean	26.01	0.01025
South Pacific Ocean	49.85	0.00531
Equatorial Pacific Ocean	88.81	−0.00323
North Pacific Ocean	32.45	0.00846
South Indian Ocean	49.63	0.00538
Equatorial Indian Ocean	32.85	−0.00592
Arctic Ocean	7.47	0.00131
The World Ocean	*348.23*	*0.00154*

ocean systems to a marked increase of anthropogenic loads on the atmosphere remains stable. A detailed analysis of such reactions shows that almost a 60% increase of the anthropogenic carbon flux during next 20 years will result in a 4.8% increase of CO_2 flux from the tropical water basins of the World Ocean to the atmosphere, but incidentally, the absorptivity of the ecosystems in the northern latitudes will increase by 12.3%. The role of land ecosystems in assimilation of excess carbon from the atmosphere will increase by 11.4%, and the role of the Arctic waters by 13.2%. As a result, in 2020 the following amounts of carbon will be assimilated from the atmosphere: $25.7 \cdot 10^6$ t C by the south-taiga forests; $35.1 \cdot 10^6$ t C by the mid-taiga forests; and $10.7 \cdot 10^6$ t C by the tundra ecosystems.

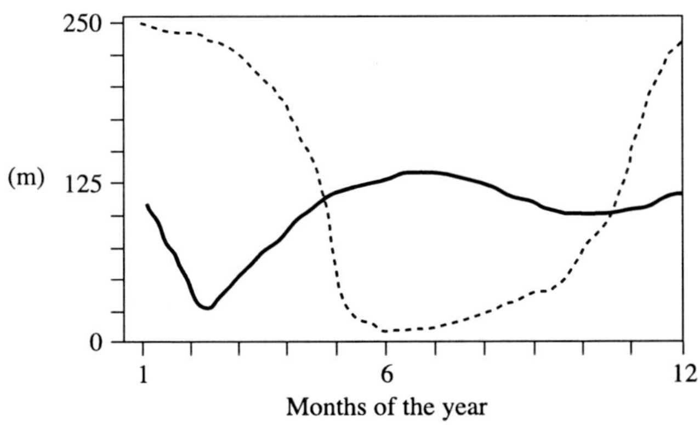

Figure 6.3. Distribution of the depth of the upper quasi-homogeneous layer of the World Ocean in the latitudinal zones 0°–10° N (solid curve) and 60°–70° N (dashed curve).

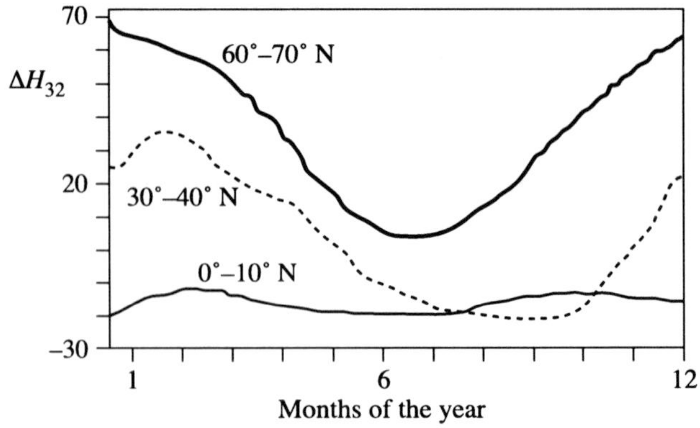

Figure 6.4. The annual distribution of carbon flux across the atmosphere–ocean border in different latitudinal zones.

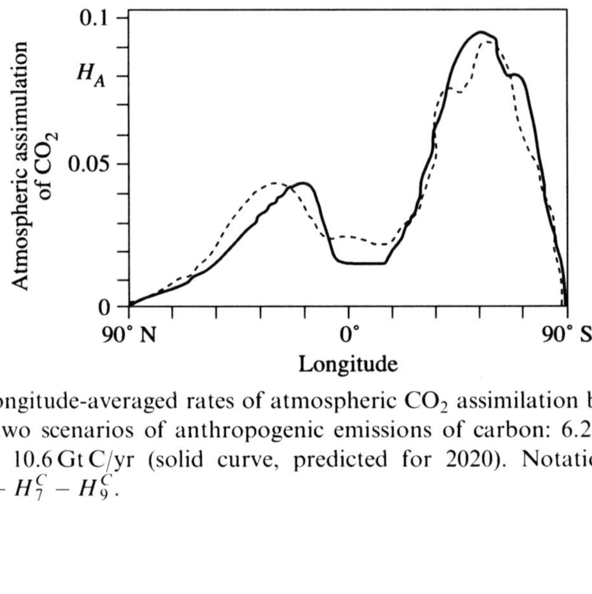

Figure 6.5. The longitude-averaged rates of atmospheric CO_2 assimilation by land and ocean ecosystems with two scenarios of anthropogenic emissions of carbon: 6.26 Gt C/yr (dashed curve, 2000) and 10.6 Gt C/yr (solid curve, predicted for 2020). Notation: $H_A = \Delta H_{32} + H_6^C + H_4^C - H_8^C - H_7^C - H_9^C$.

6.5 FUTURE DEVELOPMENT OF THE GLOBAL CARBON CYCLE MODEL

As has been mentioned repeatedly, the attention to the problem of carbon dioxide and other minor gas components (MGCs) of the atmosphere is highlighted by the alarm of scientists, economists, and politicians to the possible consequences of an expected climate change. Since fossil fuel combustion and deforestation of land surface are the most significant sources of the anthropogenic increase of concentrations of CO_2 and CH_4 in the atmosphere, these gases are being studied very thoroughly. Both these gases absorb IR radiation from the Earth surface and therefore influence the world's climate. The problem of reducing this influence was placed before politicians at the end of the 20th century and was "resolved" in a poor way with the advent of the Kyoto Protocol. In fact, a real and objective solution to this problem is to thoroughly study the biogeochemical cycles of all gases, somehow related to the carbon cycle. The GCP is concentrated on this approach, aimed both at a complex analysis of available data on the carbon cycle in the environment and at the required relevant measurements. The Project combines efforts of the IGBP, International Human Dimensions Programme of Global Changes (IHDP), and World Climate Research Programme (WCRP). The carbon cycle is considered a central cycle of the terrestrial system connected with climate, water cycle, cycles of nutrient salts, and biomass production through photosynthesis on land and in the oceans. The human participation in carbon cycle is determined by food production and habitat formation.

6.5.1 The international programme on studies of the carbon cycle in the environment

The GCP foresees studies of three subjects: (1) models and variability; (2) processes, control, and interactions; and (3) future dynamics of the carbon cycle. Such a structure of studies of the global carbon cycle should create databases and knowledge bases, which would make it possible to specify the elements of the scheme in Figure 2.8 and to increase thereby the accuracy of estimates of the distribution of carbon fluxes between the biospheric reservoirs.

The basic idea of the scheme in Figure 2.8 is that the base components of the global carbon cycle are its reservoirs: the atmosphere (mainly CO_2), oceans (different layers of water masses and bottom deposits), land biosphere (vegetation, litter, and soil), and carbon of fossil fuels (activated during the last 200 years due to human activity). The required resolution of the description of these reservoirs and fluxes of carbon between them can be found only by modelling the global carbon cycle with a gradual complication of this scheme and respective accumulation of knowledge (Schimel, 1995, 1998; Schimel *et al.*, 1996, 2001; Sun and Pethick, 2002; Trenberth *et al.*, 2002; Valiela and Bowen, 2002; Vàrallyay, 2002; Veizer, 2001; von Storch and Flöser, 1999; Wu *et al.*, 2001; Yi *et al.*, 2001; Zaragoza *et al.*, 2000).

The present ideas about carbon flux variability are based on the global scale space and ground-based observations of the concentrations of atmospheric CO_2 and other gases, on modelling the dynamics of the atmosphere, oceans, and biogeochemical processes, as well as on application of the principles of mass balance. As a result, the following information of the global balance of carbon in the present biosphere have been formed (IPCC, 2001a):

- About half the CO_2 emitted to the atmosphere from fuel burning is assimilated both by the land and oceanic sinks of carbon.
- Places of fossil fuel consumption are mainly concentrated in the northern hemisphere.
- The concentration of atmospheric CO_2 has increased considerably due to changing land cover in tropical latitudes.

- The amplitude of the intra-annual variability of the atmospheric CO_2 concentration is compatible with carbon emissions due to fossil fuel combustion.
- The inter-annual variability of carbon fluxes between the atmosphere and land biocoenoses is greater compared to fluxes between the atmosphere and the oceans.
- The complete pattern of the oceanic sources and sinks of atmospheric CO_2 are known: the tropical waters emit CO_2 and the high-latitude water bodies assimilate CO_2.
- Maximum inter-annual variations of CO_2 flux on the atmosphere–ocean border are observed in the equatorial Pacific (Sabine and Feely, 2001).

As a result, the following questions arise:

- What are the laws following which the sources and sinks of carbon vary in time?
- What are the principles of the inventory of all carbon fluxes in the environment?
- What is the contribution of human activity, including fuel burning and land use, into the distribution of sources and sinks of carbon?
- What are the social aspects of the impact on carbon cycle?
- What are the mechanisms for the global manifestation of regional and local processes of changing carbon fluxes?

The GCP foresees a search of answers to these and other questions through choosing the first-priority directions of studies:

- To determine the role of the oceans in controlling the seasonal and decadal changes of carbon fluxes in the atmosphere–ocean system. To assess the role of land biosphere in the formation of the inter-annual variability of the atmospheric CO_2 concentration.
- To determine the meridional distribution of carbon sinks in the northern hemisphere, marking out the role of North America, Eurasia, Asia, and Europe.
- To assess an importance of carbon fluxes between the oceans and land for the carbon balance, as well as carbon supplies in the inland water bodies.
- To assess the role of sediments and horizontal transport of matter in the coastal zones and open waters in the formation of the structure of the sources and sinks of carbon.
- To understand the spatial pattern and the amplitudes of the sources and sinks of carbon in the southern latitudes of the World Ocean.
- To determine the role of the Earth's tropics in the formation of carbon fluxes in the environment, with an emphasis on changing surface cover.
- To understand differences in the ways fossil fuels are used in industry and between individual regions within certain countries.
- To explain the leading forces and the consequences of various ways of regional development of the reservoirs and fluxes of carbon.
- To analyse the social and regional patterns of vulnerability and adaptability to changes in the carbon cycle.

- To determine the role of biological pumps and solubility processes in the global and regional cycles of carbon.
- To understand differences and to remove contradictions in measurements of velocities and patterns of changes in land use and transformation of land cover.
- To develop methodologies of transition from results of studies of local and regional carbon cycles to global scales.

Clearly, the carbon cycle between the biospheric reservoirs is controlled by many processes still poorly studied. Among them, the most significant processes can be selected:

- physical processes in the atmosphere, oceans, and terrestrial hydrosphere;
- biological and eco-physiological processes in the oceans and on land;
- biogeochemical transformations;
- natural and anthropogenic changes in the terrestrial biosphere; and
- processes of use by humans of fossil fuel carbon.

To understand these and other processes of the formation of the carbon cycle in the dynamical environment, new paradigms are needed on a totality of natural and anthropogenic events. Such views have been covered in Batijes (1999), Beese *et al.* (1995), Beven (2002), Ivanov-Rostovtsev *et al.* (2001), Kozoderov (1994, 1998), Gorshkov (1995), Kondratyev (1993, 1996, 1999c), Krapivin and Kondratyev (2002), Kondratyev and Krapivin (2002). Of course, a problem-oriented database is needed, which would correspond to the base of knowledge clarifying the cause-and-effect feedbacks in the nature–society system. There is such a base of knowledge, which explains in detail many processes in this system, revealing mechanisms of their occurrence and recommending a strategy of data accumulation and modelling:

- Direct exchange of CO_2 on the atmosphere–ocean border is regulated by the physical processes of the vertical mixing of water layers above the thermocline with deep waters and by additional impacts of biological processes, which pump out carbon from the water and move it down to sediments.
- The key biological mechanisms of the formation of carbon fluxes in the ocean medium are known. In the general case, these mechanisms are connected with the functioning of the carbonate system, and in a simpler presentation they are provided with the vital functions of autotrophic and heterotrophic organisms, processes of decomposition of dead organic matter, and respiration of living organisms. This spectrum should be supplemented with a circulation of nutrient elements and photosynthesis (Krapivin and Kondratyev, 2002).
- It is established that with an increasing partial pressure of CO_2 in the atmosphere the reaction of the World Ocean is non-linear both in space and in time (Orr *et al.*, 2001).
- The light, water, biogenic elements, and regimes of roughness interact following the non-linear laws described by numerical models (Krapivin, 1996).

- Many significant feedbacks in the system of the global carbon cycle have been well parameterized. Among them are the physiological responses of plants to changes of temperature and humidity of air and soil, the aerodynamics of exchange processes between land surface and the atmosphere, and dynamic and thermodynamic processes within the atmospheric boundary layer (Raupach, 2000).
- At higher concentrations of atmospheric CO_2, without restricting the influence of light, water, and nutrients, the photosynthetical assimilation by land plants becomes physiologically saturated.
- Within regional and continental scales the CO_2 exchange between land ecosystems and the atmosphere is regulated by several factors by:
  (1) extreme climatic events such as drought, appreciable changes in seasonal temperatures, and changes in solar radiation flux;
  (2) changes in the frequency of occurrence of fires, by cloud-free skies, and other large-scale fluctuations in the environment, which lead to sensible short-term losses of carbon in numerous biospheric reservoirs;
  (3) the changing boundaries of biomes and their types (e.g., change of coniferous forests for deciduous forests, of forested areas for grass, and meadows for forested areas); and
  (4) losses of biodiversity and intrusion of exotic species into the territories of sustainable existence of vegetation communities.

All studies foreseen by the plan of the GCP are aimed at obtaining reliable predictions of the consequences of human activity both in the near future and in long-term. For this purpose, it is necessary to develop an approach to estimate the functions of soil–plant formations and water bodies of the World Ocean in assimilation of CO_2 from the atmosphere. Observations show that during the 500 year period up to 1860 the climatic system functioned with a relatively narrow range of temperature and concentrations of CO_2 and CH_4 (during the last 100,000 years the concentration of CO_2 in the atmosphere varied between 180 to 280 ppmv). The last century has considerably broadened the spectrum of processes of transformation of matter and energy in the environment due to numerous new kinds of impacts on the biospheric components. The human factor in these processes has entailed an uncertainty of the future dynamics of these components, which required scenarios of the development of economy, science, and politics. The level of uncertainty is so high that the predictions cannot be realistic. According to the IPCC report (2001a), the level of carbon emissions by 2100 may reach 30 Gt C/yr, which, with the preserved rate of anthropogenic impact on the environment, may lead to CO_2 concentrations in the atmosphere reaching 950 ppm. All these estimates are obtained on the basis of expert opinion and sufficiently primitive models. Therefore, according to the GCP, an accumulation of a large base of knowledge, formation and testing of scenarios, development of models, and analysis of the present trends are first-priority directions of studies of the global changes of the environment. To synthesize the scenarios, the following set of considerations and reasons are suggested:

- Concentrations of GHGs will be increasing during some time period (IPCC, 2001a–c).
- The present and expected increase of the concentrations of GHGs, especially of CO_2, will probably be followed by the global warming and by other climate changes.
- Apart from physical factors, such as the oceanic circulation, an assimilation of CO_2 by the oceans from the atmosphere will increase in the future with growing concentrations of atmospheric CO_2.
- The response of the oceanic thermohaline and circulation systems to climate changes (especially a temperature increase and change of the hydrological cycle) will probably be non-linear and will be accompanied by an increased stratification of the upper layer of the oceans, which can reduce the ventilation of deep layers.
- An increased stratification of the upper layer of the ocean will change the provision of surface layers with nutrient salts, which will change the role of phytoplankton in carbon assimilation and cause large-scale fluctuations in the atmosphere–ocean interaction. Of special concern are the Arctic basin, the Okhotsk Sea, and other seas in northern latitudes.
- There is a difference in the estimates of direction of changes of the surface sinks of carbon due to transformation of land biosphere (Cox *et al.*, 2000; Cramer *et al.*, 2001; Friedlingstein *et al.*, 2001).
- Careless and ill-considered changes on the Earth surface will lead, in the near future, to changes of the climate and carbon cycle in nature.
- Instabilities and multiple states of equilibrium can turn out to be bound with biosphysical and biogeochemical systems, containing climatic and hydrological processes and cycles of carbon and nutrient elements. These phenomena can be caused by non-linear feedbacks in re-distribution of energy and matter between the atmosphere, land, oceans, and ice.
- Even without dynamic phenomena, such as instabilities and increasing states of equilibrium, an increase of the atmospheric CO_2 concentration during the period 1860–2100 will cause certain shifts in the climate–carbon system, changing the former distribution of the regions.
- The anthropogenic changes in carbon cycle interact with the biophysical aspects of the formation of this cycle, and hence, have the potential to lead to variations of the global climate system.
- People also react to undesirable climate changes threatening to their well-being, though these reactions are highly uncertain (e.g., the Kyoto Protocol).
- There is a certain spectrum of potential reactions of human society to undesirable climate changes, including geographical, social, economic, and nature–science directions:
  - changes in CO_2 emissions either due to transition from traditional to non-traditional types of fuel, or caused by changing the technology of its application;
  - improvement of the processes of production and use of energy, changes in life style;

- transition to renewable energy sources (bio-energy, solar energy, wind);
- increased role of forests in assimilation of excess carbon;
- increase of carbon sink from the biosphere into geological formations and deep ocean due to engineering surveys; and
- development of new technologies in agriculture, leading to a decrease of GHG emissions.

Eventually, all this requires a universal complex approach to studies of the consequences of realization of alternative scenarios of human civilization development. Clearly, a search for a safe sustainable functioning of the nature–society system is impossible without modelling the alternative scenarios. And for this purpose a global model of the functioning of this system should be derived. Such a model and a scheme of its use have been described in Krapivin and Kondratyev (2002) and in Kondratyev and Krapivin (2002).

6.5.2 Structure of the global model of the nature–society system

Synthesis of the global nature–society system is based on the idea that it is a self-organizing and self-structured system, whose elements agree in space and time through the process of natural evolution. The anthropogenic impact on this process violates this integrity. Attempts to formally parameterize the process of co-evolution of nature and humans as biospheric elements are connected with a search of a single simulation of all processes in the nature–society system, which would combine the efforts of all the environmental sciences. Such a synergetic basis is observed in many studies of global modelling.

The global modelling refers to not always appreciated spheres of knowledge, whose intrusion into non-traditional spheres of human life have been actively discussed during the last decades. Despite an existing disagreement on the role of models in studies of the nature–society system, the fact that a model makes it possible to predict consequences of an accomplishment of large-scale experiments on the transformation of the environment, as well as to provide a synthesis of a single nature–society sample on the basis of the fragments of the environment.

A combination of units within the global model simulating the development of various processes in the nature–society system requires a flexible technology to co-ordinate their spatial and temporal scales, which in turn, necessitates a synthesis of the object-oriented databases. As a result, the structure of the global model should permit sustainable development of the established hierarchy of the units. This is possible with the use of the semantic mechanisms of adaptation of the model structure to changing simulations of the configuration of the nature–society inter-connections. Figure 6.1 shows a scheme outlining the organization of such a structure. The information central line coordinates the inputs and outputs of individual units and ensures compatibility between the units and the database. This is the mechanism for synthesis.

Observations and studies of global processes in the environment according to the GCP have been aimed at overcoming the difficulties appearing in global

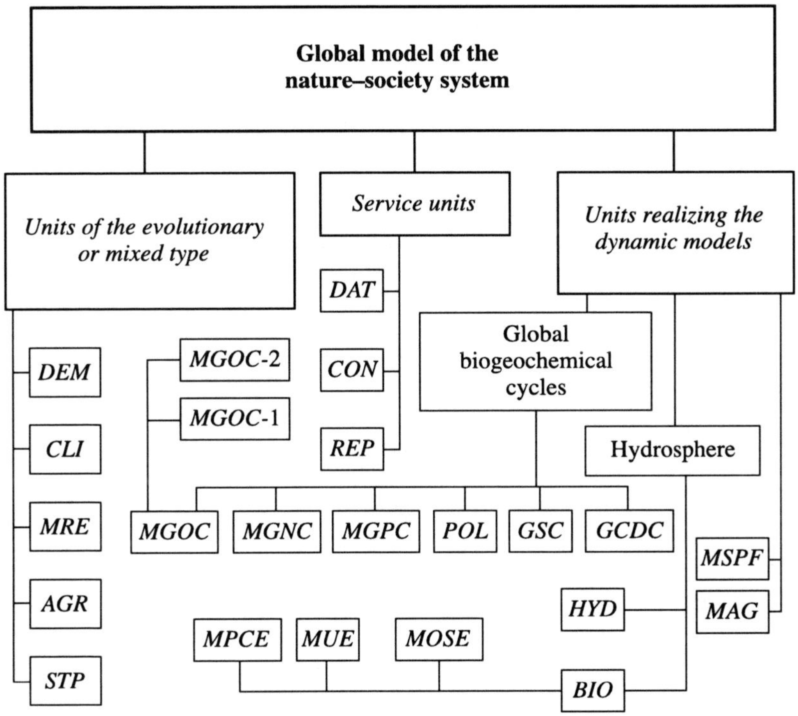

Figure 6.6. The block-scheme of the global model of the nature–society system.

modelling and in attempts to adapt an accumulated database and knowledge to the structures of available global models. One of the possible ways to solve these problems is the geoinformation monitoring system (GIMS)-technology, which makes it possible to use the global model in the regime of its constant adaptation to the current global monitoring of the environment, providing thereby a combined parameterization of the spatial and temporal variability of biospheric and geospheric processes with the procedure of observations planning.

Figure 6.6 shows a block-scheme of the global model built into the GIMS structure via the DAT unit, which controls the interface between other model units and database renewal in the monitoring regime. The CON unit adjusts the information base of the global model and its internal connections to the conditions of the simulation experiment and controls the coordination of information fluxes between the model units. The REP unit forms the report documents of the simulation experiment and connects the global model with other information systems. The remaining units of the global model are divided into two classes with different modelling approaches. The units responsible for the evolutionary modelling are the units parameterizing the demographic (DEM) and climatic (CLI) processes creating the scenarios of managing the mineral resources (MRE) and development of the scientific-technical progress (STP) as well as modelling the agricultural production

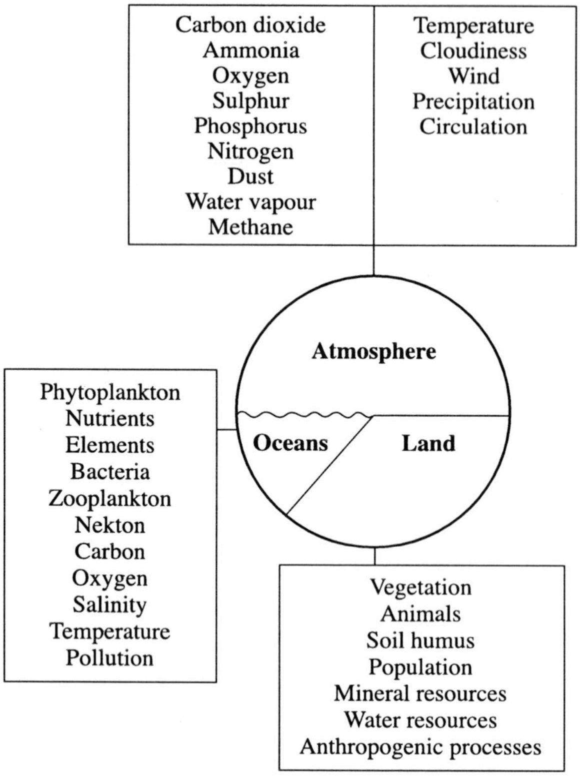

Figure 6.7. The elemental composition of the global simulation model describing the dynamics of the nature–society system.

(AGP). The units relevant to the dynamic models are based on a realization of balance relationships. The units parameterizing the global biogeochemical cycles cover the whole phase space of the nature–society system presented in Figure 6.7. Here the models of the global cycle of carbon dioxide (GCDC), sulphur (GSC), phosphorus (MGPC), nitrogen (MGNC), oxygen and ozone (MGOC), and general pollutants (POL). To simulate the totality of the processes in the hydrosphere, the global model includes the units simulating the hydrodynamic processes (HYD) and the functioning of water ecosystems in different climatic zones (BIO). The global model also includes the ecosystems of the Okhotsk Sea (MOSE), the Peruvian current (MPCE), and the equatorial upwelling (MUE). To broaden the adaptive possibilities of the global model, the structure of many units foresees various versions of parametric simulations of the processes in the nature–society system.

The scheme in Figure 6.7 shows the basic components taken into account in the global simulation model (GSM) described here. As seen from Figure 6.6, a concrete realization of each unit of the GSM is determined by a level of knowledge of the processes reflected in the unit. So, the units responsible for modelling the biogeochemical and biogeocoenotic processes are described by balance equations.

Let $\psi_s(t)$ be the content of the element ψ in the medium S at a time moment t. Then, by the law of conservation of matter and energy, write the balance equation:

$$\frac{d\psi_s}{dt} = \sum_j H_{js} - \sum_i H_{si}$$

where H_{js} and H_{si} are, respectively, incoming and outgoing fluxes relative to the medium S. Summation is made over external media i and j, interacting with the medium S. In fact, the medium S is elements of digitization of the environment by latitude φ, longitude λ, depth z, and height h. A diversity of the functional parameterizations of fluxes H_{pq} is determined by knowledge of the physical, chemical, and biological features of the element ψ.

A parameterization of the processes of photosynthesis, dying, and respiration of plants in land ecosystems, is based on knowledge of phytoceonology, which includes information about external and internal system connections of the vegetation community, including the temperature dependences of photosynthesis and evapotranspiration of plants, gas exchange between plants and the atmosphere, impacts of solar radiation on the exchange and growth processes, correlations between vegetation and soil processes, and interaction of vegetation cover with the hydrological cycle.

The GSM units responsible for parameterization of climatic and anthropogenic processes are of mixed character, that is, partially they are described by equations of motion and balance, and partially an evolutionary model which is derived based only on an observation series. Some aspects of these units are partially considered in previous chapters of the book, and some descriptions of the units have been given in Krapivin and Kondratyev (2002) and in Kondratyev and Krapivin (2002).

6.5.3 Evolution and survivability of the nature–society system

The nature–society system can be presented as a combination of nature (N) and human society (H), which constitute a single planetary system. Therefore their separation in deriving global and regional models should be considered conditional. The systems N and H have the hierarchical structures $|N|$ and $|H|$, goals $\underline{N}$ and $\underline{H}$, and modes of behaviour $\bar{N}$ and $\bar{H}$, respectively. From the mathematical point of view, an interaction of the systems N and H can be considered as a random process $\eta(t)$ with an unknown law of distribution, representing the level of intensity in relationships of these systems or estimating the state of one of them. The goals and modes of behaviour of the systems are functions of the index η. The η index can vary in such a way that the modes of behaviour of the systems can be antagonistic, indifferent, and cooperative.

The main goal of the system H is to achieve a high standard of living with guaranteed prolonged survival. The goal of the system N can be determined in terms of survival. The system N behaves itself following the objective laws of co-evolution. In this sense the selection of H and N is conditional and can be interpreted as a division of a multitude of natural processes into controlled and uncontrolled. Without dwelling upon the philosophy of this division, consider the systems H and N to be open and symmetrical in their description. The system H includes technologies, science, economic potential, industrial and agricultural production, sociological structure, size of population, etc. The H–N interaction causes changes in η, the level of which affects the structure of vectors $\bar{N}$ and $\bar{H}$. Of course, there exists a threshold of $\eta_{\max}$, beyond which humankind stops existing, but nature survives. The asymmetric nature of the systems H and N causes changes in the goal and strategy of the system H.

Apparently, in the present conditions of interaction between these systems $\eta \to \eta_{\max}$ at a rapid pace, and therefore individual components of vector H can be attributed to the class of cooperative behaviour. Since the present socio-economic structure of the world is presented by a totality of the states, we shall consider a country as a functional element of the system H. The $\eta(t)$ function reflects the result of interaction of the countries both with each other and with nature. Describe the sum of the results of their interaction by the matrix $\mathbf{B} = \|b_{ij}\|$, each element of which has a meaning of its own:

$$b_{ij} = \begin{cases} +1 & \text{for cooperative behaviour} \\ -1 & \text{for antagonistic relationships} \\ 0 & \text{for indifferent situations} \end{cases}$$

A country H_i has m_i possible ways to reach the goal H_i, in other words, it uses a series of strategies $\{\bar{H}_i^1, \ldots, \bar{H}_i^{m_i}\}$. The weight of each strategy $\bar{H}_i^j$ is prescribed by the value $p_{ij}(\sum_{j=1}^{m_i} p_{ij} = 1)$. The resulting value of the η parameter is a function of the indicated characteristics, and each time moment of the situation is described by a game-theoretic model (Krapivin, 1978).

An objective estimate of the dynamics of the natural medium $N = (N_1, N_2)$ is possible with some assumptions using the models of the biosphere N_1 and climate N_2. An accumulated experience of numerous scientists in deriving such models includes the point, regional, box, combined, and spatial models (Marchuk and Kondratyev, 1992; Krapivin, 1993). This experience enables us to move on to a synthesis of a new type of global model, covering the key bonds between the hierarchical levels of natural and anthropogenic processes.

In general, the condition of the systems H and N can be described by vectors $x_H(t) = \{x_H^1, \ldots, x_H^n\}$ and $x_N(t) = \{x_N^1, \ldots, x_N^m\}$, respectively. A combined trajectory of these systems in the $n+m$-dimensional space is described by the function $\eta(t) = F(x_H, x_N)$ determined by solutions of the global model equations. The form of F is determined by knowledge of the laws of co-evolution, and therefore there is a possibility of investigations in different spheres of science. The available estimates of F (Krapivin, 1996) reveal a correlation between the notions "*survivability*" and "*sustainability*". According to Ashby (1956), the dynamic system is "alive" within the time interval (t_a, t_b), if its determining phase coordinates are within "admissible limits" $x_{H,\min}^i \leq x_H^i \leq x_{H,\max}^i$; $x_{N,}^j \leq x_{N,\max}^j$. Since the systems H and N have a bbiological basis and limited resources, one of the indicated boundary conditions turns out to be unnecessary, that is, for the components of vector $x = \{x_H, x_N\} = \{x_1, \ldots, x_{n+m}\}$ the condition $x_{\min} \leq \eta = \sum_{i=1}^{n+m} x_i$ should be satisfied. This simple scheme requires that the total energy in the system be preserved and its elements be diverse.

Of course, the notion of the system's survivability is more capacious and informative. In the system ecology, many authors denote the stability and integrity of the system by this term, meaning the system's ability to resist external forcings. In other words, the vitality is measured by the tendency of the system to suppress large oscillations of its structure and elements, returning the system into its former state of equilibrium. Thus the system's vitality should be understood as its ability to actively resist an impact of external factors and to preserve for a long time its characteristics (considering possible states of the system when it still goes on working and is capable of performing its functions, with certain methods and conditions of its exploitation).

6.5.4 Modelling the climatic processes

The climatic component of the nature–society system represents the greatest difficulty in the synthesis of the global model, because it is characterized by a lot of feedbacks, most of which are unstable. Among them are ice-albedo, water vapour-radiation, cloudiness-radiation, aerosol-radiation, and many others. The functioning of the Earth climatic system is determined by the state of the atmosphere, oceans, cryosphere, surface of the continents with land biota, lakes, rivers, ground waters, and various anthropogenic structures. Therefore the construction of the model of climate requires a consideration of numerous factors, whose role in its formation in many cases has been studied inadequately. So far, attempts to derive a numerical model of the Earth climatic system have not given results that could be used in GSM. Also, it is not safe to say that a realization of the GCP plans will clarify the problem. Without a parameterization of climate the questions concerning the problem of the greenhouse effect will remain unanswered.

There are two approaches to the synthesis of the climatic system. One approach is based on inclusion of biospheric components into the existing climate models or in those under development. The other approach consists of the development of a unit within the numerical model of the biosphere, which would simulate dependences of biospheric components on climate parameters. In the former case, problems of unstable solutions of the respective systems of differential equations appear, which make it difficult to predict global environmental changes. In the latter case there is the possibility of obtaining steady prognostic estimates of the environmental changes, but their reliability depends on the accuracy of parameterization of correlations between the elements of the climate and the biosphere. The second approach is preferable, because it permits one to combine the climate models, which can be described on the level of scenario, with the numerical model of the biosphere. A detailed analysis of the problems of climate modelling and assessment of the present state of the problems can be found in Marchuk and Kondratyev (1992), Kondratyev

(1999b), Kondratyev and Johannessen (1993). Some models of individual components of the nature–society system, which correspond to the second approach, are discussed here. Among them are models of the atmospheric general circulation, atmosphere–ocean interaction, sensitivity of climate parameters to boundary conditions on the Earth's surface, interaction of biogeochemical and climatic processes, and others (Lin *et al.*, 2000; Chen *et al.*, 2000).

The climatic system is a physico–chemico–biological system with unlimited degrees of freedom. Therefore any attempts to model this complex system meet with serious difficulties. This explains a diversity of parametric descriptions of some processes in this system. For the global model with a time step of digitization up to one year, the use of two versions can be acceptable. The first version consists of a combined use of correlations between particular processes of the formation of the climatic situation on a given territory together with the climate scenarios. The second version is based on the use of the global monitoring data, which are the basis for the formation of a data series about climatic parameters with their territorial and temporal attachment and are used to retrieve a complete pattern of their spatial distribution. One of the widespread correlation functions is the dependence of variations of average temperature ΔT_g of the atmosphere on the CO_2 content within it:

$$\Delta T_g = \begin{cases} 2.5[1 - \exp\{-0.82(\xi - 1)\}] & \xi \geq 1 \\ -5.25\xi^2 + 12.55\xi - 7.3 & \xi < 1 \end{cases} \tag{6.1}$$

where ξ is the ratio between the present content of CO_2 in the atmosphere $C_a(t)$ and its pre-industrial level $C_a(1850)$.

As seen from (6.1), T_g is an increasing function of the amount of atmospheric CO_2. A 20% increase of CO_2 in the atmosphere raises the temperature by 0.3°C. A doubling of atmospheric CO_2 causes an increase of T_g by 1.3°C. A detailed analysis of the function (6.1) and a comparison of the observed combined variations of ΔT_g and ξ show that the use of model (6.1) makes it possible to simplify the climate unit of the nature–society model. In particular, if $(\Delta T_g)_{2[CO_2]}$ is calculated by (6.1) with a doubled concentration of atmospheric CO_2, then to assess the present trends in changes of T_g the following formula can be used:

$$\Delta T_g = (\Delta T_g)_{2[CO_2]} \ln \xi / \ln 2 \tag{6.2}$$

where, from the assumed estimates, the pre-industrial value of $C_a(1850) = 270$ ppm. The formula (6.2) satisfactorily approximates the known data with an error of about 50%. It follows from (6.2) with $C_a(1980) = 338$ ppm that $\Delta T_g = 1.3$°K, whereas a real warming is estimated by many authors at 0.6°C.

Of course, the results of the recent discussions on the greenhouse effect in connection with the growth of the CO_2 partial pressure in the terrestrial atmosphere should be reflected in the GSM. The formula (6.1) takes into account the impact of CO_2. According to Mintzer (1987), there is the possibility to broaden the consideration of the temperature effect of other greenhouse gases:

$$\Delta T_\Sigma = \Delta T_{CO_2} + \Delta T_{N_2O} + \Delta T_{CH_4} + \Delta T_{O_3} + \Delta T_{CFC11} + \Delta T_{CFC12}$$

where

$$\Delta T_{CO_2} = -0.677 + 3.019 \ln[C_a(t)/C_a(t_o)]$$

$$\Delta T_{N_2O} = 0.057\{[N_2O(t)]^{1/2} - [N_2O(t_o)]^{1/2}\}$$

$$\Delta T_{CH_4} = 0.19\{[CH_4(t)]^{1/2} - [CH_4(t_o)]^{1/2}\}$$

$$\Delta T_{O_3} = 0.7[O_3(t) - O_3(t_o)]/15$$

$$\Delta T_{CFC11} = 0.14[CFC11(t) - CFC11(t_o)]$$

$$\Delta T_{CFC12} = 0.16[CFC12(t) - CFC12(t_o)]$$

The value of t_o is identified with the year 1980, when the GHG concentrations were known.

Among simple formulae of calculations of the latitudinal distribution of average temperature over the globe one can indicate the schemes proposed by Sergin (1974):

$$T(\varphi) = T_g + \gamma(\sin^2 \varphi_T - \sin^2 \varphi) \tag{6.3}$$

where φ is the latitude in radians, γ is the pole–equator temperature difference, and φ_T is the latitude at which $T(\varphi) = T_g$. The latitudinal temperature variations can be described as:

$$T_a(\varphi, t) = T_e - 2\varphi(T_e - T_N)/\pi \text{ for the northern hemisphere}$$

$$T_a(\varphi, t) = T_e - 2\varphi(T_e - T_S)/\pi \text{ for the southern hemisphere}$$

where

$$T_N = \begin{cases} T_{N,\min} + 2t(T_{N,\max} - T_{N,\min})/t_\Delta & t \in [0, 0, 5t_\Delta] \\ T_{N,\min} + 2(t_\Delta - t)(T_{N,\max} - T_{N,\min})/t_\Delta & t \in (0, 5t_\Delta, t_\Delta] \end{cases}$$

$$T_S = \begin{cases} T_{S,\max} + 2t(T_{S,\min} - T_{S,\max})t_\Delta & t \in [0, 0, 5t_\Delta] \\ T_{S,\max} + 2(t_\Delta - t)(T_{S,\min} - T_{S,\max})/t_\Delta & t \in (0, 5t_\Delta, t_\Delta] \end{cases}$$

$T_{N,\min}(T_{S,\min})$ and $T_{N,\max}(T_{S,\max})$ are minimum and maximum temperatures at the North (South) Pole, respectively (°C); t_Δ is the duration of the year in units of measurements (Δ); and T_e is the equatorial temperature of the atmosphere (°C). Many authors use the estimates $T_{N,\min} = -30°C$, $T_{N,\max} = 0°C$, $T_{S,\min} = -50°C$, $T_{S,\max} = -10°C$, and $T_e = 28°C$.

Of course, these zonally averaged temperatures scatter widely, which leads to considerable errors. To more accurately reflect the role of various factors in changes of the basic climatic parameter (temperature), it is necessary to calculate the contribution of each factor separately. This can be done with an assumed additivity of the role of feedbacks: $\Delta T_{a,final} = \Delta T_a + \Delta T_{a,feedback}$. In some models the authors introduce the feedback coefficient β: $\Delta T_{a,final} = \beta \cdot \Delta T_a$. The β parameter is expressed through the amplification index g: $\beta = 1/(1-g)$. The value of the g

Table 6.7. Time periods of establishing an equilibrium state for some sub-systems of the Earth climate system.

Domain of influence of the climate system	Period of restoration of the equilibrium state
Atmosphere	
free	11 days
boundary layer	24 hours
World Ocean	
mixed layer	7–8 years
deep ocean	300 years
sea ice	from 24 hours to 100 years
Continents	
lakes and rivers	11 days
soil–plant formations	11 days
snow cover and surface ice	24 hours
Mountain glaciers	300 years
Ice sheets	3,000 years
Earth's mantle	30 million years

index is equivalent to albedo α, which on a global scale is a function of T_a. This dependence can be roughly approximated by the formula:

$$\alpha(T_a) = \begin{cases} \alpha_{ice} & \text{for } T_a \leq T_{ice} \\ \alpha_{free} & \text{for } T_a \geq T_{free} \\ \alpha_{free} + b(T_{free} - T) & \text{for } T_{ice} < T_a < T_{free} \end{cases}$$

Here T_{ice} and T_{free} are average planetary temperatures at which the whole surface of the Earth is ice-covered or ice-free, respectively and b is the coefficient of transition between critical conditions of the Earth's albedo. Usually it is assumed that $T_{ice} \in [263, 283]°K$.

Simple and rough climate models can be specified through an account of the characteristic times of feedback functioning. Some estimates of the time of establishing the equilibrium states in the interaction of climate sub-systems are given in Table 6.7. The lag time of reactions within the nature–society system varies widely, and its consideration is necessary when assessing the consequences of changes within one or several climate sub-systems. In particular, temperatures in the Antarctic ice sheet (along with its areal extent) mean that to raise its temperature to 0°C, the average temperature of the World Ocean should be lowered by 2°C, that is, to obtain $T_o = 3.7°C$ instead of $T_o = 5.7°C$. With account of the data in Table 6.7, it would take hundreds of years to reach this. The observed rate of the anthropogenic climate warming does not dispose of such energy supplies.

The mechanism of the anthropogenic impact on the climate system is manifested through GHG emissions, albedo change due to reconstruction of the Earth's cover, interference into the water cycle, and atmospheric pollution. Aerosol particles with

Table 6.8. Estimates of the fluxes of particles with radius <20 mm either emitted to the atmosphere or formed *in situ*.

Type of particles	Amount of particles (106 t/yr)
Particles of natural origin (weathering, erosion, etc.)	100–500
Particles from forest fires and forestry waste burning	3–150
Sea salt	300
Volcanic dust	25–150
Particles from gas emissions	
Natural processes	
Sulphates from H_2S	130–200
Ammonium salts from HN_3	80–270
nitrates from NO_x	60–430
hydrocarbons from vegetative compounds	75–200
Anthropogenic processes	
Sulphates from SO_2	130–200
Nitrates from NO_x	30–35
Hydrocarbons	15–90
Amounts of particles emitted to the atmosphere	
natural	773–2200
anthropogenic	185–415
Total flux of particles into the atmosphere	*958–2615*

radius $10^{-7} \pm 10^{-2}$ cm are observed almost at all altitudes of the atmosphere. Particles of non-anthropogenic origin get into the atmosphere from land or ocean surface and result from *in situ* chemical reactions. Particles of anthropogenic origin result mainly from fuel burning. Table 6.8 gives an idea about the relationships between these fluxes of particles into the atmosphere.

The mechanism behind the effect of particles on atmospheric temperature is explained in this way: solar radiation coming to the Earth, mainly within the range 0.4–4 m, is partially reflected and absorbed by the particles. The global albedo of the surface–atmosphere system varies, too. Besides, the particles affect the processes of moisture condensation in the atmosphere, since clouds, rain, and snow form around such condensation nucleii. Take the heat balance equation of the "surface–atmosphere":

$$(1-\alpha)E_o^* + E_a - \sigma T_s^4 = 0 \qquad (6.4)$$

where T_s is the average efficient temperature of the system's emission, close to the temperature of the average energy level near the surface 400 mb; $E_o^* = 0.487\,\text{cal/cm}^2/$ min is the mean hemispherical intensity of the incoming solar radiation; α is albedo; $\sigma = 8.14 \cdot 10^{-11}\,\text{cal/cm}^2/\text{min}$ is the Stefan–Bolzmann constant; and E_a is total intensity of the anthropogenic sources of energy per unit surface.

Let albedo $\alpha = \alpha_o - \Delta\alpha$, where $\alpha_o = 0.35$ is the albedo in the present conditions and $\Delta\alpha$ is a small part of the albedo determined by the impact of anthropogenic aerosols. From equation (6.4) we obtain an expression for temperature:

$$T_s = [E_o^*(1-\alpha)/\sigma]^{1/4}[1 + \Delta\alpha(1-\alpha_o)^{-1} + (E_a/E_o^*)(1-\alpha_o)^{-1}]^{1/4} \tag{6.5}$$

Assuming $\Delta\alpha \ll 1$ and $E_a/E_o^* \ll 1$, expand the function of the right-hand part of the equation (6.5) into a Taylor series by degrees $\Delta\alpha$ and E_a/E_o^* and write the first terms of the series as:

$$T_s \approx [E_o^*(1-\alpha_o)/\sigma]^{1/4}\{1 + 0.25\Delta\alpha(1-\alpha_o)^{-1}[1 + E_a/E_o^*]\} \tag{6.6}$$

It follows from (6.6) that the temperature for not so strong anthropogenic impacts is the sum of the terms describing the bonds in the "surface–atmosphere" system, without taking account of anthropogenic factors, and the terms T_1 and T_2, which represent the contribution of heat and aerosol emissions, respectively:

$$T_1 = 0.25(1-\alpha_o)^{-1}[E_o^*(1-\alpha_o)/\sigma]^{1/4}\, E_a/E_o^* \approx 96.046\, E_a/E_o^*$$

$$T_2 = 0.25(1-\alpha_o)^{-1}[E_o^*(1-\alpha_o)/\sigma]^{1/4}\, \Delta\alpha \approx 96.046\Delta\alpha$$

Note that the contribution of T_1 in the present conditions is very small. Assuming that $E_a = 4 \cdot 10^{-5}$ cal/cm^2/min and, hence, $E_a/E_o^* = 8.21 \cdot 10^{-4}$, then $T_1 = 0.0079°$C. Thus the direct impact of the global energy on the atmospheric average temperature is now negligible. It follows from the expression for T_1 that to raise the atmospheric temperature by 0.5°C due to thermal emissions, the condition $E_a/E_o^* = 0.0052$ should be satisfied, and this requires an increase of anthropogenic heat fluxes into the environment of 63.4 times. This is equivalent to an energy release generated by an annual burning of $570 \cdot 10^9$ t of oil equivalent.

If we assume that the energy production is proportional to the size of population, then $t_1 = 96.046\, k_{TG}\, G\sigma_S/E_o$, where G is the population density (people/km^2); σ_S is the land area (km^2); and k_{TG} is the per capita amount of energy produced (cal/min).

If we neglect the impact of aerosols on the atmospheric thermal regime, then the direct radiation E, its change dE, and the change of atmospheric turbidity dB will be related as: $dE/E = -k_B dB$, where $k_B = 0.1154$ km^2/t is the proportion coefficient and B is the amount of anthropogenic aerosols (t/km^2). After integrating this equation, we obtain: $E = E_o^*(1-\alpha_o)\exp(-k_B B)$. On the other hand, according to the definition of albedo, $E = E_o^*(1-\alpha) = E_o^*(1-\alpha_o+\Delta\alpha)$. Equating these expressions for E, we obtain $\Delta\alpha = -(1-\alpha_o)[1-\exp(-k_B B)]$. Hence, a temperature change due to pollution of the atmosphere by anthropogenic aerosols is equal to:

$$T_2 = -0.25[E_o^*(1-\alpha_o)/\sigma]^{1/4}[1-\exp(-k_B B)] = -62.43[1-\exp(-k_B B)]$$

Since the average emission of anthropogenic aerosols is estimated by many authors at $300 \cdot 10^6$ t/yr, and the average period of the aerosols residence in the atmosphere is estimated at three weeks, $17.262 \cdot 10^6$ tonnes of particles, on average, reside in the atmosphere. In this case it follows from the formula for T_2 that the atmospheric temperature should decrease by 0.84°C/yr.

Some experts consider the factor of atmospheric turbidity B_T, instead of the index β defining it as a ratio of the coefficient of attenuation of solar energy in the real atmosphere α_r to the coefficient of attenuation in an ideal atmosphere α_I: $B_T = \alpha_r/\alpha_I = (\alpha_I + \alpha_W - \alpha_A)/\alpha_I$, where α_W and α_A are the coefficients of attenuation by water vapour and by aerosols, respectively. The following estimates are assumed in the GSM:

$$B_T = \begin{cases} 3 & \text{in mid-latitudes} \\ 3.5 & \text{in tropical latitudes} \\ 2 & \text{at reduced contents of dust and water vapour} \end{cases}$$

An experience of the global climate modelling testifies to the fact that the desire of many experts to accurately and completely take into account all possible feedbacks and compounds of the climate system leads to complicated mathematical problems, solutions of which require a huge amount of data, and in most cases turn out to be unstable. Therefore the use of these complicated models as a unit of the global model of the nature–society system leads to a negative result (i.e., to the impossibility of constructing an efficient model). The most encouraging approach is to combine the climate models with the data of global monitoring. The scheme of this combination is very simple. The existing surface and satellite systems used for monitoring the climate-forming processes cover some part of the cells $\{\Omega_{ij}\}$ of the Earth's surface. Over these cells, measurements are made of temperature, cloudiness, water vapour content, aerosols and gases, albedo, and various parameters of energy fluxes. Use of simple climate models and of methods of spatial and temporal interpolations enables us to reconstruct, based on these measurements, a complete pattern of distribution of climatic parameters over the whole territory Ω (Boden *et al.*, 1994; Botta *et al.*, 2002; Clerbaux *et al.*, 2001; Dufresne *et al.*, 2001; Fan *et al.*, 1999; Grabowski and Moncrieff, 2002; Knapp *et al.*, 2002; Lambin, 2002; Linderholm *et al.*, 2002; Lockwood, 2001; Mabuchi *et al.*, 2000; McKenzic *et al.*, 1998; Mellilo *et al.*, 2002).

6.5.5 Some estimates of global eco-dynamics

The most important features of the present-day global eco-dynamics are changes of the environment and human society due to intensified development of the economy. These changes have accelerated compared to those even 50 years ago. In this regard the role of interdisciplinary studies of the nature–society system with the use of accumulated knowledge in the fields of mathematics, ecology, sociology, medicine, chemistry, geophysics, etc., has increased. Global numerical experiments with the use of the latest computer technologies have become an important tool for solving, for instance, the problems of land use and thus, numerical experiments are considered to be able to simulate real natural processes. Of course, the problems appear to assess an adequacy of the results of simulation experiments. These problems are being solved by eco-informatics (Kondratyev *et al.*, 2002). The end of the 20th century and the beginning of the 21st century are characterized by a new thinking based on accomplishment of numerical experiments, since it is clear that field experiments to change such natural systems as oceans, continental ecosystems, and the biosphere are impossible because such experiments are very dangerous and should be excluded from the practice of human–nature interaction.

The GSM described above makes it possible to accomplish numerous simulation experiments in the multidimensional phase space of the nature–society system and to predict its development for the near future. Longer term predictions are hardly ever possible at the present level of knowledge and global dataset availability. The resulting uncertainties have been analysed earlier and are closely connected in particular with numerical climate modelling. Unfortunately, there is no universal global model, so far, which might reliably simulate the dynamics of the nature–society system, reflecting interactions between the atmosphere, oceans, land, and human society. Therefore to use the GSM, it is necessary to make some a priori assumptions:

- during the period till the year 2100 the anthropogenic emissions of CO_2 are simulated with the use of the model developed by Demirchian *et al.* (2002), and emissions of other GHGs till 2050 are preserved at the level of the year 2000, decreasing thereafter by 25%;
- by the year 2050, the extraction and consumption of non-renewable fossil fuels increase by 0.5% annually in industrial countries and by 0.7% in developing countries, after which they stabilize;
- the per capita gross production in industrial and developing countries increase at constant rates of 2.5% and 1.5%, respectively;
- the enhanced level of public medical services in industrial and developing countries reduce the mortality due to the environmental changes by 90% and 50%, respectively, and a remedy for AIDS will be found by 2020;
- the air temperature regime is formed according to the model by Mintzer (1987);
- investments to agriculture depend on per capita food provision (Krapivin *et al.*, 1982);
- financing the control of environmental pollution in all countries increases by 0.2% annually till the year 2050 and then stabilizes;
- processes of deforestation and forest reconstruction starting from 2010 are balanced so that the total forested area as of 2000 is preserved, and the area of tropical forests decreases by no more than 5%;
- the oil pollution of the World Ocean in the 21st century follows the rate of the last decade of the 20th century; and
- the agricultural productivity increases steadily, to 80% by 2100.

With these assumptions, the nature–society system's dynamics can be characterized by the following features:

- The eco-dynamics of the nature–society system's functioning is characterized by the fact that the derivatives of the function describing the state of its components are never equal to zero in time periods exceeding the model's time step.

- The size of population by the year 2100 will reach $12.7 \cdot 10^9$, with relationships between the age groups characterized by the following averages: 0–15 years – 32%, 16–65 years – 53%, and older than 65 years – 15%. Disabled people will constitute 17%.
- Provision of the population with protein food increases by 6.8% by 2030 (the contribution of the World Ocean constitutes 12.1%) and then decreases by 3.2% compared to the year 2000 (the share of the World Ocean increases to 22.1%).
- The CO_2 concentration in the atmosphere will reach 486 ppm with the respective increase of the average planetary temperature by 0.87°C.
- The dynamics of the spatial distribution of the atmospheric CO_2 sinks are characterized by the increase of the role of the World Ocean and land ecosystems, whose shares in the mid-21st century will constitute 31% and 19%, respectively, and then by the end of the century the share of the World Ocean will decrease to 26.7%, with the role of land ecosystems increasing to 24.4%.
- A change of 5% in the area of tropical forests to urban ecosystems will lead to a 1.2% reduction of the total CO_2 sink.
- A complete substitution of wet tropical forests for grass ecosystems by 2050 results in the global climate change due to violation of the regional moisture exchange on the atmosphere–land border, increase of the outgoing longwave radiation flux from the deforested territory by 4.3 W/m^2, increase of the albedo of this territory by 5.7%, decrease of the absorbed shortwave radiation by 4.1%, increase of soil temperature by 1.3°C, and reduction of precipitation and evaporation by 0.7 and 0.9 mm/day, respectively.

For educational purposes, it is interesting to consider the hypothetical versions of the human–nature interaction. We consider only some situations, in accordance with hypotheses of the future dynamics of the most important anthropogenic factors. One has to point out that in contrast to many other models, the GSM makes it possible to calculate the size of a population as a function of the environmental parameters. Calculations show that mechanisms of formation of demographic processes substantially depend on the strategies of nature use. For instance, a 50% increase of the level of food provision, which is possible due to adoption of effective technologies in agriculture and development of aquasystems, leads to a 14% decrease in population density. On the contrary, a 10% decrease of the rate of expenditure of mineral resources causes a 3% increase in population density. The nature–society system is highly sensitive to possible changes in the global structure of soil–plant formations. For instance, if by the year 2050 the forest areas uniformly decrease by 10% compared to 2000, then the concentration of atmospheric CO_2 in 2100 will reach 611 ppm, that is, it will nearly double. However, with a 15% increase of the forest area by 2050, in 2100 the CO_2 partial pressure will constitute only 475 ppm.

The results of numerical modelling of the response of the nature–society system to possible anthropogenic changes of its parameters show that the proposed method enables us to assess the dynamics of various components of this system depending on hypothetically possible rates of their changes. The use of this method to retrieve the integral dynamics of the nature–society system due to changes in its individual parameters for the time period 1970–2000 based on the data published by Kondratyev and Losev (2002), Watson *et al.* (2000), and Houghton *et al.* (2001) has shown that the error of prediction for 10 and 20 years does not exceed 15% and 25%, respectively. Hence, the GSM-technology can be recommended as one of the constructive approaches to the solution of the problems of global eco-dynamics formulated earlier. Simulation experiments with the use of the GSM with available reliable data on the pre-history of the nature–society system will make it possible to reveal conditions, which limit the development of civilization, and to establish mechanisms to provide it with sustainable development. For this purpose, it is necessary to concentrate efforts of international and national scientific programmes on the synthesis of the GSM datasets. Mr Kofi Annan, UN Secretary-General, in his speech before the representatives of mass-media in Johannesburg on the last day of World Summit on Sustainable Development (WSSD) (4 September 2002) mentioned that the governments agreed to make joint efforts to preserve biodiversity and to solve the global problems of water supply, energetics, health service, and agriculture. Speaking about possible ways to solve the global problems of the present civilization, Mr Kofi Annan mentioned that there is a way that reduces poverty while preserving the environment. This is GSM-technology. Its realization requires the concentration of accumulated knowledge on the development of the GSM by improving its units and, first of all, revealing and formalizing the factors and dynamic strategies of social evolution. It is necessary to emphasize, of course, that what has been done for the time being is just a first step towards the development of adequate global simulation modelling.

References

ACECHEM – Atmospheric Composition Explorer for CHEMistry and climate interaction. (2001). European Space Agency, Noordwijk, The Netherlands, 103 pp. (ESA SP-1257 (4)).

Adamenko V. N. and Kondratyev K. Ya. (1999). Global climate changes and their empirical diagnostics. In: Yu. A. Izrael, Yu. P. Borisenkov, and K. Ya. Kondratyev (eds), *Anthropogenic Influence on the Nature of the North and Its Ecological Consequences*. The Kola Scientific Centre of the Russian Academy of Science, Apatity, Russia, pp. 17–34 [in Russian].

A Guide to World Resources 2000–2001. People and Ecosystems: The Frying Web of Life. (2000). World Resources Institute, Washington, DC, 389 pp.

Albrecht J. (2002). Instruments for Climate Policy. *Limited versus Unlimited Flexibility*. Edward Elgar, Cheltenham, UK, 288 pp.

Alcano J., Leemans R., and Kreileman E. (eds) (2001). *Global Change Scenarios of the 21st Century*. Elsevier, Amsterdam, 232 pp.

Alekseev V. V., Zaitsev S. I., and Kokorin A. O. (1990). The growth of the rate of the atmosphere–ocean gas exchange in the case of rough sea. *Bulletin of the USSR Academy of Science, FAO*, **26**(3), 322–325 [in Russian].

Alekseev V. V., Kryshev I. I., and Sazykina T. G. (1992). *Physical and Numerical Modeling of Ecosystems*. Gidrometeoizdat, St Petersburg, 367 pp. [in Russian].

Alexandrov G. and Oikawa T. (2002). TsuBiMo: A biosphere model of the CO_2 – fertilization effect. *Climate Research*, **19**, 265–270.

Allen D. R., Bevilacqua R. M., Nedoluha G. E., Randall C. E., and Manney G. L. (2003). Unusual stratospheric transport and mixing during the 2002 Antarctic winter. *Geophys. Res. Lett.*, in press.

Allen M., Raper S., and Mitchell J. (2001). Uncertainty in the IPCC's Third Assessment Report. *Science*, **293**(5529), 430.

Allison C. E. and Francey R. J. (1998). Maintaining a long term, high precision record of the $\delta^{13}C$ of carbon dioxide in the atmosphere. *Isot. Technical Study of Environmental Change: Proceedings of the International Symposium Isot. Technical Study Past and Current Environmental Changes Hydrosphere and Atmosphere, Vienna, 14–18 April 1997*, pp. 9–18.

Alverson K. (2000). Long-term biogeophysical controls on the carbon cycle and their relevance to human concerns. *IHDP Update*, **3**, 1–4.

Alverson K., Bradley R., and Pedersen T. (eds) (2002). *Paleoclimate, Global Change, and the Future*. Springer-Verlag, Heidelberg, 400 pp.

Anderson L. G., Olsson K., and Chierici M. (1998). A carbon budget for the Arctic Ocean. *Global Biogeochemical Cycles*, **12**(3), 455–466.

Andronova N. G. and Schlesinger M. E. (2001). Objective estimation of the probability density function for climate sensitivity. *Journal of Geophysics Research*, **106**(D19), 22605–22611.

Aota M., Shirasawa K., Krapivin V. F., and Mkrtchyan F. A. (1993). A project of the Okhotsk Sea GIMS. *Proceedings of the 8th International Symposium on Okhotsk Sea and Sea Ice and International Space Year/Polar Ice Extent Workshop, 1–5 February 1993, Mombetsu, Japan*, pp. 498–500.

Arneth A., Kelliher F. M., Gower J. T., Scott N. A., Byers J. N., and McSevery T. M. (1998). Environmental variables regulating carbon dioxide efflux following clear-cutting of a *Pinus radiata* D. Don plantation. *Journal of Geophysics Research*, **103**(D5), 5695–5705.

Arskiy Yu. M., Zakharov Yu. F., and Kalutskov V. A. (1992). *Ecoinformatics*. Gidrometeoizdat, St Petersburg, 520 pp. [in Russian].

Ashby W. R. (1956). *An Introduction to Cybernetics*. Chapman & Hall, London, 325 pp.

Asner G. P., Bateson C. A., Privette J. L., Saleous N., and Wessman C. A. (1998). Estimating vegetation structural effects on carbon uptake using satellite data fusion and inverse modeling. *Journal of Geophysics Research*, **103**(D22), 28839–28854.

Aurela M., Laurila T., and Tuovinen J.-P. (2001). Seasonal CO_2 balances of a subarctic mire. *Journal of Geophysics Research*, **106**(D2), 1623–1638.

Bacastow R. (1981). Numerical evaluation of the evasion factor. *Carbon Cycle Modelling, SCOPE-16*. John Wiley & Sons, New York, pp. 95–101

Bach V., Krein A., Berger A., and Longhetto A. (eds) (1987). *Carbon Dioxide in the Atmosphere*. Mir, Moscow, 530 pp. [in Russian].

Baldocchi D., Falge E., Gu L., Olson R., Hollinger D., Running S. *et al.* (2001). FLUXNET: A new tool to study the temporal and special variability of ecosystem-scale carbon dioxide, water vapor, and energy flux densities. *Bulletin American Meteorological Society*, **82**(11), 2415–2434.

Baldwin M., Hirooka T., O'Neill A., and Yoden S. (2003). Major stratospheric warming in the southern hemisphere in 2002: Dynamical aspects of the ozone hole split. *SPARC Newsletter*, **20**, 24–26.

Baliunas S. (2002). The Kyoto Protocol and global warming. *Imprimis*, **31**(3), 1–7.

Ball J. T. (1988). An analysis of stomatal conductance. Ph.D. Thesis, Stanford University, 89 pp.

Barford C. L., Wofsy S. C., Goulden M. L., Munger J. W., Pyle E. H., Urbanski S. P. *et al.* (2001). Factors controlling long- and short-term sequestration of atmospheric CO_2 in a mid-latitude forest. *Science*, **294**, 1688–1691.

Barker P. A., Street-Perrott F. A., Leng M. J., Greenwood P. B., Swain D. L., Perrott R. A. *et al.* (2001). 14,000-year oxygen isotope record from diatom silica in two alpine lakes on Mt. Kenya. *Science*, **292**, 2307–2310.

Bates N. R. and Merlivat L. (2001). The influence of short-term wind variability on air–sea CO_2 exchange. *Geophysics Research Letters*, **28**(17), 3281–3284.

Bates N. R., Knap A. H., and Michaels A. F. (1998). Contribution of hurricanes to local and global estimates of air–sea exchange of CO_2. *Nature*, **395**(6697), 58–61.

Batijes N. H. (1999). Management options for reducing CO_2-concentrations in the atmosphere by increasing carbon sequestration in soil. *Dutch National Research Programme on Global Air Pollution and Climate Change*. Amsterdam University Press, Amsterdam, 114 pp. (Report No. 410 200 031).

Battle M., Bender M. L., Tans P. P., White J. W. C., Ellis J. T., Conway T. *et al.* (2000). Global carbon sinks and their variability inferred from atmospheric O_2 and $\delta^{13}C$. *Science*, **287**(5462), 2467–2470.

Bazhin N. M. (2000). Methane emission from a residual layer. *Proceedings of the Second International Methane Mitigation Conference, 18–23 June 2000, Novosibirsk*. Novosibirsk State University Press, pp. 231–236.

Bazilevich N. I. and Rodin L. E. (1967). The map-schemes of productivity and of the biological cycle of the most significant types of land vegetation. *Bulletin of the All-Union Geographical Society*, **99**(3), 190–194 [in Russian].

Bazzaz F. A. (1986). Global CO_2 levels and the response of plants at the population and community levels. In: C. Rozenzweig and R. Dickinson (eds), *Climate–Vegetation Interactions*. UCAR, Boulder, CO, pp. 31–35.

Bazzaz F. A. (1998). Tropical forests in a future climate: Changes in biological diversity and impact on the global carbon cycle. *Climate Change*, **39**(2–3), 317–336.

Beck J. W., Richards D. A., Edwards R. L., Silverman B. W., Smart P. L., Donahue D. J. *et al.* (2001). Extremely large variations of atmospheric ^{14}C concentration during the Last Glacial Period. *Science*, **292**(5526), 2453–2458.

Beerling D. J. and Woodward F. I. (1996). Stomatal density responses to global environmental change. *Advances in Bioclimatology, Berlin*, **4**, 197–221.

Beese F. O., Hempel G., Klemmer P., Kruse-Graumann L., Labitzke K., Muhle H. *et al.* (1995). *Scenario for the Derivation of Global CO_2 Reduction Targets and Implementation Strategies*.WBGU, Dortmund, Germany, 39 pp.

Bekoriukov V. I. and Fedorov V. V. (1987). Empirical model of total ozone content above the southern hemisphere. *Meteorology and Hydrology*, **3**, 47–53 [in Russian].

Bender M. L. (2000). Tracer from the sky. *Science*, **288**(5473), 1977–1978.

Bergamaschi P., Hein R., Brenninkmeijer C. A. M., and Crutzen P. J. (2000a). Inverse modeling of the global CO cycle. 2: Inversion of $^{13}C/^{12}C$ and $^{18}O/^{16}O$ isotope ratios. *Journal of Geophysics Research*, **105**(D2), 1929–1946.

Bergamaschi P., Hein R., Heimann M., and Crutzen P. J. (2000b). Inverse modeling of the global CO cycle. 1: Inversion of CO mixing ratios. *Journal of Geophysics Research*, **105**(D2), 1909–1928.

Berner U. and Hollerbach A. (2001). Klimasystem Erde überschtzen wir das Kohlenoxid? *Stahl und Eisen*, **121**(10), 35–40.

Betts R. A. (2000). Offset of the potential carbon sink from boreal forestation by decreases in surface albedo. *Nature*, **408**, 187–190.

Beven K. (2002). Towards a coherent philosophy for modelling the environment. *Proceedings Royal Society London A*, **458**(2026), 2465–2484.

Bezborodov G. A., Bezborodov A. G., and Bezborodov Yu. G. (2003). Agroecology: composition of hydrocarbonate gases in irrigated soils of serozem belt and their control. *Engineering Ecology*, **1**, 50–62 [in Russian].

Bichele I., Moldau X., and Ross Yu. (1980). *A Sub-model of the Distribution of Assimilates and of the Growth of Plants in Conditions of Water Deficit*. The Tartu Astrophysical Observatory, Tartu, 22 pp. (Preprint No. A-5) [in Russian].

Bilek R. S., Tyler S. C., Kurshara M., and Yagi K. (2001). Investigation of cattle methane production and emission over a 24-hour period using measurements of $\delta^{13}C$ and δD of emitted CH_4 and rumen water. *Journal of Geophysics Research*, **106**(D14), 15405–15413.

Biutner E. K. (1986). *Planetary Gas Exchange*. Gidrometeoizdat, Leningrad, 240 pp. [in Russian].

Bjorkstrom A. A. (1979). A model of CO_2 interaction between atmosphere, ocean and land biota. *Global Carbon Cycle, SCOPE-13*. John Wiley & Sons, New York, pp. 403–458.

Boden T. A., Kaiser D. P., Speranski R. J., and Stoss F. W. (1994). *TRENDs '93: A Compendium of Data on Global Change*. Carbon Dioxide Informational Analysis Center, Oak Ridge National Laboratory, Oak Ridge, TN, 176 pp.

Bodenbender J., Wassmann R., Papen H., and Rennenberg H. (1999). Temporal and spatial variation of sulfur-gas-transfer between coastal marine sediments and the atmosphere. *Atmospheric Environment*, **33**(21), 3487–3502.

Bogatyrev B. G. (1988). Modelling the transition processes on the border of vegetation zones in conditions of anthropogenic climate changes. Ph.D. thesis, Moscow State University, Moscow, 114 pp. [in Russian].

Bohmer-Christiansen S. (2000). Who determines the policy concerning climate changes and how is it done? *Bulletin of the Russian Geographical Society*, **132**(3), 6–22 [in Russian].

Bolin B. and Sukumar R. (2000). Global perspective. In: B. Bolin, N. H. Ravindranath, D. J. Verardo, and O. J. Dokken (eds), *Land Use, Land-Use Change, and Forestry*. Cambridge University Press, Cambridge, UK, pp. 23–51.

Borisenkov E. P. and Pichugin Yu. A. (2001). Possible negative scenarios of the biospheric dynamics as a result of anthropogenic activity. *Proceedings of the Russian Academy of Science*, **378**(6), 812–814 [in Russian].

Botta A., Ramankutty N., and Foley J. A. (2002). Long-term variations of climate and carbon fluxes over the Amazon basin. *Geophysics Research Letters*, **29**(9), 33/1–33/4.

Bounoua L., Defries R., Collatz G. J., Sellers P., and Khan H. (2002). Effects of land cover conversion on surface climate. *Climate Change*, **52**(1–2), 29–64.

Bousquet P., Peylin P., Ciais P., Le Quéré C., Friedlingstein P., and Tans P. P. (2000). Regional changes in carbon dioxide fluxes of land and oceans since 1980. *Science*, **290**(5495), 1342–1346.

Boutton T. W., Nordt L. C., and Kuehn D. D. (1998). Late Quaternary vegetation and climate change in the North American Great Plains. Evidence from $\delta^{13}C$ of paleosol organic carbon. *Isot. Technical Study of Environmental Change: Proceedings International Symposium Isot. Technical Study of Past and Current Environmental Changes in the Hydrosphere and Atmosphere, Vienna, 14–18 April 1997*, pp. 653–662.

Bowling D. R., Delany A. C., Turnipseed A. A., Raldocchi D. D., and Monson R. K. (1999). Modification of the relaxed eddy accumulation technique to maximize measured scalar mixing ratio differences in updrafts and downdrafts. *Journal of Geophysics Research*, **104**(D8), 9121–9133.

Boysen M. (ed.) (2000). *Biennial Report 1998 & 1999*. Potsdam Institute for Climate Impact Research, Potsdam, 130 pp.

Bras R. L. (1990). *Hydrology*. Addison-Wesley, New York, 643 pp.

Braswell B. H., Schimel D. S., Privette J. L., Moore B., Emery W. J., Sulzman E. W. *et al.* (1996). Extracting ecological and biophysical information from AVHRR optical data: An integrated algorithm based on inverse modeling. *Journal of Geophysics Research*, **101**(D18), 23335–23348.

Braswell B. H., Schimel D. S., Linder E., and Moore B. (1997). The response of global terrestrial ecosystems to interannual temperature variability. *Science*, **278**, 870–872.

Bratus A., Mescherin A., and Novozhilov A. (2000). Mathematical models of interaction between pollutant and environment. *Proceedings of the Conference on Control of Oscillations and Chaos, St Petersburg, July 2000*. St Petersburg University Press, Vol. 3, pp. 569–572.

Bratus A. S., Meshcherin A. S., and Novozhilov A. S. (2001). Numerical models of the pollution–environment interaction. *Herald of Moscow State University Series 15 – Computational Mathematics and Cybernetics*, **1**, 23–28 [in Russian].

Braunlich M., Aballain O., Marik T., Jockel P., Brenninkmeijer C. A. M., Chappellaz J. *et al.* (2001). Changes in the global atmospheric methane budget over the last decades inferred from ^{13}C and D isotopic analysis of Antarctic firn air. *Journal of Geophysics Research*, **106**(D17), 20465–20482.

Brenninkmeijer C. A. M. and Röckmann T. (1998). Using isotope analysis to improve atmospheric CO budget calculations. *Isot. Technical Study of Environmental Change: Proceedings International Symposium Isot. Technical Study of Past and Current Environmental Changes in the Hydrosphere and Atmosphere, Vienna, 14–18 April 1997*, pp. 69–77.

Brenninkmeijer C. A. M. and Röckmann T. (1999). Anomalous (or not strictly mass dependent) isotope variations observed in important atmospheric trace gases. *IGACtivities Newsletter*, **16**, 13–18.

Brombach T., Hunziker J. C., Chiodini G., Cardellini C., and Marini L. (2001). Soil diffuse degassing and thermal energy fluxes from the southern Lakke plain, Nisyros (Greece). *Geophysics Research Letters*, **28**(1), 69–72.

Brown L. R., Flavin C., and French H. (eds) (2001). *State of the World 2001*. Earthscan, London, 275 pp.

Bui T. L. and Nguyen M. N. (1997). Simulation model of pollutant transport and diffusion in the mathematical problem of setting up monitoring system for air pollution. *Meteorology and Hydrology*, **442**(10), 38–48.

Burroughs W. J. (2001). *Climate Change: A Multidisciplinary Approach*. Cambridge University Press, Cambridge, UK, 298 pp.

Byakola T. (2000). Technological options and policy measures for methane mitigation in Uganda – possibilities and limitations. *Proceedings of the 2nd International Methane Mitigation Conference, 18–23 June 2000, Novosibirsk*. Novosibirsk State University Press, pp. 95–100.

Caldeira K. and Duffy P. B. (2000). The role of the Southern Ocean in uptake and storage of anthropogenic carbon dioxide. *Science*, **287**, 620–622.

Cannel M. G. R., Milne R., Hargreaves K. J., Brown T. A. W., Cruickshank M. M., Bradley R. I. *et al.* (1999). National inventories of terrestrial carbon sources and sinks: The UK Experience. *Climate Change*, **42**(2), 505–530.

Carlslaw K. S., Harrison R. G., and Kirkby J. (2002). Cosmic rays, clouds, and climate. *Science*, **298**(5599), 1732–1737.

Carvalho Jr J. A., Costa F. S., Gurgel Veras C. A., Sandberg D. V., Alvarado E. C., Gielow R. *et al.* (2001). Biomass fire consumption and carbon release rates of rainforest-clearing experiments conducted in northern Mato Grosso, Brazil. *Journal of Geophysics Research*, **106**(16), 17877–17887.

Caspersen J. P., Pacala S. W., Jenkins J. C., Hurtt G. C., Moorcroft P. R., and Birdsey R. A. (2000) Contributions of land-use history to carbon accumulation in U. S. forests. *Science*, **290**(5494), 1148–1151.

Chagas J. C. S., Newnham D. A., Smith K. M., and Shine K. P. (2002). Impact of new measurements of oxygen collision-induced absorption on estimates of short-wave atmospheric absorption. *Quarterly Journal of the Royal Meteorological Society*, **128**(585), Part A, 2377–2396.

Chahine M. T. (1992). The hydrological cycle and its influence on climate. *Nature (UK)*, **359**(6394), 373–380.

Charette M. A., Moran S. B., Pike S. M., and Smith J. N. (2001). Investigating the carbon cycle in the Gulf of Main using the natural tracer thorium 234. *Journal of Geophysics Research*, **106**(C6), 11553–11580.

Charlton A. J., O'Neill A, Lahoz W. A., Berrisford P., van Velthoven P., Eskes H., and Kelder, H. (2003) The splitting of the stratospheric polar vortex in the southern hemisphere in September 2002: Dynamical evolution and impacts. *J. Atmos. Sci.*, submitted.

Chase T. N., Pielke Sr R. A., Kittel T. G. F., Nemani R. R., and Running S. W. (2000). Simulated impacts of historical land cover changes on global climate in northern winter. *Climate Dynamics*, **16**, 93–105.

Chate D. M. and Kamra A. K. (1997). Collection efficiencies of large water drops collecting aerosol particles of various densities. *Atmospheric Environment*, **31**(11), 1631–1635.

Chen J. M., Leblanc S. G., Cihlar J., Desjardins R. L., and MacPherson J. I. (1999). Extending aircraft- and tower-based CO_2 flux measurements to a boreal region using a Landsat thematic mapper land cover map. *Journal of Geophysics Research*, **104**(D14), 16859–16878.

Chen W., Chen J., and Cihlar J. (2000). An integrated terrestrial ecosystem carbon-budget model based on changes in disturbance, climate, and atmospheric chemistry. *Ecological Modelling*, **135**(1), 55–79.

Chisholm S. W., Falkowski P. G., and Cullen J. J. (2001). Discrediting ocean fertilization. *Science*, **294**(5541), 309–311.

Chock D. P. and Winkler S. L. (2000). A trajectory-grid approach for solving the condensation and evaporation equations of aerosols. *Atmospheric Environment*, **34**(18), 2957–2973.

Choudhury B. and Kukla G. (1979). Impact of CO_2 on cooling of snow and water surfaces. *Nature*, **280**(5724), 668–671.

Christy J. R., Spencer R. W., and Lobl E. S. (1998). Analysis of the merging procedure for the MSU daily temperature time series. *J. Climate*, **11**, 2016–2041.

Ciais P. (1999a). Isotopomers of CO_2 and their use in understanding the global carbon cycle. *IGACtivities Newsletter*, **16**, 18–19.

Ciais P. (1999b). Restless carbon pools. *Nature*, **398**(6723), 111–112.

Ciais P. (2002). Taking the pulse of the carbon cycle. *IHDP Update*, **2**, 11–12.

Cihlar J., Denning S., Ahern F., Arino O., Belward A., Bretherton F. *et al.* (2002). Initiative to quantify terrestrial carbon sources and sinks. *Earth Observations from Space Transactions*, **83**(1), 1, 6–7.

Clerbaux C., Hadji-Lazaro J., Hauglustaine D., Megie G., Khattatov B., and Lamarque J.-F. (2001). Assimilation of carbon monoxide measured from satellite in a three-dimensional chemistry-transport model. *Journal of Geophysics Research*, **106**(14), 15385–15394.

Coconea G., Pintican L., and Sarbu S. (2000). Methane gas emissions from the Romanian natural gas transport system. *Proceedings of the Second International Methane Mitigation Conference, 18–23 June 2000, Novosibirsk*. Novosibirsk State University Press, pp. 297–302.

Collatz G. J., Berry J. A., Farquhar J. A., and Pierce J. (1990). The relationship between the rubisco reaction mechanism and models of leaf photosynthesis. *Plant Cell Environment*, **13**, 219–225.

Collatz G. J., Ball G. J., Grivet J. T., and Berry J. A. (1991). Physiological and environmental regulation of stomatal conductance, photosynthesis and transpiration: A model that includes a laminar boundary layer. *Agriculture and Forestry Meteorology*, **54**, 107–136.

Collatz G. J., Ribas-Carbo M., and Berry J. A. (1992). Coupled photosynthesis stomatal conductance model for leaves of C_4 plants. *Australian Journal of Plant Physiology*, **19**, 519–538.

Connors V. S., Gormsen B. B., Nolf S., and Reichle Jr H. G. (1999). Spaceborne observations of the global distribution of carbon monoxide in the middle troposphere during April and October 1994. *Journal of Geophysics Research*, **104**(D17), 21455–21470.

Cox P. M., Betts R. A., Jones C. D., Spall S. A., and Totterdell I. J. (2000). Acceleration of global warming due to carbon-cycle feedbacks in a coupled climate model. *Nature*, **408**, 184–187.

Cox P. M., Friedlingstein P., and Rayner P. (2002). Modelling climate – carbon cycle feedbacks: A cross disciplinary collaboration priority. *Global Change Newsletter*, **49**, 12–14.

Craig S. G., Holmén K. J., Bonan G. B., and Rasch P. J. (1998). Atmospheric CO_2 simulated by the National Center for Atmospheric Research Community Climate Model. 1: Mean fields and seasonal cycles. *Journal of Geophysics Research*, **103**(D11), 13213–13235.

Cramer W., Bondeau A., Woodward F. I., Prentice I. C., Betts R. A., Brovkin V. *et al.* (2001). Global response of terrestrial ecosystem structure and function to CO_2 and climate change: Results from six dynamic global vegetation models. *Global Change Biology*, **7**, 357–373.

Crutzen P. and Ramanathan V. (2000). The ascent of atmospheric sciences. *Science*, **290**(5490), 299–304.

Danilov-Danilyan V. I. and Losev K. S. (2000). *Ecological Challenge and Sustainable Development*. Progress-Tradition, Moscow, 416 pp. [in Russian].

De Lucia E. H., Hamilton J. G., Naidu S. L., Thomas R. B., Andrews J. A., Finzi A. *et al.* (1999). Net primary production of a forest ecosystem with experimental CO_2 enrichment. *Science*, **284**(5417), 1177–1179.

Dementyeva T. V. (2000). Emission of gases from peat-bog ecosystems. *Proceedings of the 2nd International Methane Mitigation Conference, 18–23 June 2000, Novosibirsk*. Nauka, Novosibirsk, pp. 223–226.

Demirchian K. S. and Kondratyev K. Ya. (1998). Development of energetics and the environment. *Bulletin of Russian Academy of Science, Energetics*, **6**, 3–27 [in Russian].

Demirchian K. S. and Kondratyev K. Ya. (1999). Scientific validity of predicted impacts of energetics on climate. *Bulletin of Russian Academy of Science, Energetics*, **6**, 3–46 [in Russian].

Demirchian K. S., Demirchian K. K., Danilevich Ya. B., and Kondratyev K. Ya. (2002). Global climate warming, energetics and geopolicy. *Annals of Russian Academy of Science, Energetics*, **3**, 221–235 [in Russian].

Dimitroulopoulou C. and Marsh A. R. W. (1997). Modelling studies of NO_3 night-time chemistry and its effects on subsequent ozone formation. *Atmospheric Environment*, **31**(18), 3041–3057.

Dolman H., Nabuurs G. J., Kuikman P., Kruijt B., and Brinkman S. (2001). *Terrestrial Carbon Sinks and the Kyoto Protocol: The Scientific Issues*. Amsterdam University Press, Dutch National Research Programme on Global Air Pollution and Climate Change, Global Change Report No. 410 200 068, 31 pp.

Doskey P. V. and Gaffney J. S. (1992). Non-methane hydrocarbons in the Arctic atmosphere at Barrow, Alaska. *Geophysics Research Letters*, **19**(4), 381–384.

Dufresne J. L., Fredlingstein P., Fairhead L., Le Treut H., Ciais P., and Monfray P. (2001). Feedback processes between carbon cycle and climate in the IPSLM coupled model. *Abstracts of the 8th Science Assembly of IAMAS, Innsbruck, 10–18 July 2001.*

Eastman J. L., Coughenour M. B., and Pielke Sr R. A. (2001). The regional effects of CO_2 and landscape change using a coupled plant and meteorological model. *Global Change Biology*, **7**, 797–815.

Egan W. G., Hogan A. W., and Zhu H. (1991). Physical variation of water vapor, and the relation with carbon dioxide. *Geophysics Research Letters*, **18**(12), 2245–2248.

Eilrich B., Burns S. J., and Steinmann P. (2000). Origin and circulation of CH_4 and CO_2 in peatlands. *Proceedings 2nd International Methane Mitigation Conference, 18–23 June 2000, Novosibirsk*. Novosibirsk University Press, pp. 227–230.

Elderfield H. (2002) Carbonate mysteries. *Science*, **296**(5573), 1618–1621.

Eliasson B., Riemer P., and Wokaun A. (1999). *Greenhouse Gas Control Technologies*. Elsevier, Hardenberg, The Netherlands, 1095 pp.

Elliott S., Lackner K. S., Ziock H. J., Dubley M. K., Hanson H. P., Barr S. *et al.* (2001). Compensation of atmospheric CO_2 buildup through engineered chemical sinkage. *Geophysics Research Letters*, **28**(7), 1235–1238.

Ellsaesser H. W. (2002). The current status of global warming. *Energy and Environment*, **13**(1), 125–129.

Engelen R. J., Denning A. S., Gurney K. R., and Stephens G. L. (2001). Global observations of the carbon budget. 1: Expected satellite capabilities for emission spectroscopy in the *Earth Observations from Space* and NPOESS eras. *Journal of Geophysics Research*, **106**(D17), 20055–20068.

Eriksson E. (1963). The yearly circulation of sulfur in nature. *Journal of Geophysics Research*, **68**, 4001–4003.

Ernst W. G. (ed.) (2000). *Earth Systems. Processes and Issues*. Cambridge University Press, Cambridge, UK, 566 pp.

Ervin D., Kahn J. R., and Livingston M. L. (eds) (2002). *Does Environmental Policy Work? The Theory and Practice of Outcomes Assessments*. Edward Elgar, Cheltenham, UK, 256 pp.

Eskes, H., Segers, A., and van Velthoven, P. (2003). Ozone forecasts of the stratospheric polar vortex splitting event in September 2002. *J. Atmos. Sci.*, submitted.

Essex C. and McKitrick R. (2002). *Taken by Storm. The Troubled Science, Policy and Politics of Global Warming*. Key Porters Books, Toronto, 320 pp.

Esteve P. (1998.) Tropical dry ecosystems modelling and monitoring from space. *Ecological Modelling*, **108**(1–3), 175–188.

Falkowski P., Scholes R. J., Boyle E., Canadell D., Elser J., Gruber N. *et al.* (2000). The global carbon cycle: A test of our knowledge of Earth as a system. *Science*, **290**, 291–296.

Faluvegi G. S., Alapaty K., Reichle Jr H. G., Mathur R., Raman S., and Connors V. S. (1999). Simulation of carbon monoxide transport during April 1994. *Journal of Geophysics Research*, **104**(D17), 21471–21485.

Fan S., Gloor M., Mahlman J., Pacala S., Sarmiento J., Takahashi T. *et al.* (1998). A large terrestrial carbon sink in North America implied by atmospheric and oceanic carbon dioxide data and models. *Science*, **282**, 442–446.

Fan S.-M., Wofsy S. C., Bakwin P. S., Jacob D. J., Anderson S. M., Kebabian P. L. *et al.* (1992). Micrometeorological measurements of CH_4 and CO_2 exchange between the atmosphere and subarctic tundra. *Journal of Geophysics Research*, **97**(D15), 16627–16644.

Fan S.-M., Sarmiento J. L., Gloor M., and Pacala S. W. (1999). On the use of regularization techniques in the inverse modeling of atmospheric carbon dioxide. *Journal of Geophysics Research*, **104**(D17), 21503–21512.

Fang J., Chen A., Peng C., Zhao S., and Ci L. (2001). Changes in forest biomass carbon storage in China between 1949 and 1998. *Science*, **292**(5525), 2320–2322.

Farquhar G. D., von Caemmerer S., and Berry J. A. (1980). A biochemical model of photosynthetic CO_2 fixation in leaves of C_3 species. *Planta*, **149**, 78–90.

Faure H. (1990). Changes in the global continental reservoir of carbon. *Paleogeography, Paleoclimatology, Paleoecology*, **82**(1–2), 47–52.

Fearnside P. M. (1997). Greenhouse gases from deforestation in Brazilian Amazonia: Net committed emission. *Climate Change*, **35**, 321–360.

Ferm M. and Hultberg H. (1999). Dry deposition and internal circulation of nitrogen, sulfur and base cations to a coniferous forest. *Atmospheric Environment*, **33**(27), 4421–4430.

Field C. B. (2001). Sharing the garden. *Science*, **294**(5551), 2490–2491.

Field C. B. and Fung I. Y. (1999). Biogeochemical cycles: The not-so-big U. S. carbon sink. *Science*, **285**(5427), 544–545.

Fleming J. R. (1998). Arrhenius and current climate concerns: Continuity or a 100-year gap? *Earth Observations from Space*, **79**(34), 406–410.

Florkowski T., Korus A., Miroslaw J., Necki J. M., Neubert R., Schmidt M., and Zimnoch M. (1998). Isotopic composition of CO_2 and CH_4 in a heavily polluted atmosphere and in a remote mountain area (Southern Poland). *Isot. Technical Study of Environmental Change: Proceedings International Symposium Isot. Technical Study of Past and Current Environmental Changes in the Hydrosphere and Atmosphere, Vienna, 14–18 April 1997*, pp. 37–48.

Fonteyn D., Berrisford P., Bonjean S., Chabrillat S., Charlton A., Daerden F., Errera Q., and Lahoz, W. (2003). Chemical and dynamical data assimilation for the Antarctic winter 2002. *J. Atmos. Sci.*, submitted.

Forest C. E., Stone P. H., Sokolov A. P., Allen M. R., and Webster M. D. (2002). Quantifying uncertainties in climate system properties with the use of recent climate observations. *Science*, **295**(5552), 113–114.

Francey R. J., Trudinger C. M., and Levchenko V. A. (2000). ^{14}C variations in the atmosphere: A southern perspective. Paper given at the *12th National Congress of the Australian Institute of Physics, Hobart, 1–5 July, 1966. ANARE Res. Notes*, **102**, pp. 193–209.

Frantz B. R., Kashgarian M., Coale K. H., and Foster M. S. (2000). Growth rate and potential climate record from a rhodolith using ^{14}C accelerator mass spectrometry. *Limnology and Oceanography*, **45**(8), 1773–1777.

Friedlingstein P., Bopp L., Ciais P., Dufresne J. L., Fairhead L., LeTreut H. *et al.* (2001). Positive feedback between future climate change and the carbon cycle. *Geophysical Research Letters*, **28**, 1543–1546.

Friend, A. D. (1998). Parametrization of a global daily weather generator for terrestrial ecosystem modelling. *Ecological Modelling*, **109**(2), 121–140.

Fung I. (2000). Variable carbon sinks. *Science*, **290**(5495), 1313–1314.

Gale J. and Freund P. (2000). Reducing methane emissions to combat global climate change: The role Russia can play. *Proceedings of the 2nd International Methane Mitigation Conference, 18–23 June 2000, Novosibirsk*, pp. 73–80.

Galitsky V. V. (1985). Horizontal structure and dynamics of an aged vegetation community: Numerical modeling. *Numerical Modeling of Biogeocenotic Processes*. Science, Moscow, pp. 59–69 [in Russian].

Gasse F. and Lin R. (1998). Lacustrine isotope archives of past climate and environmental changes in the tropics and subtropics. Some examples from Africa. *Isot. Technical Study of Environmental Change: Proceedings International Symposium Isot. Technical Study of Past and Current Environmental Changes in the Hydrosphere and Atmosphere, Vienna, 14–18 April 1997*. IIASA Press, Vienna, pp. 519–531.

Gbondo-Tugbawa S. S. and Driscoll C. T. (2002). Evaluation of the effect of future controls on sulfur dioxide and nitrogen oxide emissions on the acid-base status of a northern forest ecosystem. *Atmospheric Environment*, **36**(10), 1631–1643.

Gerstengarbe F.-W. (2002). *Angewandte statistik*. PIK, Potsdam, Germany, 100 pp. (Report No. 75).

Gildor H. and Tziperman E. (2001). Physical mechanisms behind biogeochemical glacial-interglacial CO_2 variations. *Geophysics Research Letters*, **28**(12), 2421–2424.

Gillon J. and Yakir D. (2001). Influence of carbonic anhydrase activity in terrestrial vegetation on the ^{18}O content of atmospheric CO_2. *Science*, **291**(5513), 2584–2587.

Gleick P. H. (1993). *A Guide to the World's Fresh Water Resources*. Oxford Science, Oxford, UK, 473 pp.

Gloor M., Bakwin P., Hurst D., Lock L., Draxler R., and Tans P. (2001). What is the concentration footprint of a tall tower? *Journal of Geophysics Research*, **106**(D16), 17831–17840.

Goddéris Y., François L. M., and Veizer J. (2001). The early Paleozoic carbon cycle. *Earth and Planetary Science Letters*, **190**(3–4), 181–196.

Gorshkov V. G. (1990). Biospheric dynamics and environmental sustainability. *Science and Technical Results: Theoretical and General Problems of Geography*. VINITI Press, Moscow, 237 pp. [in Russian].

Gorshkov V. G. (1995). *Physical and Biological Basis of Life Stability*. VINITI Press, Moscow, 472 pp. [in Russian].

Gorshkov V. G. and Makaryeva A. M. (1999). Impact of the virgin and mastered biota on the global environment. *Research of the Earth from Space*, **5**, 3–11 [in Russian].

Gorshkov, V. G., Gorshkov, V. V., and Makarieva, A. M. (2000). *Biotic Regulation of the Environment*. Springer–Praxis, Chichester, UK, 364 pp.

Gorshkov V., Makarieva A., and Mackey B. (2002). Biological theory and global change science. *Global Change Newsletter*, **48**, 11–14.

Goudrian J., van Keulen H. and van Laar H.H. (eds) (1990). *The Greenhouse Effect and Primary Productivity in European Agro-ecosystems*. Pudoc, Wageningen, The Netherlands, 90 pp.

Grabowski W. W. and Moncrieff M. W. (2002). Large-scale organization of tropical convection in two-dimensional explicit simulations: Effects of interactive radiation. *Quarterly Journal of the Royal Meteorological Society*, **128**(585), Part A, 2349–2376.

Grant R. F., Gouldren M. L., Wofsy S. C., and Berry J. A. (2001). Carbon and energy exchange by a black spruce-moss ecosystem under changing climate: Testing the mathematical model ECOSYS with data from the BOREAS experiment. *Journal of Geophysics Research*, **106**(D24), 33605–33621.

Grigoryev Al. A. and Kondratyev K. Ya. (2001a). *Ecological Disasters*. Russian Academy of Science, St Petersburg, 206 pp. [in Russian].

Grigoryev Al. A. and Kondratyev K. Ya. (2001b). *Ecological Catastrophes*. Russian Academy of Science, St Petersburg, 661 pp. [in Russian].

Grogan P., Illeris L., Michelsen A., and Jonasson S. (2001). Respiration of recently-fixed plant carbon dominates mid-winter ecosystem CO_2 production in subarctic heath tundra. *Climate Change*, **50**(1–2), 129–142.

Grooss J.-U., Konopka P., and Mueller R. (2003). Ozone chemistry during the 2002 Antarctic vortex split. *J. Atmos. Sci.*, submitted.

Grossman D. (2001). Dissent in the maelstrom. *Scientific American*, **285**(5), 38–39.

Gruber N., Keeling C. D., and Bates N. R. (2002). Interannual variability in the North Atlantic ocean carbon sink. *Science*, **298**(5602), 2374–2378.

Guaduong F. Q. C. and Masao M. (2001). The carbon cycle of sandy lands in China and its global significance. *Climate Change*, **48**(4), 535–549.

Gyalistras D. (2002). *How Uncertain Are Regional Climate Change Scenarios? Examples for Europe and the Alps*. PIK, Potsdam, Germany, pp. 85–100 (Report No. 75).

Haan D., Zuo Y., Gros V., and Brenninkmeijer C. A. M. (2001). Photochemical production of carbon monoxide in snow. *Journal of Atmospheric Science*, **40**(3), 217–230.

Habib S. (2000). Technology thrusts for future Earth science applications. *Proceedings SPIE*, **2150**, 124–137.

Hackler J. L., Houghton R. A., and Lawrence K. T. (1999). The U. S. carbon budget: Contributions from land-use change. *Science*, **285**(5427), 574–577.

Hadley A. and Toumi R. (2002). A simple model which predicts some non-linear features between atmospheric sulphur and sulphur emissions. *Environmental Pollution*, **119**(3), 365–374.

Hansell D. A. and Carlson C. A. (eds) (2002). *Biogeochemistry of Marine Dissolved Organic Matter*. Elsevier, Amsterdam, 469 pp.

Hansen J. E. and Sato M. (2001). Trends of measured climate forcing agents. *Proceedings National Academy of Science USA*, **98**(26), 14778–14783.

Hansen J. E., Sato M., Ruedy R., Lacis A., and Oinas V. (2000). Global warming in the twenty-first century: An alternative scenario. *Proceedings U.S. National Academy of Science*, **97**(18), 9875–9880.

Harnik N., Scott R. K., and Perlwitz J. (2003). Wave reflection and focusing prior to the major stratospheric warming of September 2002. *J. Atmos. Sci.*, submitted.

Harris R. C., Sachse G. W., Hill G. F., Wade L., Bartlett K. B., Collins J. E. *et al.* (1992). Carbon monoxide and methane in the North American Arctic and subarctic troposphere: July–August 1988. *Journal of Geophysics Research*, **97**(D15), 16589–16600.

Hartmann D. L. (2002). Tropical surprises. *Science*, **295**(5556), 811–812.

Harvey L. D. D. (2001). A quasi-one-dimensional coupled climate-change cycle model. 2: The carbon cycle component. *Journal of Geophysics Research*, **106**(15), 22355–22372.

Harvey L. D. D. and Huang Z. (2001). A quasi-one-dimensional coupled climate-change cycle model. 1: Description and behavior of the climate component. *Journal of Geophysics Research*, **106**(C10), 22339–22353.

Haszpra L., Bercza P. S., Bakwin P. S., Berger B. W., Davis K. J., and Weidinger T. (2001). Measuring system for the long-term monitoring of biosphere/atmosphere of carbon dioxide. *Journal of Geophysics Research*, **106**(D3), 3057–3069.

Heans K. A. (2001). Assessment of pre-industrial carbon dioxide content in the atmosphere using hydro-chemical data. *1st International Conference on Global Warming and the Next Ice Age, Dalhousie University, 19–24 August 2001*. Dalhousie University Press, Dalhousie, New Brunswick, pp. 140–144.

Hemming D., Switsur V. R., Waterhouse J. S., and Heaton T. H. E. (1998). Carbon stable isotope response of three tree species to recent climate and atmospheric changes. *Isot. Technnical Study of Environmental Change: Proceedings International Symposium Isot. Technical Study of Past and Current Environmental Changes in the Hydrosphere and Atmosphere, Vienna, 14–18 April 1997*. IIASA Press, Vienna, pp. 800–802.

Hensen A., van den Bulk W. C. M., Vermeulen A. T., and Wyers G. P. (1998). *CO_2 Exchange between Grassland and the Atmosphere. Results over a Four Year Period of CO_2 Measurements at Cabauw, The Netherlands.* Amsterdam University Press, 53 pp. (Dutch National Research Programme on Global Air Pollution and Climate Change Report No. 410 200 020).

Hernández P. A., Notsu K., Salazar J. M., Mori T., Natale G., Okada H. *et al.* (2001). Carbon dioxide degassing by advective flow from Usu Volcano, Japan. *Science*, **292**(5514), 83–87.

Hoerling M. and Kumar A. (2003). The perfect ocean for drought. *Science*, **299**(5607), 691–694.

Holtmann I., Morais J. (2000). Global change and the carbon cycle. *LUCC Newsletter*, **5**, 16–17.

Hood D. W. (1981). Preliminary observations of the carbon budget of the eastern Bering Sea shelf. In: D. W. Hood and J. A. Calder (eds), *The Eastern Bering Shelf: Oceanography and Resources*, Vol. 1. University of Washington Press, Seattle, WA, pp. 347–358.

Hoppel K., Bevilacqua R., Allen D., Nedoluha G., and Randall C. (2003) POAM III observations of the anomalous 2002 Antarctic ozone hole. *Geophys. Res. Lett.*, **30**, 1394. doi:10.1029/2003GL016899.

Houghton J. T. and Yihmi D. (eds) (2001). *Climate Change. The Scientific Basis.* Cambridge University Press, Cambridge, UK, 881 pp. (contribution of Working Group I to the 3rd Assessment Report of the Intergovernmental Panel on Climate Change).

Houghton J. T., Filho G. M., Calander B. A., Harris N., Kattenberg A., and Maskell K. (1996). *Climate Change 1995: The Science of Climate Change.* Cambridge University Press, Cambridge, UK, 572 pp. (Intergovernmental Panel on Climate Change).

Houghton J. T., Ding Y., Griggs D. J., and Noguer M. (2001). *Climate Change 2001: The Scientific Basis.* Cambridge University Press, Cambridge, UK, 892 pp. (contribution of Working Group I to the 3rd Assessment Report of the Intergovernmental Panel Group on Climate Change).

Houghton R. A. (1999). The annual net flux of carbon to the atmosphere from changes in land use 1850–1990. *Tellus*, **518**, 298–313.

Houghton R. A., Skole D. L., Nobre C. A., Hackler J. L., Lawrence K. T., and Chomentowski W. H. (2000). Annual fluxes of carbon from deforestation and regrowth in the Brazilian Amazon. *Nature*, **403**(6767), 301–304.

Howarth R. W., Stewart J. W. B., and Ivanov M. V. (eds) (1992). *Sulphur Cycling on the Continents: Wetlands, Terrestrial Ecosystems, and Associated Water Bodies.* John Wiley & Sons, Chichester, UK, 372 pp.

Huber M. and Caballero R. (2003). Eocene El Niño: Evidence for robust tropical dynamics in the "Hohouse". *Science*, **299**(5608), 877–881.

Huesemann M. N., Skillman A. D., and Crecelius E. A. (2002). The inhibition of marine nitrification by ocean disposal of carbon dioxide. *Marine Pollution Bulletin*, **44**(2), 142–148.

Hughes G. B., Giegengack R., and Kritikos H. N. (1999). Spectral indications of unexpected contributors to atmospheric CO_2 variability. *International Journal of Climatology*, **19**(8), 813–820.

Hulst, R. van (1979). The dynamics of vegetation: Succession in model communities. *Vegetation*, **39**(2), 85–96.

Indermühle A., Stocker T. F., Joos F., Fischer H., Smith H. J., Wahlen M. *et al.* (1999). Holocene carbon-cycle dynamics based on CO_2 trapped in ice at Taylor Dome, Antarctica. *Nature*, **398**(6723), 121–126.

IPCC (2001a). *Climate Change 2001: Impacts, Adaptation, and Vulnerability*. Cambridge University Press, Cambridge, UK, 1032 pp. (contribution of Working Group II to the 3rd Assessment Report of the Intergovernmental Panel on Climate Change).

IPCC (2001b). *Climate Change 2001: Mitigation*. Cambridge University Press, Cambridge, UK, 752 pp. (contribution of Working Group III to the 3rd Assessment Report of the Intergovernmental Panel on Climate Change).

IPCC (2001c) *Climate Change 2001: The Scientific Basis*. Cambridge University Press, Cambridge, UK, 881 pp. (contribution of Working Group I to the 3rd Assessment Report of the Intergovernmental Panel on Climate Change).

Irion R. (2001) Fathoming the chemistry of the deep blue sea. *Science*, **293**(5531), 790–793.

Isidorov V. A., and Kondratyev K. Ya. (2001). On global carbon cycle problems. *Problems of Environmental and Natural Resources*, **5**, 56–83.

Ito A. and Oikawa T. (2002). A simulation model of the carbon cycle in land ecosystems (Sim-CYCLE): A description based on dry-matter production theory and plot-scale validation. *Ecological and Modelling*, **151**(2–3), 143–176.

Itoh H. (2002). True versus apparent Arctic oscillation. *Geophysics Research Letters*, **19**(8), 109/1–109/4.

Ivanov A. (1978). *Introduction to Oceanography*. Mir, Moscow, 392 pp. [in Russian].

Ivanov-Rostovtsev A. G., Kolotilo L. G., Tarasiuk Yu. F., and Sherstiankin P. P. (2001). *Self-organization and Self-regulation of Natural Systems*. Russian Geographical Society, St Petersburg, 216 pp. [in Russian].

Jadin E. A., and Vargin P. N. (2003) Interannual variations of the Antarctic ozone hole, stratospheric circulation and sea surface temperature in the south Oceans. *J. Atmos. Sci.*, submitted.

Jagovkina S. V., Karol I. L., Zubov V. A., Lagun V. E., Reshemikov A. I., and Rosanov E. V. (2000). The estimation of the gas deposits leakage input into the total methane flux from the West Siberian region. *Proceedings of the 2nd International Methane Mitigation Conference, 18–23 June 2000, Novosibirsk*. Novosibirsk University Press, pp. 263–267.

Jenkins R., Barton J., Bartzokas A., Hesselberg J., and Knutsen H. M. (2002). *Environmental Regulation in the New Global Economy*. Edward Elgar, Cheltenham, UK, 360 pp.

Jepma C., Meijer H., Uiterkamp T. S., and Weesie P. (2001). Editorial. *Energy and Environment*, **12**(5–6), iii–ix.

Johnson D. E. and Ulyatt M. I. (2000). Variations in the proportion of methane of total greenhouse gas emissions from US and NZ dairy production systems. *Proceedings of the 2nd International Methane Mitigation Conference, 18–23 June 2000, Novosibirsk*. Novosibirsk State University Press, pp. 249–254.

Jones C. D. and Cox P. M. (2001). Modelling the volcanic signal in the atmospheric CO_2 record. *Global Biogeochemical Cycles*, **15**(2), 453–466.

Joos F., Plattner G.-K., Stocker T. F., Marchal O., and Schmittner A. (1999). Global warming and marine carbon cycle feedbacks on future atmospheric CO_2. *Science*, **284**(5413), 464–468.

Kabat P. and Nobre C. A. (2000). Brazilia – European carbon research studies in LBA. *BAMS News*, **8**, 15–18.

Kaiser J. (2000). Soaking up carbon in forests and fields. *Science*, **290**(5493), 922–923.

Kamall R. (2000). Capture and storage of CO_2. *Greenhouse Issues*, **51**, 2–4.

Kaminski T., Rayner P. J., Heimann M., and Enting I. (2001). On aggregation errors in atmospheric transport inversions. *Journal of Geophysics Research*, **106**(D5), 4703–4715.

Karley M. J., Beven K. J., and Oliver H. R. (1993). A method for predicting spatial distribution of evaporation using simple meteorological data. *Proceedings of International*

Symposium "Land Surface Range Space and Timescales", Yokohama, 13–16 July 1993. IAHS, Yokohama, pp. 619–626 (Publication No. 212).

Karol I. L. (2000). Impact of the flights of world transport aircraft on the ozonosphere and climate. *Meteorology and Hydrology*, **7**, 17–32 [in Russian].

Kashapov R. Sh. (2002). On the balance of organic carbon in the nature–economic system of Bashkortostan. *Bulletin of Russian Geographic Society*, **134**(3b), 39–42 [in Russian].

Katul G. G. and Albertson J. D. (1999). Modeling CO_2 sources, sinks, and fluxes within a forest canopy. *Journal of Geophysics Research*, **104**(D6), 6081–6092.

Katz M. E., Pak D. K., Dickens G. R., and Miller K. G. (1999). The source and fate of massive carbon input during the latest Paleocene thermal maximum. *Science*, **286**(5444), 1531–1533.

Katz R. W. (2002). Techniques for estimating uncertainty in climate change scenarios and impact studies. *Climate Research*, **20**, 167–185.

Keeling R. F. and Visbeck M. (2001). Antarctic stratification and glacial CO_2. *Nature*, **412**(6847), 605–606.

Keeling R. F., Manning A. C., McElroy E. M., and Shertz S. R. (1998). Methods for measuring changes in atmospheric O_2 concentration and their application in southern hemisphere air. *Journal of Geophysics Research*, **103**(D3), 3381–3397.

Keller M., Rocha H., Trumbora S., and Kruijt B. (2001). Investigating the carbon cycle of the Amazon forests. *Global Change Newsletter*, **45**, 14–19.

Kelley J. J. (1987). Carbon dioxide in the Arctic environment. *Journal of Earth Science*, **35**(2), 341–354.

Kennedy D. (2001). An unfortunate U-turn on carbon. *Science*, **291**(5513), 2115–2116.

Kennedy D. (2003). The policy drought on climate change. *Science*, **299**(5605), 309–310.

Kennedy M. J., Pevear D. R., and Hill R. J. (2002). Mineral surface control of organic carbon in black shale. *Science*, **295**(5555), 657–660.

Kerr R. A. (2001). How grasses got the upper hand. *Science*, **293**, 1572–1573.

Kerr R. A. (2002). Reducing uncertainties of global warming. *Science*, **295**(5552), 29–31.

Khalil M. A. K., Rasmussen R. A., Ren L., Wang M. X., Shearer M. J., Dalluge R. W. *et al.* (2000). Methane emissions from rice fields. *Proceedings of the 2nd International Methane Mitigation Conference, 18–23 June 2000, Novosibirsk*. Novosibirsk State University Press, pp. 13–30.

Kheshgi H. S., Jain A. K., Kotamarthi V. R., and Wuebbles D. J. (1999a). Future atmospheric methane concentrations in the context of the stabilization of greenhouse gas concentrations. *Journal of Geophysics Research*, **104**(D16), 19183–19190.

Kheshgi H. S., Jain A. K., and Wuebbles D. S. (1999b). Model-based estimation of the global carbon budget and its uncertainty from carbon dioxide and carbon isotope records. *Journal of Geophysics Research*, **104**(D24), 31127–31143.

Khotuntsev Yu. L. (2001). *Human Beings, Technologies, Environment*. Sustainable World, Moscow, 224 pp. [in Russian].

Kirilenko A. P. (1990). Numerical modeling of the production process and of the water cycle of the forest ecosystems. Ph.D. thesis, Computing Center of Russian Academy of Science, Moscow, 151 pp. [in Russian].

Kiseleva S. V. (1990). Study of the processes of CO_2 transport under different conditions of the water–air interface. Ph.D. thesis, Moscow State University, Moscow, 171 pp. [in Russian].

Klubov S. V., Grigorian S. V., and Davydova S. L. (2000). Natural and technogenic distributions of the elements of the pollutants within the marine coastal zone. *Ecological Systems and Devices*, **5**, 42–45 [in Russian].

Knapp A. K., Fay P. A., Blair J. M., Collins S. L., Smith M. D., Carlisle J. D. *et al.* (2002). Rainball variability, carbon cycling, and plant species diversity in a mosaic grassland. *Science*, **298**(5601), 2202–2205.

Kojima T. (1998). *The Carbon Dioxide Problem. Integrated Energy and Environmental Policies for the 21st Century*. Gordon & Breach, London, 240 pp.

Kok M. T. J. and Verweij W. (1999). The Kyoto Protocol: Implications for research. *The NRP Workshop: The Road after Kyoto, 20 January 1998*. Amsterdam University Press, 22 pp. (Dutch National Research Programme on Global Air Pollution and Climate Change Report No. 410 200 018).

Kolbe A. E. (1999). Geographic extension of an uneven-aged, multi-species matrix growth model for northern hardwood forests. *Ecological Modelling*, **121**(2–3), 235–253.

Kolber Z. S., Plumley F. G., Lang A. S., Beatty J. T., Blankenship R. E., van Dover C. L. *et al.* (2001). Contribution of aerobic photoheterotrophic bacteria to the carbon cycle in the ocean. *Science*, **292**(5526), 2492–2495.

Kondragunta S., Flynn L. E., Neuendorffer A., Miller A. J., Long C., Nagatani R., Zhou S., Beck T., Beach E., McPeters R., Stolarski R., Bhartia P. K., DeLand M. T., and Huang, L.-K. (2003). Vertical structure of the anomalous 2002 Antarctic ozone hole. *J. Atmos. Sci.*, submitted.

Kondratyev K. Ya. (1990). *Key Problems of Global Ecology*. VINITI, Moscow, 454 pp. [in Russian].

Kondratyev K. Ya. (1991). Priorities in global ecology. *Bulletin of USSR Academy of Science, Geography*, **6**, 21–30 [in Russian].

Kondratyev K. Ya. (1992). *Global Climate*. Science, St Petersburg, 359 pp. [in Russian].

Kondratyev K. Ya. (1993). Ecology and politics. *Bulletin of Russian Geographic Society*, **125**(2), 78–91 [in Russian].

Kondratyev K. Ya. (1996). Global changes and demographic dynamics. *Bulletin of Russian Geographic Society*, **128**(3), 1–12 [in Russian].

Kondratyev K. Ya. (1997). Key issues of global change at the end of the second millennium. *Newsletter of the European Geophysical Society*, **63**, 4–8.

Kondratyev K. Ya. (1998a). Ecological risk: Real and hypothetical. *Bulletin of Russian Geographic Society*, **130**(3), 13–24 [in Russian].

Kondratyev, K. Ya. (1998b). *Multidimensional Global Change*. Wiley–Praxis, Chichester, UK, 771 pp.

Kondratyev, K. Ya. (1999a). *Atmospheric Ozone Variability*. Springer–Praxis, Chichester, UK, 592 pp.

Kondratyev, K. Ya. (1999b). *Climatic Effects of Aerosols and Clouds*. Springer–Praxis, Chichester, UK, 264 pp.

Kondratyev K. Ya. (1999c). *Ecodynamics and Geopolicy. Vol. 1: Global Problems*. Russian Academy of Science, St Petersburg, 1036 pp. [in Russian].

Kondratyev K. Ya. (2000a). Global changes on the border of two millennia. *Herald of Russian Academy of Science*, **70**(9), 788–796 [in Russian].

Kondratyev K. Ya. (2000b). Studying the Earth from space: The Earth Observations from Space scientific plan. *Research of the Earth from Space*, **3**, 82–91 [in Russian].

Kondratyev K. Ya. (2001a). Impact of NATO's bombardment on the environment of Yugoslavia. *Bulletin of Russian Academy of Science*, **1**, 63–67 [in Russian].

Kondratyev K. Ya. (2001b). On the problem of global carbon cycle. *Problems of the Environment and Natural Resources*, **6**, 56–83 [in Russian].

Kondratyev K. Ya. (2002). Global climate changes: Reality, hypotheses, and fantasies. *Research of the Earth from Space*, **1**, 3–23 [in Russian].

Kondratyev K. Ya. (2003a). High-latitude environmental dynamics in the context of global change. *Idöjàràs*, **107**(1), 1–29.

Kondratyev K. Ya. (2003b). Radiative forcings due to aerosols. *Optics of the Atmosphere and Ocean*, **16**(1), 5–18.

Kondratyev K. Ya. and Demirchian K. S. (2000). Global climate changes and carbon cycle. *Bulletin of Russian Academy of Science*, **132**(4), 1–20 [in Russian].

Kondratyev K. Ya. and Demirchian K. S. (2001a). Global climate and the Kyoto Protocol. *Problems of the Environment and Natural Resources*, **6**, 2–15 [in Russian].

Kondratyev K. Ya. and Demirchian K. S. (2001b). Climate of the Earth and the Kyoto Protocol. *Herald of the Russian Academy of Science*, **71**(11), 100–107 [in Russian].

Kondratyev K. Ya. and Galindo I. (2001). *Global Change Situations: Today and Tomorrow*. Universidad de Colima, Colima, Mexico, 164 pp.

Kondratyev K. Ya. and Grigoryev Al. A. (2000). Natural and anthropogenic ecological catastrophes: Meteorological disasters and catastrophes. *Research of the Earth from Space*, **4**, 3–19 [in Russian].

Kondratyev K. Ya. and Grigoryev Al. A. (2002). *Environmental Disasters. Anthropogenic and Natural*. Springer–Praxis, Chichester, UK, 484 pp.

Kondratyev K. Ya. and Isidorov V. A. (2001). Global carbon cycle. *Optics of the Atmosphere and Ocean*, **14**(1), 1–10 [in Russian].

Kondratyev K. Ya. and Johannessen O. (1993). *The Arctic and Climate*. PROPO, St Petersburg, 140 pp. [in Russian].

Kondratyev K. Ya. and Krapivin V. F. (2001a). Biocomplexity and the global ecological monitoring. *Research of the Earth from Space*, **1**, 1–8 [in Russian].

Kondratyev K. Ya. and Krapivin V. F. (2001b). An expert system to control the greenhouse effect. *Problems of the Environment and Natural Resources*, **6**, 23–39 [in Russian].

Kondratyev K. Ya. and Krapivin V. F. (2003a). Climate change and carbon cycle. *Problems of Environmental and Natural Resources*, **1**, 2–26.

Kondratyev K. Ya. and Krapivin V. F. (2003b). Global carbon cycle and climate. *Research of the Earth from Space*, **1**, 3–15.

Kondratyev K. Ya. and Losev K. S. (2002). Illusions and reality of sustainable developments strategy. *Herald Russian Academy of Science*, **72**(7), 592–681 [in Russian].

Kondratyev K. Ya. and Varotsos C. A. (1995). Atmospheric ozone variability in the context of global change. *International Journal of Remote Sensing*, **16**(10), 1851–1881.

Kondratyev K. Ya. and Varotsos C. A. (2000). *Atmospheric Ozone Variability: Implications for Climate Change, Human Health, and Ecosystems*. Springer–Praxis, Chichester, UK, 758 pp.

Kondratyev, K. Ya. , Ortner, J., and Preining, O. (1992). Priorities of global ecology now and in the next century. *Space Policy*, **8**(1), 39–48.

Kondratyev, K. Ya., Moreno Pena F., and Galindo I. (1997). *Sustainable Development and Population Dynamics*. Universidad de Colima, Colima, Mexico, 128 pp.

Kondratyev K. Ya., Krapivin V. F., and Pshenin E. S. (2000). Conception of regional geoinformational monitoring. *Research of the Earth from Space*, **6**, 1–8 [in Russian].

Kondratyev K. Ya., Demirchian K. S., Baliunas S., Adamenko V. N., Bohmer-Christiansen S., Idso Sh. B. *et al.* (2001a). *Global Climate Changes: Conceptual Aspects*. Russian Academy of Science, St Petersburg, 125 pp. [in Russian].

Kondratyev K. Ya., Losev K. S., Ananicheva M. D., and Chesnokova I. V. (2001b). Some problems of the landscape history and ecology in the context of biotic regulation. *Bulletin of Russian Geographic Society*, **133**(5), 22–29 [in Russian].

Kondratyev K. Ya., Krapivin V. F., and Phillips G. W. (2002a). *Global Environmental Change: Modelling and Monitoring*. Springer-Verlag, Berlin, 319 pp.

Kondratyev K. Ya., Krapivin V. F., and Phillips G. V. (2002b). *Problems of High-latitude Environmental Pollution*. St Petersburg State University, St Petersburg, 280 pp. [in Russian].

Kondratyev K. Ya., Krapivin V. F., and Savinykh V. P. (2003a). *Perspectives of Civilization Development: Multidimensional Analysis*. Logos, Moscow, 546 pp. [in Russian].

Kondratyev K. Ya., Losev K. S., Ananicheva M. D., and Chesnokova I. V. (2003b). The cost of Russia's environmental services. *Bulletin of Russian Academy of Science*, **73**(1), 3–13 [in Russian].

Konopka P., Grooß J-U., and Steinhorst H-M. (2003). Mixing and chemical ozone loss during and after the Antarctic polar vortex major warming in September 2002. *J. Atmos. Sci.*, submitted.

Kontorovich A. E., Dobretsov N. L., Lanerov I. P., Korzhubayev A. G., and Livshits V. R. (1999). The energy strategy of Russia in the 21st century. *Herald of Russian Academy of Science*, **69**(9), 771–789 [in Russian].

Korner C. (1998). Tropical forests in a CO_2-rich atmosphere. *Climate Change*, **39**(2–3), 297–315.

Kozoderov V. V. (1994). Problems of global changes in the context of spaceborne observations and modeling the biogeophysical processes. *Research of the Earth from Space*, **6**, 51–67 [in Russian].

Kozoderov V. V. (1998). Development of the theory of biospheric global changes from data of the spaceborne observations of the Earth. *Research of the Earth from Space*, **6**, 25–39 [in Russian].

Kraabol A. G. and Stordal F. (2000). Modelling chemistry in aircraft plumes. 2: The chemical conservation of NO_x to reservoir species under different conditions. *Atmospheric Environment*, **34**(23), 3951–3962.

Kraabol A. G., Konopka P., Stordal F., and Schlager H. (2000). Modelling chemistry in aircraft plumes. 1: Comparison with observations and evaluation of a layered approach. *Atmospheric Environment*, **34**(23), 3939–3950.

Krapivin V. F. (1978). *On the Theory of Complex Systems' Survivability*. Soviet Radio, Moscow, 248 pp. [in Russian].

Krapivin V. F. (1993). Mathematical model for global ecological investigations. *Ecological Modelling*, **67**(2–4), 103–127.

Krapivin V. F. (1996). The estimation of the Peruvian current ecosystem by a mathematical model of biosphere. *Ecological Modelling*, **91**(1), 1–14.

Krapivin V. F. (2000). The greenhouse effect and the global biogeochemical cycle of carbon dioxide. *Problems of the Environment and Natural Resources*, **2**, 2–18 [in Russian].

Krapivin V. F. and Kondratyev K. Ya. (2002). *Global Changes of the Environment*. St Petersburg University, St Petersburg, 724 pp. [in Russian].

Krapivin V. F. and Marenkin D. S. (1999). Co-evolution of man and nature. *Problems of the Environment and Natural Resources*, **1**, 2–10 [in Russian].

Krapivin V. F. and Nazarian N. A. (1997). A numerical model to study the global sulfur cycle. *Numerical Modeling*, **9**(8), 36–50 [in Russian].

Krapivin V. F. and Potapov I. I. (2002). *Methods of Ecoinformatics*. VINITI, Moscow, 496 pp. [in Russian].

Krapivin V. F. and Vilkova L. P. (1990). Model estimation of excess CO_2 distribution in biosphere structure. *Ecological Modelling*, **50**, 57–78.

Krapivin V. F., Svirezhev Yu. M., and Tarko A. M. (1982). *Mathematical Modeling of the Global Biospheric Processes*. Science, Moscow, 272 pp. [in Russian].

Krapivin V. F., Vilkova L. P., Rochon G. L., and Hicks D. R. (1996). Model estimation of the role of urban areas in global CO_2 dynamics. *Proceedings of Eco-Informa '96, 4–7 November 1996, Florida*, pp. 17–22.

Krueger K., Naujokat B., and Labitzke, K. (2003). The unusual midwinter warming in the southern hemisphere stratosphere 2002: A comparison to northern hemisphere phenomena. *J. Atmos. Sci.*, submitted.

Krupchatnikov V. N. (1998). Simulation of CO_2 exchange processes in the atmosphere–surface biomass system by the climate model ECSSib. *Russian Journal of Numerical, Analytical and Mathematical Modelling*, **13**(6), 479–492.

Kuck Jr L. R., Smith T., Balsley B. B., Helmig D., Conway T. J., Tans P. P. *et al.* (2000). Measurements of landscape-scale fluxes of carbon dioxide in the Peruvian Amazon by vertical profiling through the atmospheric boundary layer. *Journal of Geophysics Research*, **105**(D17), 22137–22146.

Kukla G. J. (2000). The last interglacial. *Science*, **287**, 987–988.

Kump L. R., Brantley S. L., and Arthur M. A. (2000). Chemical weathering, atmospheric CO_2, and climate. *Annual Reviews of Earth and Planetary Sciences*, **28**, 611–667.

Kuznetsov O. L. and Bolshakov B. E. (2002). *Sustainable Development: Scientific Basis of Planning the Nature–Society–Man System*. Gumanistika, St Petersburg, 615 pp. [in Russian].

La Deau S. and Clark J. S. (2001). Rising CO_2 levels and the fecundity of forest trees. *Science*, **292**(5514), 95–97.

Lafleur P. M. (1999). Growing season energy and CO_2 exchange at a subarctic boreal woodland. *Journal of Geophysics Research*, **104**(D8), 9571–9580.

Lafleur P. M., Roulet N. T., and Admiral S. W. (2001). Annual cycle of CO_2 exchange at a bog peat-land. *Journal of Geophysics Research*, **106**(D3), 3071–3082.

Lai Ch. T., Katul G., Oren R., Ellsworth D., and Schfer K. (2000). Modeling CO_2 and water vapor turbulent flux distributions within a forest canopy. *Journal of Geophysics Research*, **105**(D21), 26333–26351.

Lal R. (2001). Potential of desertification control to sequester carbon and mitigate the greenhouse effect. *Climate Change*, **51**, 35–72.

Lal R., Follett R. F., and Stewart B. A. (eds) (1998). *Soil Processes and the Carbon Cycle*. Springer–CRC Press, Berlin, 609 pp.

Lambin E. F. (2002). New global-scale datasets on land-use/cover change. *LUCC Newsletter*, **8**, 1–2.

Langmann B. (2000). Numerical modelling of regional scale transport and photochemistry directly together with meteorological processes. *Atmospheric Environment*, **34**(21), 3585–3598.

Lavorel S., Lambin E. F., Flannigan M., and Scholes M. (2002). Fires in the Earth system: The need for integrated research. *Global Change Newsletter*, **48**, 7–10.

Leach G. (2002). Think before you sink. *Tiempo Issled*, **46**, 18–22.

Ledley T. S., Sundquist E. T., Schwartz S. E., Hall D. K., Fellows J. D., and Killeen T. L. (1999). Climate change and greenhouse. *Earth Observations from Space*, **80**(39), 453–474.

Lee K., Wannikhof R., Takahashi T., Doney S. C., and Feely R. A. (1998). Low interannual variability in recent oceanic uptake of atmospheric carbon dioxide. *Nature*, **396**(6707), 155–159.

Lee X., Fuentes J. D., Staebler R. M., and Neumann H. H. (1999). Long-term observation of the atmospheric exchange of CO_2 with a temperate deciduous forest in southern Ontario, Canada. *Journal of Geophysics Research*, **104**(D13), 15975–15984.

Legendre L. and Krapivin V. F. (1992). Model for vertical structure of phytoplankton community in Arctic regions. *Proceedings of the 7th International Symposium on the Okhotsk Sea and Sea Ice, 2–5 February 1992, Mombetsu, Japan*. Okhotsk Sea and Cold Ocean Research Association, Mombetsu, Japan, pp. 314–316.

Legendre P. and Legendre L. (1998). *Numerical Ecology*. Elsevier, Amsterdam, 853 pp.

Lenaz R., Giovanelli G., Ori C., Colombo T., Masarie K., Ciattaglia L. *et al.* (2000). Shipboard report on atmospheric CO_2 concentrations recorded continuously from the Mediterranean Sea to Antarctica. *Il Nuovo Cimento*, **23C**, Ser. 2, No. 5, 507–514.

Leslie P. H. (1995). On the use of matrices in certain population mathematics, *Biometrica*, **33**(3), 183–212.

Lieth H. (1985). A dynamic model of the global carbon flux through the biosphere and its relations to climatic and soil parameters. *International Journal of Biometeorology*, **29**(2), 17–31.

Lin D.-L., Sakoda A., Shibasaki R., Goto N., and Suzuki M. (2000). Modelling a global biogeochemical nitrogen cycle in terrestrial ecosystems. *Ecological Modelling*, **135**(1), 89–110.

Linderholm H. W., Moberg A., and Grudd H. (2002). Peatland pines as climate indicators? A regional comparison of the climate influence on Scots pine growth in Sweden. *Canadian Journal of Forest Research*, **32**(8), 1400–1410.

Liping G., Erda L., and Zhongpei L. (2000). Methane emission flux and mitigation options and its relationship with N_2O emission from paddy soils. *Proceedings of the 2nd International Methane Mitigation Conference, 18–23 June 2000, Novosibirsk*. Novosibirsk State University Press, pp. 217–222.

Lischke H., Lotter A. F., and Fischlin A. (2002). Untangling a Holocene pollen record with forest model simulations and independent climate data. *Ecological Modelling*, **150**(1–2), 1–21.

Lockwood J. G. (2001). Abrupt and sudden climatic transitions and fluctuations: A review. *International Journal of Climatology*, **21**(9), 1153–1180.

Loehle C. (1993). Geological methane as a source for post-glacial CO_2 increases: The hydrocarbon pump hypothesis. *Geophysics Research Letters*, **20**(4), 1415–1418.

Logofet D. O. (2002). Matrix population models: Construction, analysis and interpretation. *Ecological Modelling*, **148**(3), 307–310.

Lomborg B. (2001). *The Sceptical Environmentalist. Measuring the Real State of the World.* Cambridge University Press, Cambridge, UK, 515 pp.

Losev K. S. (2001). *Ecological Problems and Perspectives of Sustainable Development of Russia in the 21st Century*. Kosmosinform, Moscow, 400 pp. [in Russian].

Lovett R. A. (2002). Rain might be leading carbon sink factor. *Science*, **296**(5574), 1787–1788.

Lowe D. C., Manning M. R., Brailsford G. W., Bromley A. M., and Ferretti D. (1998). Latitudinal gradients of ^{13}C in atmospheric methane in the Pacific Ocean region. *Isot. Technnical Study of Environmental Change: Proceedings International Symposium Isot. Technical Study of Past and Current Environmental Changes in the Hydrosphere and Atmosphere, Vienna, 14–18 April 1997*. IIASA Press, Vienna, pp. 43–56.

Lowry D., O'Brien P., Nisbet E. G., and Rata N. D. (1998). $\delta^{13}C$ of atmospheric methane. *Isot. Technnical Study of Environmental Change: Proceedings International Symposium Isot. Technical Study of Past and Current Environmental Changes in the Hydrosphere and Atmosphere, Vienna, 14–18 April 1997*. IIASA Press, Vienna, pp. 57–67.

Lueker T. J., Keeling R. F., and Dubey M. K. (2001). The oxygen to carbon dioxide ratios observed in emissions from a wildfire in Northern California. *Geophysics Research Letters*, **28**(12), 2413–2416.

Mabuchi K., Sato Y., and Kida H. (2000). Numerical study of the relationships between climate and the carbon dioxide cycle on a regional scale. *Journal of the Meteorological Society of Japan*, **78**(1), 25–46.

Macdonald A. M., O'Neil Baringer M., Lee K., Wallace D., and Wanninkhof R. (2002a). Subtropical Atlantic carbon transport. *International WOCE Newsletter*, **42**, 14–19.

Macdonald A. M., Wanninkhof R., O'Neil Baringer M., Robbins P. E., and Wallace D. W. R. (2002b). Oceanic biogeochemical fluxes: A summary of the JGOFS portion of the WOCE/JGOFS Transport Workshop, Southampton, 25–29 June 2001. *International WOCE Newsletter*, **42**, 20–21.

Manabe S. (1983). Carbon dioxide and climate change. *Advances in Geophysics*, **25**, 39–82.

Mander Ü. and Jongman R. (2000). *Landscape Perspectives of Land Use Changes*. WIT Press, Southampton, UK, 224 pp. (Advances in Ecological Sciences Vol. 6).

Manney G. L., Sabutis J. L., Allen D. R., Lahoz W. A., Scaife A. A., Randall C. E., Pawson S., Swinbank R., and Naujokat B. (2003). Simulations of dynamics and transport during the September 2002 Antarctic major warming. *J. Atmos. Sci.*, submitted.

Mannion A. M. (1997). *Global Environmental Change*, 2nd edn. Longman, London, 387 pp.

Marchuk G. I. and Kondratyev K. Ya. (1992). *Priorities in Global Ecology*. Science, Moscow, 264 pp. [in Russian].

Marland G., McCarl B. A., and Schneider U. (2001). Soil carbon: Policy and economics. *Climate Change*, **51**(1), 101–117.

Matsumura H. (2000). *Japan and the Kyoto Protocol Conditions for Ratification*. Royal Institute of International Affairs, London, 66 pp. (Energy and Environment Programme).

McCarl B. A. and Schneider U. A. (2001). Greenhouse gas mitigation in U. S. agriculture and forestry. *Science*, **294**(5551), 2481–2482.

McGillis W. R., Edson J. B., Hare J. E., and Fairrall C. W. (2001). Direct covariance air–sea CO_2 fluxes. *Journal of Geophysics Research*, **106**(C8), 16729–16746.

McIntyre, A. D. (1999). The environment and the oil companies. *Marine Pollution Bulletin*, **38**(3), 155–156.

McKenzie C., Schiff S., Aravena R., Kelly C. J. T., and Lons V. (1998). Effect of temperature on production of CH_4 and CO_2 from peat in a natural and flooded boreal forest wetlands. *Climate Change*, **40**(2), 247–266.

McNeil B. I., Matear R. J., Key R. M., Bullister J. L., and Sarmiento J. L. (2003). Anthropogenic CO_2 uptake by the ocean based on the global chlorofluorocarbon data set. *Science*, **299**(3604), 235–239.

McNulty S. G. (2002). Hurricane impacts on US forest carbon sequestration. *Environmental Pollution*, **116**, S17–S24.

Meadows M. (2000). United Kingdom methane emissions: Trends, projections and mitigation options. *Proceedings of the 2nd International Methane Mitigation Conference, 18–23 June 2000, Novosibirsk*. Novosibirsk State University Press, pp. 37–44.

Meijer E. W. and Velthoven P. (1997). The effect of the conversion of nitrogen oxides in aircraft exhaust plumes in global models. *Geophysics Research Letters*, **24**(23), 3013–3016.

Mejer H. A. J. (2001). The science of greenhouse gases: Uncertainties in sources and sinks, and implications for verification. *Energy and Environment*, **12**(5–6), 425–446.

Mellilo J. M., Stendler P. A., Aber J. D., Newkirk K., Lux H., Bowles F. P. *et al.* (2002). Soil warming and carbon-cycle feedbacks to the climate system. *Science*, **298**(5601), 2173–2176.

Mészàros E. (2002). Atmospheric wet deposition as a nutrient supply for the vegetation. *Idöjàràs*, **106**(3–4), 103–112.

Metting F. B., Smith J. L., Amthor J. S., and de Izaurral R. C. (2001). Science needs and new technology for increasing soil carbon sequestration. *Climate Change*, **51**(1), 11–34.

Metzger R. A. and Benford G. (2001). Sequestering of atmospheric carbon through permanent disposal of crop residue. *Climate Change*, **49**(1–2), 11–19.

Milford C., Hargreaves K. J., Sutton M. A., Loubet B., and Cellier P. (2001). Fluxes of CO_2 over upland moorland in the vicinity of agricultural land. *Journal of Geophysics Research*, **106**(D20), 24169–24181.

Mintzer I. M. (1987). *A Matter of Degrees: The Potential for Controlling the Greenhouse Effect*. World Resources Institute, Tokyo, 70 pp. (Research Report No. 15).

Mirkin B. M. (1986). *Vegetation Communities: What Are They?* Science, Moscow, 160 pp. [in Russian].

Mirtskhoulava Ts. E. (2003). Geoecology problems: Quantitative assessment of the state and risk dangers for geographic systems and their sustainable development. *Engineering Ecology*, **1**, 25–34 [in Russian].

Moe A. and Tanger K. (2000). *The Kyoto Mechanisms and Russian Climate Politics*. Royal Institute of International Affairs, London, 110 pp. (Energy and Environment Programme).

Monahan E. C. and Dam H. G. (2001), Bubbles: An estimate of their role in the global oceanic flux of carbon. *Journal of Geophysics Research*, **106**(C5), 9377–9384.

Monin A. S. and Krasitsky V. P. (1985). *Phenomena on the Ocean Surface*. Gidrometeoizdat, Leningrad, 375 pp. [in Russian].

Monnin E., Indermühle A., Dällenbach A., Flückiger J., Stauffer B., Stocker T. F. *et al.* (2001). Atmospheric CO_2 concentrations over the last glacial termination. *Science*, **291**(5501), 112–114.

Morgan J. A. (2002). Looking beneath the surface. *Science*, **298**(5600), 1903–1904.

Morgan M. G. (2000). Managing carbon from the bottom up. *Science*, **289**(5480), 2285–2286.

Nabuurs G. J. and Sikkema R. (2001). International trade in wood products: Its role in the land use change and forestry carbon cycle. *Climate Change*, **49**(4), 377–395.

Nabuurs G. J., Dolman A. J., Verkaik E., Whitmore A. R., Daamen W. P., Oenema O. *et al.* (2000). *Resolving Issues on Terrestrial Biospheric Sinks in the Kyoto Protocol*. Dutch National Research Programme on Global Air Pollution and Climate Change, The Netherlands, 10 pp. (Report No. 410 200 030).

Nakicenovic N., Grübler A., and McDonald A. (eds) (1998). *Global Energy Perspectives*. Cambridge University Press, Cambridge, UK, 299 pp.

NASA/NOAA (2002a). Ozone Hole Press Release, 30 September 2002 (`http://toms.gsfc.nasa.gov`).

NASA/NOAA (2002b). Ozone Hole Press Release, 6 December 2002 (`http://www.gsfc.nasa.gov/topstory/20021206ozonehole.html`).

Nedoluha G. E., Bevilacqua R. M., Fromm M. D., Hoppel K. W., and Allen, D. R. (2003). POAM measurements of PSCs and water vapor in the 2002 Antarctic vortex. *Geophys. Res. Lett.*, submitted.

Nefedova E. I. (1994). Mathematical modeling of the global carbon cycle in the atmosphere–ocean system. Ph.D. thesis, Computing Center of the Russian Academy of Science, 112 pp. [in Russian].

Nefedova E. I. and Tarko A. M. (1993). Study of the global carbon cycle using the zonal model in the atmosphere–ocean system. *Proceedings of the Russian Academy of Science*, **333**(5), 645–647 [in Russian].

Nepstad D., McGrath D., Alencar Barros A. C., Carvalho G., Santilli M., and Vera Diaz M. del C. (2002). Frontier governance in Amazonia. *Science*, **295**(5555), 621–631.

Newell R. and Pizer W. (2002). Discounting the benefits of climate change policies using uncertain rates. *Resources*, **146**, 15–20.

Newman P. A., and Nash E. R. (2003). The unusual southern hemisphere stratosphere winter of 2002. *J. Atmos. Sci.*, submitted.

Ni J. (2001). Carbon storage in terrestrial eco-systems of China: Estimates at different spatial resolutions and their responses to climate change. *Climate Change*, **49**(3), 339–358.

Nicolis C. and Nicolis G. (1995). From short-scale atmospheric variability to global climate dynamics: Toward a systematic theory of averaging. *Journal of Atmospheric Science*, **52**(11), 1903–1913.

Nielsen T. T. (1999). Characterization of fire regimes in the Experiment for Regional Sources and Sinks of Oxidants (EXPRESSO) study area. *Journal of Geophysics Research*, **104**(D23), 30713–30723.

Niles J. O. and Schwarze R. (2001). The value of careful carbon accounting in wood products. Editorial. *Climate Change*, **49**(4), 371–376.

Nilsson S., Jonas M., and Obersteiner M. (2002). COP-6: A healing shock? An editorial essay. *Climate Change*, **52**(1–2), 25–28.

Nitu C., Krapivin V. F., and Bruno A. (2000a). *Intelligent Techniques in Ecology*. Printech, Bucharest, 150 pp.

Nitu C., Krapivin V. F., and Bruno A. (2000b). *System Modelling in Ecology*. Printech, Bucharest, 260 pp.

Nordhaus W. D. (2001). Global warming economics. *Science*, **294**, 1283–1284.

Novozhilov A. S. (2002). Identification of the parameters of a dynamic system simulating the pollution–environment interaction. *Bulletin of the Russian Academy of Science (Series: Theory of Systems and Management)*, **3**, 123–127 [in Russian].

Oechel W. C., Vourlitis G. L., Hastings S. J., Zulneta R. C., Hinzman L., and Kane D. (2000). Acclimation of ecosystem CO_2 exchange in the Alaskan Arctic in response to climate warming. *Nature*, **406**(6799), 978–981.

O'Neill B. C. and Oppenheimer M. (2002). Dangerous climate impacts and the Kyoto Protocol. *Science*, **296**(5575), 1971–1972.

Orr J. C, Maier-Reimer E., Mikolajewicz U., Monfray P., Sarmiento J. L., Toggweiler J. R. *et al.* (2001). Estimates of anthropogenic carbon uptake from four three-dimensional global ocean models. *Global Biogeochemical Cycles*, **15**, 43–60.

Orsolini Y. J., Randall C. E., Manney G. L., and Allen, D. R. (2003). An observational study of the final breakdown of the southern hemisphere stratospheric vortex in 2002. *J. Atmos. Sci.*, submitted.

Pacala S. W., Hurtt G. C., Baker D., Peylin P., Houghton R. A., Birdsey R. A. *et al.* (2001). Consistent land- and atmosphere-based U. S. carbon sink estimates. *Science*, **292**, 2316–2320.

Papakyriakou T. N. and McCaughey J. H. (1991). An evolution of evapotranspiration for a mixed forest. *Canadian Journal of Forestry Research*, **21**(11), 1622–1631.

Patin S. (1999). *Environmental Impact of the Offshore Oil and Gas Industry*. EcoMonitor, New York, 425 pp.

Patra P. K. and Maksyutov S. (2002). Incremental approach to the optimal network design for CO_2 surface source inversion. *Geophysics Research Letters*, **29**(10), 97/1–97/4.

Peng C. (1999). Modelling the response of net primary productivity (NPP) of boreal forest ecosystems to changes in climate and disturbance regime. *Ecological Modelling*, **122**(3), 175–193.

Peng C., Jiang H., Apps M. J., and Zhang Y. (2002a). Effects of harvesting regimes on carbon and nitrogen dynamics of boreal forests in central Canada: A process model simulation. *Ecological Modelling*, **155**(2–3), 177–189.

Peng C., Liu J., Dang O., Apps M. J., and Jiang H. (2002b). TRIPLEX: A generic hybrid model for predicting forest growth and carbon and nitrogen dynamics. *Ecological Modelling*, **153**(1–2), 109–130.

Pennise D. M., Smith H. R., Kithinji J. P., Rezende M. A., Raad T. J., Zhang J. *et al.* (2001). Emissions of greenhouse gases and other airborne pollutants from charcoal making in Kenya and Brazil. *Journal of Geophysics Research*, **106**(D20), 24143–24155.

Peritunen J. (1998). LIGNUM: A model combining the structure and the functioning of trees. *Ecological Modelling*, **108**(1–3), 189–198.

Pervaniuk V. S. (2001). A spatial numerical model of the global biogeochemical cycles of carbon and nitrogen in the atmosphere–ocean system. Ph.D. thesis, Computing Center of the Russian Academy of Science, 118 pp. [in Russian].

Peterson B. J., Holmes R. M., McClelland J. W., Vörösmarty C. J., Lammers R. B., Shiklomanov A. U. *et al.* (2002). Increasing river discharge to the Arctic Ocean. *Science*, **298**(5601), 2171–2175.

Philander G. (2001). Why global warming is controversial. *Science*, **294**(5549), 2105–2106.

Pielke Sr R. A. (2001a). Carbon sequestration – the need for an integrated climate system approach. *Bulletin of the American Meteorological Society*, **82**(11), 20–21.

Pielke Sr R. A. (2001b). Earth system modeling – An integrated assessment tool for environmental studies. In: T. Matsuno and H. Kida (eds), *Present and Future of Modeling Global Environmental Change: Toward Integrated Modeling*. TERRA Publishing, Tokyo, pp. 311–337.

Pielke Sr R. A. (2002). Overlooked issues in the U. S. national climate and IPCC assessments. *Climate Change*, **52**(1–2), 1–11.

Pittock A. B. (2002). What we know and don't know about climate change: Reflections on the IPCC TAR. *Climate Change*, **53**(4), 393–411.

Polishchuk Yu. M. (1992). *Simulation Modelling of the Systems with Natural Components*. Science, Novosibirsk, 228 pp. [in Russian].

Popovicheva O. B., Starik A. M., and Favorsky O. N. (2000). Impacts of aviation on the gas and aerosol composition of the atmosphere. *Bulletin of the Russian Academy of Science: FAO*, **2**, 163–176 [in Russian].

Porte A. and Bartelink H. H. (2002). Modelling mixed forest growth: A review of models for forest management. *Ecological Modelling*, **150**(1–2), 141–188.

Post W. M., Izaurralde R. C., Mann L. K., and Bliss, N. (2001). Monitoring and verifying changes of organic carbon in soil. *Climate Change*, **51**(1), 73–99.

Potosnak M. J., Wofsy S. C., Denning A. S., Conway T. J., Munger J. W., and Barnes D. H. (1999). Influence of biotic exchange and combustion sources on atmospheric CO_2 concentrations in New England from observations at a forest flux tower. *Journal of Geophysics Research*, **104**(D8), 9561–9569.

Potter C. S. and Klooster S. A. (1999). Detecting a terrestrial biosphere sink for carbon dioxide: Interannual ecosystem modeling for the mid-1980s. *Climate Change*, **42**(3), 489–503.

Potter C. S., Klooster S. A., de Carvalho C. R., Genovese V. B., Torregrosa A., Dungan J. *et al.* (2001). Modeling seasonal and interannual variability in ecosystem carbon cycling for the Brazilian Amazon region. *Journal of Geophysics Research*, **106**(D10), 10423–10446.

Prather M., Derwent R., Ehhalt D., Fraser P., Sanhaeza E., and Zhou X. (1994). Other trace gases and atmospheric chemistry. In: J. T. Houghton, L. G. Meira Filho, and J. Bruce (eds), *Climate Change 1994: Radiative Forcing of Climate Change and an Evaluation of the IPCC IS92 Emission Scenarios.* Cambridge University Press, Cambridge, UK, pp. 73–126.

Qi F., Guoduong C., and Masao M. (2001). The carbon cycle of sandy lands in China and its global significance. *Climate Change*, **48**(4), 535–549.

Qi Y., Xu M., and Wu J. (2002). Temperature sensitivity of soil respiration and its effects on ecosystem carbon budget: Nonlinearity begets surprises. *Ecological Modelling*, **153**(1–2), 131–142.

Quay P. (2002). Ups and downs of CO_2 uptake. *Science*, **298**(5602), 2344–2345.

Rabotnov T. A. (1987). *Experimental Phytocenology*. Moscow State University, Moscow, 160 pp. [in Russian].

Ragas A. M. J. and R. S. E. W. Leuven (1999). Modelling of water quality-based emission limits for industrial discharges in rivers. *Water Science and Technology*, **39**(4), 185–192.

Risnen J. (2001). CO_2-induced climate change in GMIP2 experiments: Quantification of agreement and role of internal variability. *Journal of Climate*, **14**(9), 2088–2104.

Ramanathan V. and Coakley J. A. (1978). Climate modelling through radiative–convective models. *Reviews of Geophysical Space Physics*, **16**, 465–489.

Randall C. E., Manney G. L., Allen D. R., Bevilacqua R. M., Hornstein J., Trepte C., Lahoz W. A., Ajtic J., and Bodeker G., (2003) Reconstruction and simulation of stratospheric ozone distributions during the 2002 austral winter. *J. Atmos. Sci.*, submitted.

Raschke E., Meywerk J., and Warrack K. (2001). The Baltic Sea experiment (BALTEX): A European contribution to the investigation of the energy and water cycle over a large drainage basin. *Bulletin of the American Meteorological Society*, **82**(11), 2389–2414.

Rau G. H. and Caldeira K. (2002). Minimizing effects of CO_2 storage in ocean. *Science*, **295**, 275–276.

Raupach M. R. (2000). Equilibrium evaporation and the convective boundary layer. *Boundary Layer Meteorology*, **96**, 107–141.

Raupach M. R., Barrett D., Briggs P., and Kirby M. (2001). Water, carbon and nutrient cycles on the Australian continent. *Global Change Newsletter*, **48**, 15–22.

Raymond C. (2002). Ice sheets on the move. *Science*, **298**(5601), 2147–2148.

Reichle Jr H. G., Anderson B. E., Cannors V. S., Denkins T. C., Forbes D. A., Gormsen B. B. *et al.* (1999). Space shuttle based global CO measurements during April and October 1994, MAPS instrument, data reduction, and data validation. *Journal of Geophysics Research*, **104**(D17), 21443–21454.

Reid P. C. (2001). Climate change and the Continuous Plankton Recorder survey. *Marine Observations*, **71**(353), 118–123.

Reilly J., Stone P. H., Forest C. E., Webster M. D., Jacoby H. D., and Prinn R. G. (2001). Climate change: Uncertainty and climate change assessments. *Science*, **293**(5529), 430–431.

Riedo M., Gyalistras D., and Fuhrer J. (2000). Net primary production and carbon stocks in differently managed grasslands: Simulation of site-specific sensitivity to climate change. *Ecological Modelling*, **134**(2–3), 207–227.

Rojstaczer S., Sterling S. M., and Moore N. J. (2001). Human appropriation of photosynthesis products. *Science*, **294**, 2549–2552.

Romankevich E. A. and Vetrov A. A. (2001). *The Carbon Cycle in the Arctic Seas of Russia.* Science, Moscow, 303 pp. [in Russian].

Romashkin P. A., Hurst D. F., Elkins J. W., Dutton G. S., and Wamsley P. R. (1999). Effect of the tropospheric trend on the stratospheric tracer–tracer correlations: Methyl chloroform. *Journal of Geophysics Research*, **104**(D21), 26643–26652.

Roo-Zeilinska E. and Solon J. (1997). Effect of geographical location on species composition, vegetation diversity and phytoindicative characteristics in pine forests. *Environmental Pollution*, **98**(3), 347–360.

Rosa L. P. and Ribeiro S. K. (2001). The present, past, and future contributions to global warming of CO_2 emissions from fuels. *Climate Change*, **48**(2–4), 289–308.

Roscoe H. K., Shanklin J. D., and Colwell S. R. (2003) Has the Antarctic vortex split before? *J. Atmos. Sci.*, submitted.

Rosenberg N. J. and Izaurralde R. C. (2001). Storing carbon agricultural soils to help head off a global warming. *Climate Change*, **51**(1), 1–10.

Royer D. L., Wing S. L., Beerling D. J., Jolley D. W., Koch P. L., Hockey L. J. *et al.* (2001). Paleobotanical evidence for near present-day levels of atmospheric CO_2 during part of the Tertiary. *Science*, **292**(5525), 2310–2313.

Sabine C. L. and Feely R. A. (2001). Comparison of recent Indian Ocean anthropogenic CO_2 estimates with a historical approach. *Global Biogeochemical Cycles*, **15**, 31–42.

Safonov V. S., Sobolevskaya Yu. S., and Ivanova N. A. (1999). Emission of CO_2 by conifers with increased concentrations of O_3 and CO_2. *Optics of the Atmosphere and Ocean*, **12**(12), 1157–1160 [in Russian].

Sarma V. V. S. S., Kumar M. D., and Manerikar M. (2001). Emission of carbon dioxide from a tropical estuarial system, Goa, India. *Geophysics Research Letters*, **28**(7), 1239–1242.

Sarmiento J. L. and Wofsy S. C. (1999). *A U. S. Carbon Cycle Science Plan*. US Global Change Research Program, Washington, DC, 69 pp. (a report of the Carbon and Climate Working Group).

Scaife A. A., Jackson D. R., Swinbank R., Butchart N., Thornton H. E., Keil M., and Henderson, L. (2003) Stratospheric vacillations and the major warming over Antarctica in 2002. *J. Atmos. Sci.*, submitted.

Scanlon T. M. and Albertson J. D. (2001). Turbulent transport of carbon dioxide and water vapor within a vegetation canopy during unstable conditions: Identification of episodes using wavelet analysis. *Journal of Geophysics Research*, **106**(D7), 7251–7262.

Schafer A. and Victor D. G. (1999). Global passenger travel: Implications for carbon dioxide emissions. *Energy*, **24**(8), 657–679.

Schimel D. S. (1995). Terrestrial biogeochemical cycles: Global estimates with remote sensing. *Remote Sensing and Environment*, **51**(1), 49–56.

Schimel D. S. (1998). The carbon equation. *Nature*, **393**(6682), 208–209.

Schimel D. S., Alves D., Enting I., Heimann M., Joos F., Raynand D. *et al.* (1996). CO_2 and carbon cycle. *Climate Change*. Cambridge University Press, Cambridge, UK, pp. 76–86.

Schimel D. S., House J. I., Hibbard K. A., Bousquet P., Ciais P., Peylin P. *et al.* (2001). Recent patterns and mechanisms of carbon exchange by terrestrial ecosystems. *Nature*, **414**, 169–172.

Schlesinger W. H. (1999). Carbon and agriculture: Carbon sequestration in soils. *Science*, **284**(5423), 2095–2096.

Scholes R. J. and Noble I. R. (2001). Climate change: Storing carbon on land. *Science*, **294**(5544), 1012–1014.

Schönwiese C.-D. (2002). Klima in der Diskussion. *AFZ/Wald*, **57**(8), 386–389.

Schulze E.-D., Wirth C., and Heimann M. (2000). Managing forests after Kyoto. *Science*, **289**, 2058–2059.

Sedjo R. A. (2001). Forest sinks as a tool for climate-change policy-making: A look at the advantages and challenges. *Resources*, **143**, 21–23.

Seibel B. A. and Walsh P. J. (2001). Potential impacts of CO_2 injections on deep-sea biota. *Science*, **294**(5541), 319–321.

Seidov D. G. (1987). Numerical models of the ocean circulation. *Earth and Universe*, **5**, 28–34 [in Russian].

Sellers P. J. (1985). Canopy reflectance, photosynthesis and transpiration. *International Journal of Remote Sensing*, **6**, 1335–1372.

Sellers P. J. (1987). Canopy reflectance, photosynthesis and transpiration. II: The role of biophysics in the linearity of their interdependence. *Remote Sensing of the Environment*, **21**, 143–183.

Sellers P. J., Berry J. A., and Collatz G. J. (1992). Canopy reflectance, photosynthesis, and transpiration. III: A reanalysis using improved leaf models and a new canopy integration scheme. *Remote Sensing of the Environment*, **42**, 187–216.

Sellers P. J., Meeson B. W., Hall F. G., Asrar G., Murphy R. E., Schiffer R. A. *et al.* (1995). Remote sensing of the land surface for studies of global change: Models – algorithms – experiments. *Remote Sensing of the Environment*, **51**(1), 3–26.

Sellers P. J., Bounoua L., Collatz G. L., Randall D. A., Dazlich D. A., Los S. O. *et al.* (1996a). Comparison of radiative and physiological effects of doubled atmospheric CO_2 on climate. *Science*, **271**, 1402–1406.

Sellers P. J., Los S. O., and Tucker C. J. (1996b). A revised land surface parametrization (SiB2) for atmospheric GCMs. Part II: The generation of global fields of terrestrial biophysical parameters from satellite data. *Journal of Climate*, **9**(4), 708–737.

Sellers P. J., Randall D. A., and Collatz G. J. (1996c). A revised land surface parametrization (SiB2) for atmospheric GCMs. Part I: Model formulation. *Journal of Climate*, **9**(4), 676–705.

Sellers P. J., Dickinson R. E., Randall D. A., Betts A. K., Hall F. G., Berry J. A. *et al.* (1997). Modeling the exchanges of energy, water, and carbon between continents and the atmosphere. *Science*, **275**, 502–509.

Seppelt R. and Voinov A. (2002). Optimization methodology for land use pattern using spatially explicit landscape models. *Ecological Modelling*, **151**(2–3), 125–142.

Sergin S. Ya. (ed.) (1974). *Study of the Glaciers–Ocean–Atmosphere System.* USSR Far-East Science Centre, Vladivostok, 138 pp. [in Russian].

Shaw M. R., Zavaleta E. S., Chiariello N. R., Cleland E. E., Mooney K. A., and Field C. B. (2002). Grassland responses to global environmental changes suppressed by elevated CO_2. *Science*, **298**(5600), 1987–1990.

Sherman K. (2003). Assessment and recovery of large marine ecosystems. *Proceedings of the PAME Workshop, Protection of the Arctic Marine Environment, 25–27 February 2003, Stockholm, Sweden.* Amsterdam University Press, pp. 1–36.

Shimji M., Nakazawa T., Higurachi K., and Aoki S. (2000). Latitudinal distribution of atmospheric CO_2 sources and sinks inferred by $\delta^{13}C$ measurements from 1985 to 1991. *Journal of Geophysics Research*, **105**(D19), 24315–24326.

Shindell D. (2003). Whither Arctic climate? *Science*, **299**(5604), 215–216.

Siani G., Paterne M., Michel E., Sulpizio R., Sbrana A., Arnold M. *et al.* (2001). Mediterranean Sea surface radiocarbon reservoir age changes since the last glacial maximum. *Science*, **294**(5548), 1917–1920.

Simmons A., Hortal M., Kelly G., McNally A., Untch A., and Uppala S. (2003) ERA-40 Project Report Series – analyses and forecasts of stratospheric winter polar vortex break-up: September 2002 in the southern hemisphere and related events from ECMWF operations and ERA-40. *J. Atmos. Sci.*, submitted.

Sinnhuber B.-M., Weber M., Amankwah A. and Burrows J. P. (2003) Total ozone during the unusual Antarctic winter of 2002. *Geophys Res. Lett.*, in press.

Sleep N. H. and Zahnle K. (2001). Carbon dioxide cycling and implications for climate on ancient Earth. *Journal of Geophysics Research*, **106**(E1), 1373–1399.

Smith K. A., Thomson P. E., Clayton H., McTaggart I. P., and Conen F. (1998). Effects of temperature, water content and nitrogen fertilisation on emissions of nitrous oxide by soils. *Atmospheric Environment*, **32**(19), 3301–3309.

Smith S. V. (2002). Carbon–nitrogen–phosphorus fluxes in the coastal zone: The global approach. *Global Change Newsletter*, **49**, 7–11.

Smith T. M. and Shugart H. H. (1993). The transient response of terrestrial carbon storage to a perturbed climate. *Nature*, **361**, 523–526.

Soon W. and Baliunas S. (2003). Proxy climatic and environmental changes of the past 1000 years. *Climate Research*, **23**, 89–110.

Soon W., Baliunas S., Demirchian, K. S., Kondratyev K. Ya., Idso Sh. B., and Postmentier E. S. (2001). Impact of anthropogenic CO_2 emissions on climate: Unsolved problems, *Bulletin of the Russian Geographic Society*, **133**(2), 1–19 [in Russian].

Sorokhtin O. G. and Ushakov S. A. (1996). Adiabatic theory of the atmospheric greenhouse effect. *Herald of Moscow State University (Ser. 5: Geography)*, **5**, 27–37 [in Russian].

Sorokhtin O. G. and Ushakov S. A. (1998). An accumulation of carbon dioxide in the atmosphere: Damage or benefit? *Natural Gas Industry*, **6**, 44–48 [in Russian].

Sorokhtin O. G. and Ushakov S. A. (1999). *Global Evolution of the Earth*. Moscow State University, Moscow, 446 pp. [in Russian].

Spero H. J. and Lea D. W. (2002). The cause of carbon isotope minimum events on glacial terminations. *Science*, **296**(5567), 522–525.

Staudt A. C., Jacob D. J., Logan J. A., Bachiochi D., and Krishnamurti T. N. (2001). Continental sources, transoceanic transport, and interhemispheric exchange of carbon monoxide over the Pacific. *Journal of Geophysics Research*, **106**(D23), 32571–32589.

Stauffer B., Blunier T., Dällerbach A., Indermühle A., Schwander J., Stocker T. F. *et al.* (1998). Atmospheric CO_2 concentration and millennial-scale climate change during the last glacial period. *Nature*, **392**(6671), 59–62.

Steffen W. and Tyson P. (eds) (2001). *Global Change and the Earth System: A Planet under Pressure*. IGBP Science 4, Stockholm, 32 pp.

Stein A. F. and Lamb D. (2000). The sensitivity of sulphur wet deposition to atmospheric oxidants. *Atmospheric Environment*, **34**(11), 1681–1690.

Stempell D. (1985). *Weltbevolkerung*. Springer-Verlag, Berlin, 205 pp.

Stephens S. M., Frankling S. C., Stagg R. M., and Brown J. A. (2000). Sub-lethal effects of exposure of juvenile turbot to oil produced water. *Marine Pollution Bulletin*, **40**(11), 928–937.

Stephens B. B. and Keeling R. F. (2000). The influence of Antarctic sea ice on glacial–interglacial CO_2 variations. *Nature*, **404**(6774), 171–174.

Stern L. A., Baisden W. T., Amundson R., and Wang Y. (1998). The influence of soils on oxygen isotope ratio of atmospheric CO_2. *Isot. Technnical Study of Environmental Change: Proceedings International Symposium Isot. Technical Study of Past and Current Environmental Changes in the Hydrosphere and Atmosphere, Vienna, 14–18 April 1997*. IIASA Press, Vienna, pp. 27–36.

Sternberg L. and De Angells D. (2002). Carbon isotope composition of ambient CO_2 and recycling: A matrix model. *Ecological Modelling*, **154**(1–2), 179–192.

Stockwell D. Z., Giannakopoulos C., and Plantevin P. H. (1999). Modelling NO_x from lightning and its impact on global chemical fields. *Atmospheric Environment*, **33**(27), 4477–4493.

Stolarski R. S., McPeters R. D., and Newman P. A. (2003). The 2002 Antarctic ozone, *J. Atmos. Sci.*, submitted.

Streets D. G., Jiang K., Hu X., Sinton J. E., Zhang Z.-Q., Xu D. *et al.* (2001). Recent reductions in China's greenhouse gas emissions. *Science*, **294**(5548), 1835–1837.

Strich S. (2002). Ergebnisse der Klimakonferenz von Marrakesch. *AFZ/Wald*, **57**(4), 278–279.

Sturm M., Racine C., and Tape K. (2001). Increasing shrub abundance in the Arctic. *Nature*, **411**, 546–547.

Sun Z. and Pethick D. (2002). Comparison between observed and modeled radiative properties of stratocumulus clouds. *Quarterly Journal of the Royal Meteorological Society*, **128**(586), Part B, 2691–2712.

Svirezhev Yu. M. (2002). Simple spatially distributed model of the global carbon cycle and its dynamic properties. *Ecological Modelling*, **155**(1), 53–69.

Tangley L. (2001). High CO_2 levels may give fast-growing trees an edge. *Science*, **292**(5514), 36–37.

Tenn A. and Davidesin F. (1998). Carbon isotope composition of atmospheric CO_2 in Romania. *Isot. Technnical Study of Environmental Change: Proceedings International Symposium Isot. Technical Study of Past and Current Environmental Changes in the Hydrosphere and Atmosphere, Vienna, 14–18 April 1997.* IIASA Press, Vienna, pp. 19–26.

Terziev F. S., Zatuchnoy B. M., and Gershanovich D. E. (eds) (1993). *The Okhotsk Sea.* Gidrometeoizdat, St Petersburg, 168 pp. [in Russian].

Thompson D. W. J., Baldwin M. P., and Solomon S. (2003) Stratospheric precursors of tropospheric climate anomalies in the southern hemisphere: 1979–2002. *J. Atmos. Sci.*, submitted.

Tian H., Helillo J. M., Kicklighter D. W., McGuire A. D., Helfrich III J. V. K., Moore III B. *et al.* (1998). Effect of interannual climate variability on carbon storage in Amazonian ecosystems. *Nature*, **396**(6712), 664–667.

Tiezen L. L. (1973). Photosynthesis and respiration in arctic tundra grasses: Field light intensity and temperature responses. *Arctic and Alpine Research*, **5**, 239–251.

Tikunov V. S. (1997). *Classifications in Geography*. Institute of the Arts, Moscow, 364 pp. [in Russian].

Todd M. C., and Mackay A. W. (2003) Large scale climatic controls on Lake Baikal ice cover. *J. Climate*, in press.

Tol R. S. J. (2000). International climate policy: An assessment. *IHDP Update*, **3**, 11–12.

Trenberth K. E., Karl T. R., and Spence T. W. (2002). The need for a systems approach to climate observations. *Bulletin of the American Meteorological Society*, **83**(11), 1593–1602.

Tsushima Y. and Manabe S. (2001). Influence of cloud feedback on annual variation global mean surface temperature. *Journal of Geophysics Research*, **106**(D19), 22646–22655.

Ugolnitsky G. A. (1999). *Management of the Ecologo-economic Systems*. High School Book, Moscow, pp. 79–89 [in Russian].

Ugolnitsky G. A., Demianenko Ya. M., and Dubrov D. V. (2000). Methods of the hierarchical optimization in the problems of forestry. *Computer Modeling: Ecology*. High School Book, Moscow, pp. 79–89 [in Russian].

Valiela I. and Bowen J. L. (2002). Nitrogen sources to watersheds and estuaries: Role of land cover mosaics and losses within watersheds. *Environmental Pollution*, **118**(2), 239–248.

van Oijen T. (1998). *Effect of UV Radiation on the Global Carbon Cycle: The Influence of Marine Phytoplankton's Response to UV-B Increase*. Dutch National Research Programme on Global Air Pollution and Climate Change, Amsterdam University Press, 25 pp. (Report No. 410 200 021).

van Wesemael B. and Lambin E. F. (2001). Carbon sinks and conserving biodiversity. *Science*, **294**(5549), 2094–2095.

Vàrallyay G. (2002). Climate change and soil processes. *Idöjàràs*, **106**(3–4), 113–122.

Varotsos C. A. (2002a). Climate change problems and carbon dioxide emisions: Expecting Rio + 10 (Editorial). *ESPR – Environmental Science and Pollution Research*, **9**(2), 97–98.

Varotsos C. A. (2002b). The southern hemisphere ozone hole split in 2002. *ESPR – Environmental Science and Pollution Research*, **9**(6), 375–376.

Varotsos C. A. (2002c) Atmospheric pollution and remote sensing: Implications for the southern hemisphere ozone hole split in 2002 and the northern mid-lattitude ozone trend. *COSPAR Symposium on Natural Hazards Using Space Technology, Houston, TX, 9–19 October, 2002*. Proceedings to be published in *Advances in Space Research* (in press).

Varotsos C. A. (2003a). What is the lesson from the unprecedented event over Antarctica in 2002. *ESPR – Environmental Science and Pollution Research*, **10**(2), 81–82.

Varotsos C. A. (2003b). No ozone hole episode in Antarctica. *Earth Observations from Space* (in press).

Varotsos C. A. (2003c) Major sudden warming and strange twist of the ozone hole over Antarctica in 2002. *European Physics News* **34**(2), 66–67.

Varotsos C. A. (2003d) Why did a "no-ozone-hole" episode occur in Antarctica. *EOS*, **84**(19), 183.

Varotsos C. A. (2003e) A note on "What caused Antarctica to exhibit quite strong planetary waves in 2002?" *J. Atmos. Sci.*, submitted.

Varotsos C. A. and Deligiorgi D. G. (1991). Sea surface temperature and Southern Oscillation signal in the upper stratosphere–lower mesosphere. *International Journal of Climatology*, **11**, 77–83.

Varotsos C. A. and Kondratyev K. Ya. (1998a). Total ozone dynamics in mid-latitudes of the northern hemisphere. *Doklady Akademij Nauk*, **359**(6), 821–822.

Varotsos C. A. and Kondratyev K. Ya. (1998b). The impact of lower stratospheric circulation on the vertical distribution of ozone concentration: A case study of Athens, Greece. *Doklady Akademij Nauk*, **360**(4), 594–599.

Varotsos C. A. and Kondratyev K. Ya. (1998c). The potential interconnection between the vertical ozone concentration distribution in mid-latitudes and processes in the polar stratosphere. *Doklady Akademij Nauk*, **360**(4), 544–553.

Varotsos C. A. and Kondratyev K. Ya. (2003). The unusual split in the Antarctic ozone hole in September 2002. *Studying the Earth from Space*, **1**(4), 78–81, 92–93 [in Russian].

Varotsos C., Kondratyev K. Ya., and Chronopoulos G. (1998) On total ozone content decrease in middle latitudes of the northern hemisphere from satellite observations data in Athens region (Greece). *Doklady Akademij Nauk*, **355**(1), 104–105.

Vasilevich V. I. (1983). *Essays on Theoretical Phytocenology*. Science, Leningrad, 226 pp. [in Russian].

Vasiliev A. S. (1991). Hydrodynamic model for seasonal variability of physical fields in the sea of Okhotsk. *Proceedings of the 6th International Symposium on Okhotsk Sea and Sea Ice, 3–6 February 1991, Mombetsu, Japan*, pp. 294–297.

Vay S. A., Anderson B. E., Conway T. J., Sachse G. W., Collins Jr J. E., Blake D. R. *et al.* (1999). Airborne observations of the tropospheric CO_2 distribution and its controlling factors over the South Pacific Basin. *Journal of Geophysics Research*, **104**(D5), 5663–5676.

Vega M., Pardo R., Barrado E., and Deban L. (1998). Assessment of seasonal and polluting effects of the quality of river water by exploratory data analysis. *Water Research*, **32**(12), 3581–3593.

Veizer J. (2001). Four billion years of carbon cycle: What's the message for us? *Proceedings of the 1st International Conference on Global Warming and the Next Ice Age, Dalhousie University, 19–24 August 2001*, Dalhousie University Press, Dalhousie, New Brunswick, pp. 28–29.

Vernadsky V. I. (1944). A few words about the noosphere. *Successes of Modern Biology*, **18**(2), 49–93 [in Russian].

Vilkova L. P., Novichikhin E. P., Santalov N. P., and Yakovleva G. D. (1998). Estimation of the parameters of the biogeochemical cycle of carbon on global and regional scales. *Problems of the Environment and Natural Resources*, **7**, 24–37 [in Russian].

Vinogradov B. V. (1983). A quantitative estimation of the potential of the soil moisture remote sounding. *Proceedings of the USSR Academy of Science*, **272**(1), 247–250 [in Russian].

von Storch H. and Flöser G. (eds) (1999). *Anthropogenic Climate Change*. Springer-Verlag, Berlin, 351 pp.

Wacker U. (1995). Competition of precipitation particles in a model with parametrized cloud microphysics. *Journal of Atmospheric Science*, **52**(14), 2577–2589.

Walter, H. and Breckle, S-W. (1985). *Ecological Systems of the Geosphere*. Springer-Verlag, Berlin, 242 pp.

Wang F., Chen J., and Forsling W. (1997). Surface properties of natural aquatic sediments. *Water Research*, **31**(7), 1796–1800.

Watson, R. T., Noble, I. R., Bolin, B., Ravindranath, N. H., Verardo, D. J., and Dokken, D. J. (eds) (2000). *Land Use, Land-use Change, and Forestry*. Cambridge University Press, Cambridge, UK, 377 pp.

Wauben W. M. F., Van Velthoven P. F. J., and Kelder H. (1997). 3D chemistry transport model study of changes in atmospheric ozone due to aircraft – NO emissions. *Atmospheric Environment*, **31**(12), 1819–1836.

Weber M., Dhomse S., Wittrock F., Richter A., Sinnhuber B.-M., and Burrows J. P. (2003). Dynamical control of NH and SH winter/spring total ozone from GOME observations in 1995–2002. Geophys. Res. Lett., in press.

Weisenstein D. K., Ko M. K. W., Dyominov I. G., Pitarui G., Riccardully L., Visconti G. *et al.* (1998). The effect of sulfur emissions from HSCT aircraft: A 2D model intercomparison. *Journal of Geophysics Research*, **103**(ND1), 1527–1547.

White A., Cannel M. G. R., and Friend A. D. (1999). Climate change impacts on ecosystems and the terrestrial carbon sink: A new assessment. *Global Environmental Change*, **9**(1001), S21–S30.

Wirtz, K. W. (2000). Second order up-scaling: Theory and an exercise with a complex photosynthesis model. *Ecological Modelling*, **126**(1), 59–71.

WMO (2003). *The Antarctic Ozone Hole Split*. World Meteorological Organization, Geneva.

Wofsy S. C. (2001). Climate change: Where has all the carbon gone? *Science*, **292**(5525), 2261–2262.

Wofsy S. C., Goulden M. L., Munger J. W., Pyle E. H., Urbanski S. P., Hutyra L. *et al.* (2001). Factors controlling long- and short-term sequestration of atmospheric CO_2 in a mid-latitude forest. *Science*, **294**(5547), 1688–1691.

Woodbury P. B., Beloin R. M., Swaney D. P., Gollands B. E., and Weinstein D. A. (2002). Using the ECLIPS software environment to build a spatially explicit component-based model of ozone effects on forest ecosystem. *Ecological Modelling*, **150**(3), 211–238.

Wu Z., Schneider E. K., Hu Z.-Z., and Cao L. (2001). *The Impact of Global Warming on ENSO Variability in Climate Records*. COLA, Calverton, MD, No. 110, 22 pp. (technical report).

Wuebbles D. (2002). Oversimplifying the greenhouse. An editorial essay. *Climate Change*, **52**(4), 3240–3245.

Xu Y. and Carmichael G. R. (1999). An assessment of sulfur deposition pathways in Asia. *Atmospheric Environment*, **33**(21), 3473–3486.

Yakovlev O. I. (1998). *Space Radiophysics*. RFFI Press, Moscow, 432 pp. [in Russian].

Yakovlev O. I. (2001). *Space Radio Science*. Taylor & Francis, London, 320 pp.

Yang X., Wang M., Huang Y., and Wang Y. (2002). A one-compartment model to study soil carbon decomposition rate at equilibrium situation. *Ecological Modelling*, **151**(1), 63–73.

Yanovsky R. G. (1999). *Global Changes and Social Safety*. Academia, Moscow, 358 pp. [in Russian].

Yemshanov D. and Perera A. H. (2002). A spatially explicit stochastic model to simulate boreal forest cover transitions: General structure and properties. *Ecological Modelling*, **150**(1–2), 189–209.

Yi C., Davis K. J., Bakwin P. S., Berger B. W., and Marr L. C. (2001). Influence of advection on measurements of the net ecosystem–atmosphere exchange of CO_2 from a very tall tower. *Journal of Geophysics Research*, **105**(D8), 9991–9999.

Yokozawa, M. (1998). Effects of competition mode on spatial pattern in plant communities. *Ecological Modelling*, **106**(1), 1–16.

Young O. R. and Wefering D. M. (2002). The Global Carbon Project. *IHDP Update*, **1**, 20–21.

Zaitsev S. I. (1988). Laboratory modeling of the processes of the ocean–atmosphere CO_2 exchange. Ph.D. thesis, Moscow State University, Moscow, 170 pp. [in Russian].

Zaragoza G., Lopez-Puertas M., Lopez-Valverde M. A., and Taylor F. W. (2000). Global distribution of CO_2 in the upper mesosphere as derived from UARS/ISAMS measurements. *Journal of Geophysics Research*, **105**(D15), 19829–19840.

Zavarzin G. A. (Ed.) (1999). *Carbon Cycle on the Territory of Russia*. Russian Academy of Science, Moscow, 329 pp. [in Russian].

Zavarzin G. A. and Kolotilova N. N. (2001). *Introduction to Natural Microbiology*. Moscow University, Moscow, 256 pp. [in Russian].

Zerchaninova I. L. and Potapov I. I. (2001). Study of technologies considered by the Intergovernmental Group of Experts with respect to climate change in the context of progress achieved in Russia. *Problems of the Environment and Natural Resources*, **4**, 3–22 [in Russian].

Zonneveld C. (1998). A cell-based model for the chlorophyll *a* to carbon ratio in phytoplankton. *Ecological Modelling*, **113**(1–3), 55–70.

Zor'kin L. M., Subbota M. I. and Stadnik E. V. (1986). *The Methane in Our Life*. NEDRA Publishing, Moscow, 150 pp. [in Russian].

Zuev V. V. (2000). Destruction of the ozone layer – an ecological catastrophe or the greatest mystery of the century? *Proceedings of the 2nd Symposium on the Ecology of Siberian Rivers and the Arctic, Tomsk, 24–26 November 2000*. Tomsk University Press, Tomsk, Russia, pp. 36–40 [in Russian].

Index

Lightning Source UK Ltd.
Milton Keynes UK
UKOW040846190712

196229UK00001B/4/A

9 783540 008095